2nd International Conference on Green Energy Technology (ICGET 2017)

IOP Conference Series: Earth and Environmental Science
Volume 83

Rome, Italy
18 - 20 July 2017

Editors:

Rizzo Renato
Allen Barnett
Tariq Shamim
Kei Eguchi

ISBN: 978-1-5108-4773-6
ISSN: 1755-1307

Printed from e-media with permission by:

Curran Associates, Inc.
57 Morehouse Lane
Red Hook, NY 12571

Some format issues inherent in the e-media version may also appear in this print version.

Printed by Curran Associates, Inc. (2017)

For permission requests, please contact the Institute of Physics at the address below.

Institute of Physics
Dirac House, Temple Back
Bristol BS1 6BE UK

Phone: 44 1 17 929 7481
Fax: 44 1 17 920 0979

techtracking@iop.org

Additional copies of this publication are available from:

Curran Associates, Inc.
57 Morehouse Lane
Red Hook, NY 12571 USA
Phone: 845-758-0400
Fax: 845-758-2633
Email: curran@proceedings.com
Web: www.proceedings.com

Table of contents

Volume 83

2nd International Conference on Green Energy Technology (ICGET 2017)
18–20 July 2017, Rome, Italy

Accepted papers received: 18 August 2017
Published online: 7 September 2017

Preface

Power System Management

Energy and Power Engineering

Renewable Energy and Electrochemistry

Energy Conservation and Emission Reduction

Environmental Engineering and Management

IOP Conf. Series: Earth and Environmental Science **83** (2017) 011001 doi:10.1088/1755-1315/83/1/011001

PREFACE

It is our great pleasure to welcome you to 2017 2nd International Conference on Green Energy Technology (ICGET 2017) which will be held in SAPIENZA University of Rome, Rome, Italy during 18-20 July, 2017. ICGET 2017 is dedicated to issues related to Green Energy Technology.

The major goal and feature of the conference is to bring academic scientists, engineers, industry researchers together to exchange and share their experiences and research results, and discuss the practical challenges encountered and the solutions adopted. Professors from Italy, Canada, USA, Poland are invited to deliver keynote speeches and plenary speeches regarding latest information in their respective expertise areas. It will be a golden opportunity for the students, researchers and engineers to interact with the experts and specialists to get their advice or consultation on technical matters, sales and marketing strategies.

This proceedings present a selection from papers submitted to the conference from universities, research institutes and industries. All of the papers were subjected to peer-review by conference committee members and international reviewers. The papers selected depended on their quality and their relevancy to the conference. The volume tends to present to the readers the recent advances in the field of E Green Energy Technology and various related areas, such as Novel Energy Conversion Studies for RESs, New Approaches in Lightings, Renewable Energy Systems in Smart Cities, Catalysis and Hybrid RESSs, Renewable (Green) Energy Systems and Sources (RESSs) as Wind Power, Hydropower, Solar Energy, Biomass, Biofuel, Geothermal, Energy, Wave Energy, Tidal energy, Hydrogen & Fuel Cells, Energy Storage and etc.

We would like to thank all the authors who have contributed to this volume and also to the organizing committee, reviewers, speakers, chairpersons, sponsors and all the conference participants for their support to ICGET 2017.

Prof. Marco Casini
SAPIENZA University of Rome, Italy
July 31, 2017

IOP Conf. Series: Earth and Environmental Science **83** (2017) 011002 doi:10.1088/1755-1315/83/1/011002

Conference Committee

Conference General Co-Chairs

Prof. Marco Casini, Design and Architecture Technology, SAPIENZA University of Rome, Italy

Prof. Jim P. Zheng, Florida State University, Florida, USA

Prof. Hossam A.Gabbar, University of Ontario Institute of Technology (UOIT), Canada

Program Co-Chairs

Prof. Pierluigi Siano, Department of Industrial Engineering, University of Salerno, Italy

Prof. Kei Eguchi, Fukuoka Insititute of Technology, Japan

Guest Editors

Prof. Rizzo Renato, University of Naples Federico II, Italy

Prof. Allen Barnett, The University of New South Wales, Australia

Prof. Tariq Shamim, Masdar Institute of Science & Technology, UAE

Prof. Kei Eguchi, Fukuoka Insititute of Technology, Japan

Technical Committee

Prof. Mohamed DAROUACH, University of Lorraine France, France

Prof. Rizzo Renato, University of Naples Federico II, Italy

Prof. Mohsen A. Jafari, Rutgers University, America

Prof. Mary Eshaghian-Wilner, The University of Southern California, America

Prof. Allen Barnett, The University of New South Wales, Australia

Prof. Subhransu Sekhar Dash, SRM University, India

Prof. Tariq Shamim, Masdar Institute of Science & Technology, UAE

Prof. Hyung Hee Cho, Dept. of Mechanical Engineering, Yonsei University, South Korea

Prof. Young Nam Chun, Chosun University, South Korea

Prof. Ossama Gouda, Cairo University, Egypt

Prof. Jie Yu, Institute of Nuclear Energy Safety Technology, CAS, China

Assoc. Prof. Angela Russo, Politecnico di Torino, Italy

Assoc. Prof. Farhad Shahnia, Murdoch University, Perth, Australia

Assoc. Prof. Ali Abu El Humos, Jackson State University, America

Assoc. Prof. Buford Boyd Pollett, University of Tulsa, America

Assoc. Prof. Quan Li, University of Edinburgh, UK

Assoc. Prof. Khaled H. Ahmed, University of Aberdeen, UK

Assoc. Prof. Jaroslaw Krzywanski, Jan Dlugosz University in Czestochowa, Poland

Assoc. Prof. Sima Noghanian, University of North Dakota, America

Dr. Farzaneh Hadafi, Department of Architecture, Islamic Azad University, Iran

Dr. Andrew Seagar,Griffith University, Australia

Dr. Prasad Kaparaju, Griffith University, Australia

Dr. Bailey Bubach, University of North Dakota, America

Dr. Ali Arefi, Murdoch University, Perth, Australia, Australia

Dr. Mostafa Mohamed Sayed Ahmed, Faculty of Engineering, Assiut University, Egypt

Dr. Ahmed Elnaem Elnozahy, Faculty of Engineering, Assiut University, Egypt

Peer review statement

All papers published in this volume of *IOP Conference Series: Earth and Environmental Science* have been peer reviewed through processes administered by the proceedings Editors. Reviews were conducted by expert referees to the professional and scientific standards expected of a proceedings journal published by IOP Publishing.

IOP Conf. Series: Earth and Environmental Science **83** (2017) 012001 doi:10.1088/1755-1315/83/1/012001

Practical Efficiency of Photovoltaic Panel Used for Solar Vehicles

T. Koyuncu[1]

1 University of Adiyaman, Faculty of Technology, Adiyaman, Turkey.
E-mail address: tkoyuncu@adiyaman.edu.tr

Abstract. In this experimental investigation, practical efficiency of semi-flexible monocrystalline silicon solar panel used for a solar powered car called "Firat Force" and a solar powered minibus called "Commagene" was determined. Firat Force has 6 solar PV modules, a maintenance free long life gel battery pack, a regenerative brushless DC electric motor and Commagene has 12 solar PV modules, a maintenance free long life gel battery pack, a regenerative brushless DC electric motor. In addition, both solar vehicles have MPPT (Maximum power point tracker), ECU (Electronic control unit), differential, instrument panel, steering system, brake system, brake and gas pedals, mechanical equipments, chassis and frame. These two solar vehicles were used for people transportation in Adiyaman city, Turkey, during one year (June 2010-May 2011) of test. As a result, the practical efficiency of semi-flexible monocrystalline silicon solar panel used for Firat Force and Commagene was determined as 13 % in despite of efficiency value of 18% (at 1000 W/m^2 and 25 °C) given by the producer company. Besides, the total efficiency (from PV panels to vehicle wheel) of the system was also defined as 9%.

1. Introduction

Scarcity and cost of fossil fuels combined with their greenhouse gas emissions make the development of non-fossil fuel based methods of transportation a high-priority task. Therefore, academic studies regarding renewable and clean energy like solar energy are rapidly increasing in the last decade. This regenerative, clean and free energy could be harnessed in several ways. This energy can be converted to electric power by using solar photovoltaic (PV) modules. The produced electricity from solar panels can also be stored in the batteries to operate DC motors of solar powered vehicles [1]-[5]. We believe that solar powered vehicles are smart solution for transportation of people particularly for small and flat location cities that have relatively high solar radiation along the year.

Adiyaman City (Latitude= 37,45°, Longitude= 38,17° and Altitude= 672 m) is located in Southeast of Turkey. It has a university called "Adiyaman University". The university campus is about 5 km far from the city centre. City population is about 210000 and area of the city is 1700 km^2. Most of the city roads have the gradient (slope) less than 10%. In the city the solar radiation is relatively high enough to be efficiently used. The annual total solar energy is about 1600 kWh/m^2.year.

In this experimental investigation, practical efficiency of semi-flexible monocrystalline silicon solar panel used for a solar powered car called "Firat Force" and a solar powered minibus called "Commagene" was determined.

2. Materials and methods

Firat Force has 6 solar PV modules (4x30x0.125x0.125+2x10x0.125x0.125=2.19 m^2), a maintenance free long life gel battery pack (6 batteries: 8 V, 190 Ah, maximum 250 Ah for short period, 9120 Wh), a regenerative (can also be used as generator) brushless DC electric motor (4.5 kW, maximum power 6 kW) and Commagene has 12 solar PV modules (8x14x0.125x0.125+2x12x 0.125x0.125+2x72x0.125x0.125=4.38 m^2), a main-tenance free long life gel battery pack (12 batteries: 6 V, 190 Ah, maximum 250 Ah for short period, 13680 Wh), a regenerative (can also be used as generator) brushless DC electric motor (6.3 kW, maximum power 8 kW)). In addition, each vehicle has MPPT (Maximum power point tracker), ECU (Electronic control unit), differential and external automatic charge device. Each vehicle was also equipped with a mechanical steering system, a free front and rear mechanical suspension system, the chassis constructed from stainless steel pipe covered with special paint, the frame constructed from impact-resist fiber composite material, an instrument panel, a brake and a gas pedal, all standard systems such as signaling, lighting and warning systems.
Some needed technical specifications concerned with Firat Force and Commagene are given in Table 1. In addition, schematic presentation of PV module location, electrical system and power supply for the vehicles are also given in Figure 1, 2 3, 4 and 5.
The electric energy for charging battery pack is supplied by the PV modules. The energy is converted from the solar energy to electric power through the installation of solar photovoltaic modules located on top of the vehicles (Figure 6a, 6b) [6]-[9]. The supplied electric energy is stored in batteries that feed the DC motor.

Table 1. Technical specifications of solar vehicles.

	Vehicle Type	
Specifications	**Firat Force**	**Commagene**
Class	Prototype	Prototype
Axle number	2	2
Wheel number	4	4
Powered axle	1 (rear)	1 (rear)
Unloaded mass	705 kg	1550 kg
Maximum loaded mas.	1025 kg	2600 kg
Motor type	Regenerative brushless DC	Regenerative brushless DC
Motor efficiency	88%	90%
Motor nominal power	4.5 kW	6.3 kW
Motor nominal rpm	3000 rpm ± %10	3000 rpm ± %10
Motor nominal torque	25 Nm	34 Nm
Suggested speed	40 km/h	40 km/h
PVmodule type	Semiflexible Monocrystal Si	Semiflexible Monocrystal Si
Module efficiency	18% @ 1000 W/m^2, 25 °C	18% @ 1000 W/m^2, 25 °C
Module area	[4(6x5) + 2(2x5)] x [0.125x0.125] = 2.19 m^2	[8(7x2) + 2(6x2) +2(6x12)]x [0.125 x 0.125] = 4.38 m^2
MPPT efficiency	95 %	95 %
Battery pack	6 batteries, 8 V, 190 Ah	12 batteries, 6 V, 190 Ah
Battery type	Maintenance free long life gel	Maintenance free long life gel
Battery energy	9120 Wh @ C20 discharge	13680 Wh @ C20 discharge
Battery voltage	Nominal 48V	Nominal 72 V

IOP Conf. Series: Earth and Environmental Science **83** (2017) 012001 doi:10.1088/1755-1315/83/1/012001

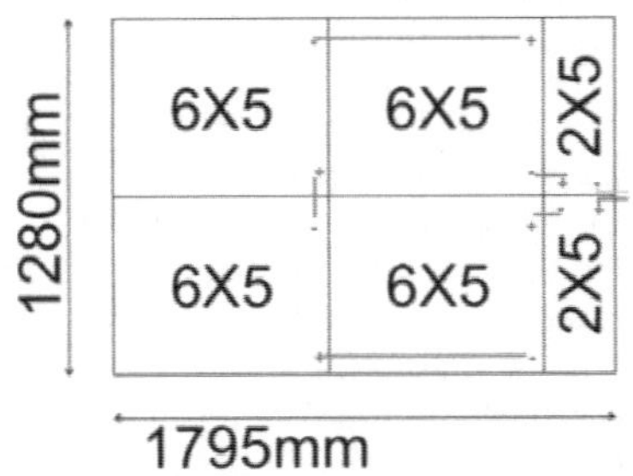

Figure 1. Schematic presentation of PV module location for Firat Force.

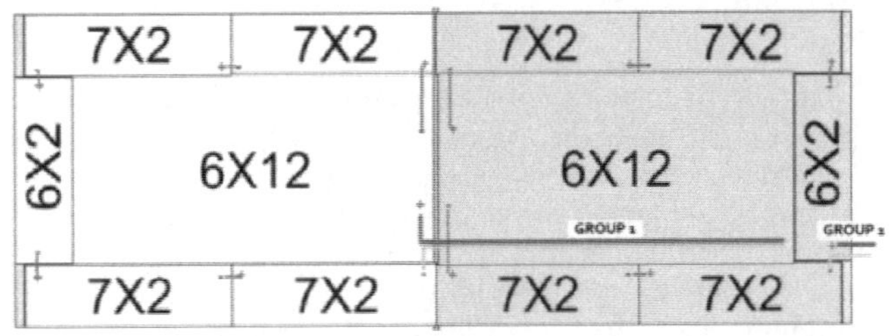

Figure 2. Schematic presentation of PV module location for Commagene.

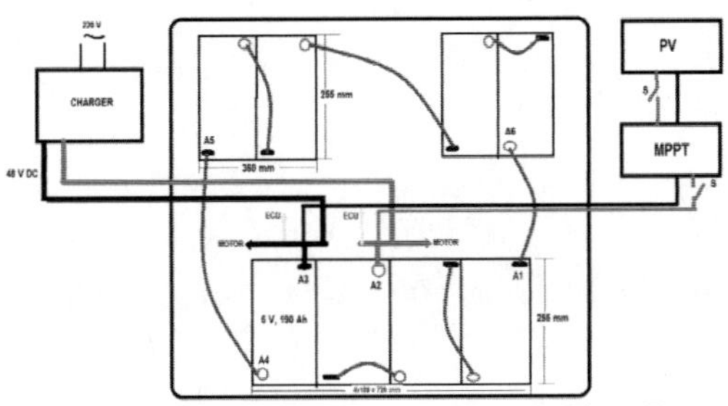

Figure 3. Schematic presentation of Firat Force electrical system.

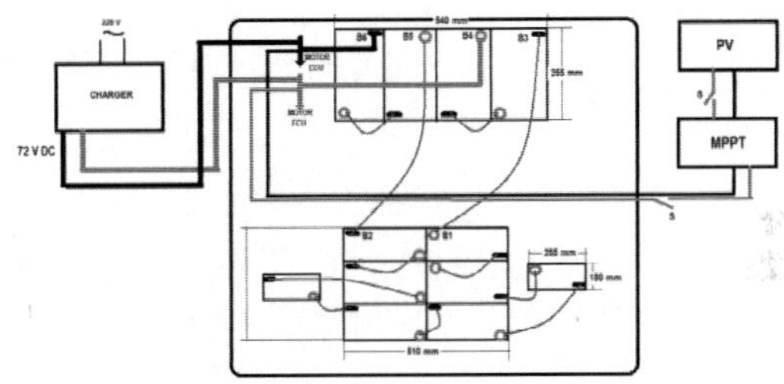

Figure 4. Schematic presentation of Commagene electrical system.

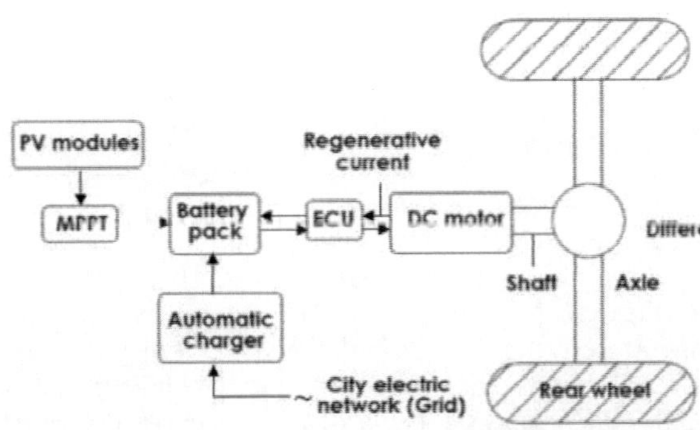

Figure 5. Schematic presentation of power supply for Firat Foce and Commagene.

Figure 6a. A view of Firat Force PV panel location.

Figure 6b. A view of Commagene PV panel location.

The needed total area of solar photovoltaic module was determined based on the DC motor power. In order to calculate the approximate power needed for any vehicle, the average incident solar radiation and daytime of the city (mean of about 40 years), vehicle speed, efficiency of the PV panels, road features and grade, carrying number of people, needed daily road distance for transportation and the efficiency of all equipments have been considered. To determine the total needed force and power for any vehicle to overcome all resistances, Equations 1, 2, 3, 4 and some coefficients were used. In addition, the power supplied by PV panels, the efficiency of PV panels (from PV panel to ECU), the efficiency of total system (from PV panel to vehicle wheel) were calculated by using Equations 5, 6, 7 and 8 [10]-[16]. As seen from Equation 7 that basically, the efficiency of PV panel equals to electric power produced by PV panels divided by solar power that comes to the total surface of the PV panel. All symbols used in equations are described in the Nomenclature.

$$F_{TT} = F_{AC} + F_{GR} + F_{AE} + F_{RR} \quad (1)$$

$$F_{TT} = \lambda ma + mgp + 0.5\rho c_w A_C v^2 + mgf_R \quad (2)$$

$$q_{TP} = F_{TT} v \quad (3)$$

$$p = \sin\alpha \cong tg\alpha \quad (4)$$

$\lambda = 1.03...1.45;\ a = 0.3...0.5 m/s^2;\ \rho \cong 1.226 Ns^2/m^4 (@1.0133 bar \& 15^0 C);\ c_w \cong 0.30;\ A_C \cong 1.85 m^2;\ f_R = 0.01...0.02$

$$q_{SR} = I_R A_{PV} \quad (5)$$

$$q_{PV} = VC \quad (6)$$

$$\eta_{PV} = \frac{q_{PV}}{q_{SR}} \quad (7)$$

$$\eta_{SY} = \eta_{PV}\eta_{MPPT}\eta_{BA}\eta_{ECU}\eta_{DC}\eta_{TR} \quad (8)$$

$\eta_{PV} \cong 0.18;\ \eta_{MPPT} \cong 0.95;\ \eta_{BA} = 0.80...0.90;\ \eta_{ECU} = 0.90...0.95;\ \eta_{DC} \cong 0.88;\ \eta_{TR} \cong 0.98;$

The solar vehicles were used for people transportation in the city to see the unexpected problems regarding design, manufacture and vehicle systems during one year (June 2010-May 2011) (Figure 7a, 7b). During this period we have made some preliminary tests. These preliminary experiments showed us that the best way to see the exact performance or efficiency of any solar powered vehicle under city normal traffic conditions is to use the solar vehicles nonstop with full charged batteries while PV panels off and PV panels on. Therefore, the vehicles were tested while PV panels off (battery pack was charged with electricity only), PV panels on (battery pack was charged with PV panels during use) on the standard road (asphalt-paved road with gradient of about 0%) and on the mixed road (asphalt-paved and unimproved roads with different gradients of 0%...20%) during 3 months (June, July and August 2011) under normal city traffic flow. All tests were made between the times of 10:00 (morning)-17:00 (afternoon). Firat Force was loaded with 2 and 5 persons (driver included) while Commagene loaded with 5 and 10 persons (driver included) during all tests for their total transportation distance. The difference of transportation distance between PV panels off and PV panels on positions was recorded as transportation distance supplied by solar energy. As known, this energy is renewable and clean energy. On this account, Firat Force and Commagene apparently can be called clean or green vehicles. However in many literatures, all electric vehicles are called green or clean vehicle. It should be noted that it completely depends on the energy sources that electric produced from. It is not possible to say that all electric vehicles are green or clean.

Figure7a. A view of Firat Force test activities. **Figure 7b.** A view of Commagene test activities.

Besides, the term "efficiency" should be used instead of term "performance" for PV cell, module or panel, in despite of given by some literatures. Because, term "efficiency" is very appropriate to explain the amount of energy conversion from one form to another (form light or solar to electricity for PV cell). However, the term "performance" is particularly suitable to clarify the speed comparision of any activity, process or work.

3. Results and Observations

Firat Force and Commagene were tested on the standard road (asphalt-paved road with gradient of about 0%) and on the mixed road (asphalt-paved and unimproved roads with different gradients of 0%...20%) during 3 months (June, July and August 2011) under normal city traffic flow for practical efficiency of PV panels. All data regarding solar radiation intensity and PV panels were given in Table 2 and 3 for Firat Force and Commagene, respectively. Average values of these data can also be seen from these tables. The average practical PV panel efficiency was seen as approximately 13% from Table 2 and 3 for Firat Force and Commagene. The total system efficiency (from PV panel to vehicle wheel) was also seen as approximately 9% (calculated from Equation 8) for both solar vehicles. In addition, the average solar radiation and daily sunshine hours were determined as 525.22 W/m^2 and 8.11 h/day for Adiyaman city, Turkey (Figure 8). The average calculated transportation distance, amount of saved fuel, amount of saved money and percentage of full charged with PV panel for one year were also given for Firat Force and Commagene (Figure 9a,9b and 10a,10b) [9], [17].

Table 2. Data regarding Firat Force

Road Fea.	P.No	S.I.(W/m^2)	A (m^2)	S.P. (W)	PV-P(W)	PV-E(%)
Stdrd. road	2	980	2.19	2146	245	11.4
Stdrd. road	5	975	2.19	2135	250	11.7
Average efficiency of PV + MPPT						***11.5***
Average efficiency of PV						***≈ 13***
System total efficiency (from PV to wheel)						***≈ 9***
PV module efficiency at 1000 W/m^2 and 25 °C						***18***

Road Feat: Road feature; P. No. : Passsenger number; S.I. :Solar radiation intensity; A : PV panel surface area; S.P. : Solar power; PV-P : PV panel power; PV-E : PV panel efficiency.

Table 3. Data regarding Commagene

Road Fea.	P.No	S.I.(W/m^2)	A(m^2)	S.P.(W)	PV-P(W)	PV-E (%)
Stdrd. road	5	980	4.38	4292	510	11.8
Stdrd. road	10	980	4.38	4292	485	11.3
Average efficiency of PV + MPPT						***11.5***
Average efficiency of PV						***≈ 13***

IOP Conf. Series: Earth and Environmental Science **83** (2017) 012001 doi:10.1088/1755-1315/83/1/012001

System total efficiency (from PV to wheel)	***≈ 9***
PV module efficiency at 1000 W/m² and 25 °C	***18***

Road Feat: Road feature; P. No. : Passsenger number; S.I. Solar radiation intensity; A: PV panel surface area; S.P. : Solar power; PV-P : PV panel power; PV-E : PV panel efficiency.

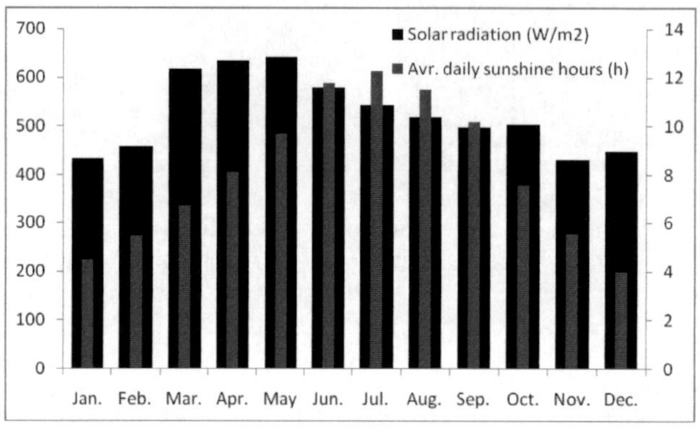

Figure 8. Solar radiation (right axis: W/m^2) and average daily sunshine duration (left axis: h) related to months.

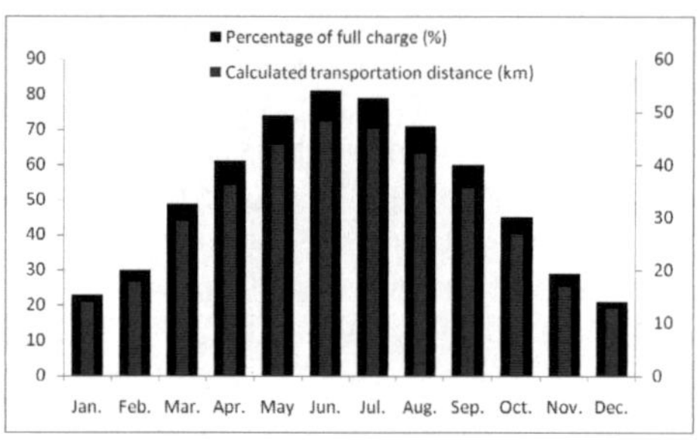

Figure 9a. Charge of solar PV panels as a percentage of full charged battery pack (right axis: %) and calculated transportation distance (left axis: km) related to months (for Firat Force).

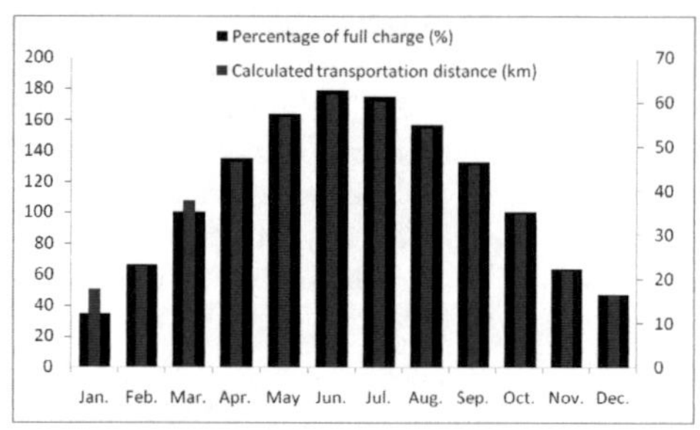

Figure 9b. Charge of solar PV panels as a percentage of full charged battery pack (right axis: %) and calculated transportation distance (left axis: km) related to months (for Commagene).

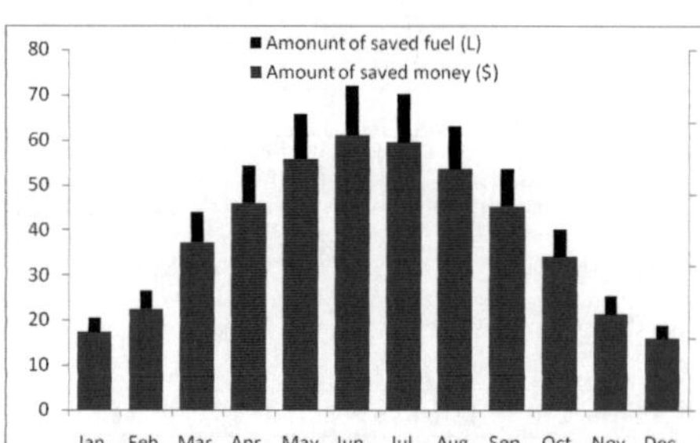

Figure 10a. Amount of saved fuel (right axis: L) and amount of saved money (left axis: $) related to months per year (The fuel consumption of an internal combustion engine (ICE) was accepted as 5L/100 km and Turkey fuel sale price (March 2013, 2.65 $/L) was considered for the calculation) (for Firat Force)

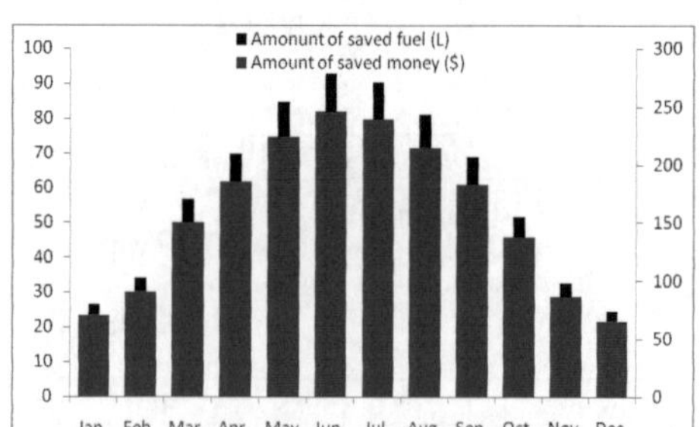

Figure 10b. Amount of saved fuel (right axis: L) and amount of saved money (left axis: $) related to months per year (The fuel consumption of an internal combustion engine (ICE) was accepted as 5L/100 km and Turkey fuel sale price (March 2013, 2.65 $/L) was considered for the calculation) (for Commagene).

4. Conclusions

The main conclusion of this experimental investigation is that solar power vehicles should be used with nonstop use under city normal traffic conditions to define the practical (most realistic) efficiency of solar panels. The results showed us that the average practical efficiency of semi-flexible monocrystalline silicon PV panel is approximately 13% in despite of 18% given by the manufacturer company for standard test conditions of 1000 W/m2 and 25 °C. The solar vehicle total system efficiency form PV panel to the vehicle wheel was also seen about 9%. Therefore, in order to design any solar vehicle with minimum mistake the practical losses of the PV panels and vehicle power supply system should be considered. Otherwise, theoretical approximation should not meet the practical needed.

5. Acknowledgement

I would like to thank too much to Adiyaman University for supporting this experimental investigation.

6. Nomenclature

a : *Acceleration,* m/s^2
A_C : *Frontal area of the vehicle,* m^2
A_{PV} : *Photovoltaic PV area,* m^2
C : *Average current, A*
c_w : *Coefficient of drag*
F_{AC} : *Needed force for acceleration, N*
F_{AE} : *Needed force for aerodynamic resistances, N*
F_{GR} : *Needed force for grade resistance, N*
f_R : *Coefficient of rolling friction*
F_{RR} : *Needed force for rolling resistance, N*
F_{TT} : *Needed total force to overcome all resistances, N*
g : *Gravitational acceleration,* m/s^2
I_R : *Incident solar radiation,* W/m^2
m : *Vehicle mass, kg*
q_{PV} : *Photovoltaic (PV) power, W*
q_{SR} : *Incident solar power, W*
q_{TP} : *Needed total power of the vehicle, W*
v : *Speed of the vehicle,* m/s
V : *Average voltage, V*

Greek

α : *Road grade*
η_{BA} : *Battery efficiency*
η_{DC} : *DC motor efficiency*
η_{ECU} : *Electronic control unit (ECU) efficiency*
η_{MPPT} : *Maximum power point tracker (MPPT) efficiency*
η_{PV} : *Photovoltaic (PV) efficiency*
η_{SY} : *System efficiency*
η_{TR} : *Transmission efficiency*
λ : *Rotational inertia factor*
ρ : *Air density,* kg/m^3

7. References

[1] Dayem, A. M. A. 2011. Set-up and performance investigation of an innovative solar vehicle. 1st International Workshop on Diffusion of New Energy Technologies in Chine, March 25-27, Shanghai, China.
[2] Garrison, M. 2008. The 2007 solar D house. 5th International Conference on Urban Regeneration and Sustainability, Skiathos Isl., Greece, September 24-26.
[3] Hazami, M., Kooli, S., Lazaar, M., Farhat, A. and Belghith, A. 2005. Performance of a solar storage collector. Desalination, (183): 167-172.
[4] Tsai, C. S. and Ting, C.H. 2010. Evaluation of a multi-power system for an electric vehicle. International Conference on Control, Automation and Systems (ICCAS 2010), p.p. 1308-1311, Gyeonggi do, South Korea, October 27-30.
[5] Ustun, O., Yilmaz, M., Gokce, C. et al., 2009. IEEE International Electric Machines and Drives Conference (IEMDC 2009), Miami, FL., USA. , May 03-06.
[6] Ageev, MD., Blidberg,,,DR., Jalbert, J. et al., 2002. Results of the evaluation and testing of the solar powered AUV and its subsystems. IEEE Workshop on Autonomous Underwater Vehicles, San Antonio, TX. , June 20-21.

[7] Egiziano, L., Giustiniani, A., Lisi, G. et al., 2007. Experimental characterization of the photovoltaic generator for a hybrid solar vehicle. IEEE International Symposium on Industrial Electronics. Vigo, Spain, June 04-07.
[8] Lu, W., Yamayee, Z. A., Melton, A. et al., 2007. A solar powered battery charger for a hybrid electric vehicle. 37th Annual Frontiers in Education Conference, Milwaukee, WI. , October 10-13.
[9] Redpath, D.A.G., Mcllveen-Wright, D., Kattakayam,T., Hewitt, N.J. and Karlowski, J. 2011. Battery powered electric vehicles charged via solar photovoltaic arrays developed for light agricultural duties in remote hilly areas in the Southern Mediterranean region. Journal of Cleaner Production, (19): 2034-2048.
[10] Koyuncu, T., Engin, K.E. and Lule, F. 2014. Design, manufacture and test of a solar powered minibus (Commagene) for transportation of people in Adiyaman city, Turkey. 29th European Photovoltaic Solar Energy Conference and Exibition (EU PVSEC 2014), Amsterdam, The Netherlands, September 22-26.
[11] Koyuncu, T. 2015. Design, manufacture and test of a solar powered car (Firat Force) for transportation of governmental officers in Adiyaman city, Turkey. The 6th International Congress of Energy and Environment Engineering and Management (CIIEM15), Paris, France, July 22-24.
[12] Cengel, Y.A. and Boles, M.A. 1994. Thermodynamics: an engineering approach (Second edition). McGraw-Hill, Inc., New York.
[13] Gibson, T.L. and Kelly, N.A. 2010. Solar photovoltaic charging of Lithium-ion batteries. Journal of Power Sources, (195): 3928-3932.
[14] Goktan, A.G., Guney, A. and Ereke, M. 1995. Tasıt frenleri (vehicle brakes: in Turkish). ITU Machine Faculty, Automotive Main Science, Istanbul.
[15] Holman, J.P. 1994. Experimental methods for engineers (Sixth edition). McGraw-Hill Inc., New York.
[16] Poulek, V. and Libra, ,M. 2006. Solar energy: photovoltaics-promossing trend for today and close future. Czech University of Agriculture in Prague, Prague, Czech Republic.
[17] Sorrentino, M., Rizzo, G. and . Arsie, I. 2011. Analysis of a rule-based control strategy for on-board energy management of series hybrid vehicles. Control Engineering Practice, (19): 1433-1441.

State-of-The-Art of Modeling Methodologies and Optimization Operations in Integrated Energy System

Zhan Zheng[1] and Yongjun Zhang[1]

1 School of Electric Power, South China University of Technology,Guangzhou, China
E-mail: zz_wallace@163.com

Abstract. Rapid advances in low carbon technologies and smart energy communities are reshaping future patterns.Uncertainty in energy productions and demand sides are paving the way towards decentralization management. Current energy infrastructures could not meet with supply and consumption challenges, along with emerging environment and economic requirements.Integrated Energy System(IES) whereby electric power,natural gas,heating couples with each other demonstrates that such a significant technique would gradually become one of main comprehensive and optimal energy solutions with high flexibility,friendly renewables absorption and improving efficiency.In these global energy trends,we summarize this literature review. Firstly the accurate definition and characteristics of IES have been presented.Energy subsystem and coupling elements modeling issues are analyzed.It is pointed out that decomposed and integrated analysis methods are the key algorithms for IES optimization operations problems,followed by exploring the IES market mechanisms.Finally several future research tendencies of IES,such as dynamic modeling,peer-to-peer trading,couple market design,are under discussion.

1. Introduction

Nowadays the energy industry is facing fierce transformation from generations to generations. Wide distribution of utility networks offers the convenience of energy consumption and variable resources integration.Web smart grid has brought the concept of ICT infrastructure construction and fast-response communication into the development of flexibility and real-time energy management[1]. With arising of distributed energy resources and regional integrated energy networks, the traditional electric energy system is witnessing the lessening impact of unified interconnected networks and the emerging influence of decentralized energy generations tendency. The double operations mode of 'Interconnection' and 'decentralization' will govern the vision of energy generation, transmission and consumption.Integrated energy system (IES) has aroused under such macro energy transients stages. It represents frontier demands for modern energy system of high efficiency, complementary medias for redundancy and variable resources consumption with systemic stability. Consequently this novel issue will contribute to academic and industrial focuses continuously.

This paper aims to provide the readers with ampliate and miniature overview in integrated energy system (IES) with the emphasis on the energy systems modeling methodologies and optimization operations.We retrospect and summarize the basic IES concepts with comparisons of former theories. Following by introductions of subsystem analysis and coupling components modeling, up-to-date optimization algorithms and energy market mechanisms are presented. Future research tendencies are also externalized.

2. IES concepts and definition

Integrated Energy System as a concept has been approached by peers from different aspects. Three different summarizing perspectives are proposed as Smart Grid 2.0, Interconnected Energy Network and Internet + Energy Network,in which the merge modes are different in physical and informative merge characteristics. Key technical constraining conditions development in time-scale and application scenarios in different regions contributes to the typical forms of Internet + Energy Net development [2].Variable perspectives reflect the up-to-date technology sharping the energy and power systems. In a strict definition, 'integration' is related to parallel energy medias coupling and restructured while 'system' is focusing on the physical main infrastructure body with deriving other minor application issues.

In 2007, energy hub [3] was proposed by G. Anderson in ETH as the linear energy transfer relationship between multi-energy flow while later the system concept has appeared. G. Strbac proposed Multi-Vector Energy System [4] enabling storage and transport, which is based on the UK situation of highly co-operation of natural gas and electricity with large wind farms integration. The Multi-Energy System [5] raised by P. Mancarella, divides the internal and external regional networks and could be under the interpretation of microgrids and virtual power plants. Besides, Wu Jianzhong is focusing on the combination scheming with coupling elements in Integrated Energy System operations[6].

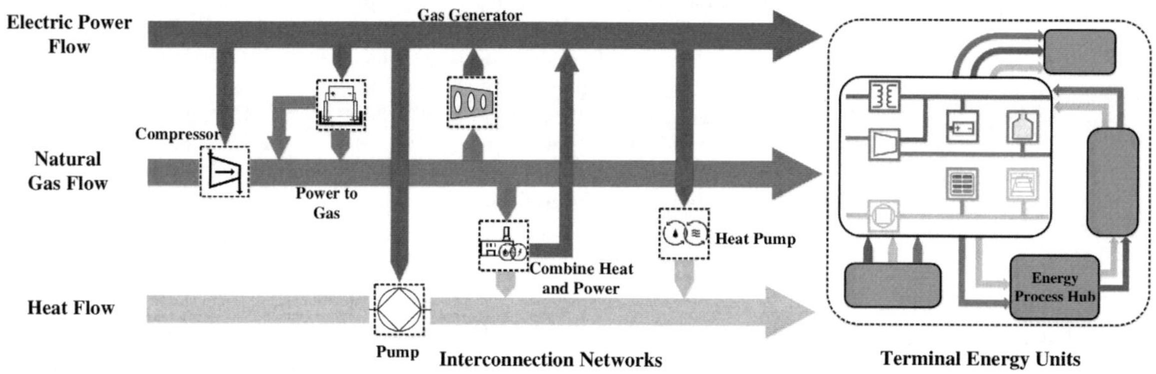

Fig. 1 A typical diagram of overall Integrated Energy System concept with multi-energy flow

As it is shown in Figure 1, an typical overall concept depiction of integrated energy system is presented:electric power flow,natural gas flow and heating flow are the majority energy carriers,which are consists of interconnection networks and terminal energy units.From the integrated perspective,system maximum benefits could be realized as follows:

- High penetration harvest of renewable and intermittent energy.
- Complementarity of mutual storage for load shifting and curtailments.
- Reliability for energy supply,changing the scenarios of single energy supply or non-switchable energy supplies.
- Efficiency improvements by providing much more available conversion sections

3. IES modeling techniques,approaches and analysis methods

3.1. Unique energy network mode issues

With Kirchhoff voltage law and Kirchhoff current law,similar analogy can be concluded from these three energy carriers.For node energy balance,the Kirchhoff current law,gas flowrate balance equation and heating mass flow continuity equation could be all included in same principle.The node variables loss can be formed into Kirchhoff voltage law,gas pressure drop equation and loop pressure equation.In some ways,IES modeling could be made into unified formulation considering node variables and branch variables.

Table 1. Comparison with three energy system modeling's variables

Energy Carrier	Node Variable	Branch Variable	Constraints
Electric Power	Voltage	Current	Kirchhoff voltage law; Kirchhoff current law
Natural Gas	Pressure	Gas flow rate	Pressure drop equation; Flowrate balance equation
Heating	Pressure, Temperature	Mass flow rate,Heat power	Flow continuity equation; Loop pressure equation; Head loss equation

3.2. Coupling components modeling in IES

The coupling components between the two energy carriers could be taken account into linear relation expression describing the numeral single input-output situation,which could be formulated as follows:

$$E_1 = \varphi E_2 \tag{1}$$

Where E_1 and E_2 represent different energy flow and is φ the energy transfer efficiency.

*3.2.1. Combined Heating and Power.*With the efficiency of up to 80%,the combined heating and power(CHP) could realize the multiple outputs in power and heating as cogeneration unit.It is also called combined heat and power district heating and sometimes works as decentralized energy.Some lower power CHP is defined as micro-CHP for inhouse usage.Commonly the ratio between heating generation and power generation is constant.Thus,the operations problem could turn into single optimal energy flow problem based another other fixed energy flow.CHP will be strongly supported when local combined network is under interconnection with outer power grid;otherwise in islanded mode,the CHP should be adjusted more accurately to fulfill demands.

3.2.2. Power to Gas. Power to Gas(P2G) facilities could promote antiparallel flow,through water electrolysis generating hydrogen or methane(synthesized by carbon dioxide and oxygen).P2G could realize long-term and wide-area storage,which is beneficial to co-planning and co-operation of power system and gas network.Seasonal storage flexibility with renewables harvest could be also improved by P2G.However, the overall process of P2G owns probably transfer efficiency of only 20%-40%.The less development of P2G commercialization and the unexpected market demands for hydrogen will be the major obstacles of its further popularization.

3.2.3. Heat Pump. Driven by electricity,the heat pump is another high potential tool to increase system's flexibility,which could be sourced from electric networks or fluctuant renewable energy.Serving as the negative load,it could support the CHP operations when CHP reaches its heating output limitation.Further more,the curtailment of renewable peaking generation is also achieved in IES or isolated microgrid equipped with heat pumps.

*3.2.4. Gas turbine.*The gas turbine is the linkage between natural gas network and electric power network.It could be considered as the gas load or power generation from two perspectives.In small IES communities such as microgrids,gas turbine is taken into account in operating as main resources with high dynamic response.In long-term's energy substitution,gas turbine plays a necessary role in system flexibility adjustment.

3.2.5. Comprehensive Energy Hub. Energy hub issues could be expressed in conversion matrix connecting the multiple inputs and outputs,describing the linear relationship.Such kind of model could be extended and characterized by self-defined components such as battery energy storage systems and solar heat exchanger [7].The energy management system containing non-dispatchable resources and dispatchable energy resources or storage could be consider as electrical hub,optimized by different levels' dispatch strategies [8].Such methodology can be used for non-linear hub problems.

4. System modeling and optimization operations

4.1. Decomposed analysis method

The decomposition problem could be described as:under hypothesis of given value in master problem,subproblem could be calculated to convergence through iterations. The decomposition techniques have been widely applied in solving possible decomposable structures such as complicating variables or complicating constraints.Different kinds of energy flows could be defined as the elements under the decomposable framework. Dual decomposition mathematical structure includes individual and coupling constraints, which is forcing the Lagrangian multipliers to relax the coupling constraints [9]. The decomposition-coordination algorithm with inner, boundary and coordinated variables,has been proposed in [10], in which sub-problems are firstly solved and then part of boundary variables are selected as coordinated variables to solve interfaced constraints.The decomposed techniques could reduce the calculation scale when the energy system is increasingly complex but it is hard to guarantee the convergence due to much more iterations.

4.2. *Integrated analysis method*

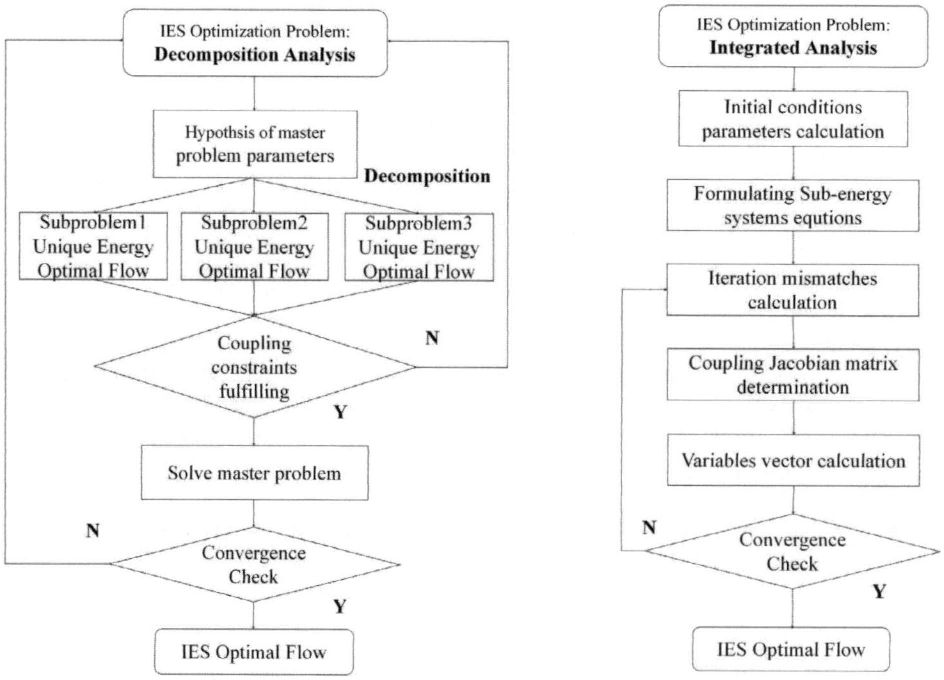

Fig.2 IES problem analysis methods comparison

Unlike the decomposed analysis methods,the integrated analysis methods are directly forging the subproblems into unique equation.Total general expression (2) is investigated as the combination of electrical network,natural gas network and heat network,covering all variables in unique vector x ,with Jacobian matrix J and energy imbalances vector $f(x^{(n)})$.

$$x^{(n+1)} = x^{(n)} - J^{-1} f(x^{(n)}) \quad (2)$$

The integrated electrical-hydraulic-thermal method [11] is explored with grid-connected and islanded modes,determining value of Jacobian submatrix elements and final optimal results. An expansive Newton-Raphson algorithm has been described in [12] to construct the coupling matrix and achieve optimal energy flow , following initially satisfying the calculation of CHP energy dispatch and other coupling equipments constraints.In summary,the integrated analysis method is to implement the non-linear optimization in unique general equation by forming the state variables vectors and coupling Jacobian matrix.Compared to decomposed analysis method,integrated analysis method takes less calculation steps but increases calculation scales and time.

5. Market mechanism for IES

New market structures with integrated approach optimization of electricity-natural gas system should be adapted by implementation of co-optimizing the day-ahead and real-time dispatch, where stochastic models are used to describe the uncertainty renewable energy integration while real-time energy balancing cost is solved by the specific realization of power production. [13] defines the market majority bodies of three different energy(gas-heat-power) participants in multi-lateral trading (MLT).In such a case, gas system is an absolute provider and heat system is an absolute consumer,making power system a prosumer to buy gas and sell electricity to heat networks,which could be turned into a bi-level problem solved by two phase methods: Rough Estimation and Elaborate Adjustment. The common solution to IES market problem is to establish the double stages or multi-stages stochastic programming with assumption of one energy market clearing being fixed.

6. Future research tendencies towards IES

Upon the complex coupling issues and various demanding energy challenges,utilities are facing more obstacles in system planning an operations.It is meaningful to continue concentrating efforts on IES. New research will be needed for the further development to guarantee the IES optimization operations .

6.1. Dynamic models modeling issue

For dynamic models issue,independent systems research are conducted. As for multi-vector energy, no current paper is covering.Drawing on the experiences of RTDS applied in power system, natural gas and heating dynamic models could be established in physical ways connected with electrical part when parts of them are under high interactions with electricity.The other models of natural gas and heating,which own slow dynamic characteristics and are not highly coupled with electricity,could be constructed in digital simulation tools[14].Further search work on such issues is worthy.

6.2. Peer-to-peer trading technology application in IES

Along with blockchain technology spread,P2P technology unlocks the decentralized decision making in regional energy market,ensuring customers freely handling energy trading.However,the current research progress is only focusing on electricity.The coupling of other energy carriers and P2P makes IES market mechanism more complex,especially in multi-energy flows P2P's multidirection trading.

6.3. Couple energy market design for IES

The different timing scales with large differences facilitate possibilities of coherent coupling energy markets.Incentive demand response cases under IES will be the most potential technology for such energy market[15,16].Coupling elements contributes to flexibility increase,which is meaningful to the pricing schemes in couple markets with high complementarity.Based on above, the coupling market should be designed under diverse markets regulations,integrated externality network limitations and optimal multi-energy flows convexity solutions.

7. Conclusions

The calls of Cleantech,whether from green energy consumption or environment protections,is existing as a hotspot.The Integrated Energy System has boosted the booming of next generations of smart infrastructures.This paper provides concepts and characteristics for IES.Through concerning systems modeling with solutions and markets mechanisms,this paper also offer insights into potential technologies in IES.

8. Acknowledgements

This paper is supported by National Natural Science Foundation of China (51377060).

9. References

[1] Zhaoyang D, Junhua Z, Fushuan W, et al 2014 From smart grid to energy internet basic concept and research framework *J. Auto. of Elec. Powe. Syst.* **38(15)** 1-11
[2] Licheng L, Yongjun Z, Zexing C, Zexiang C, Yongxia H, Ping Y 2016 Merger between smart grid and energy-net:mode and develop prospects *J. Auto. of Elec. Powe. Syst.* **40(11)** 1-9
[3] Martin G, Gaudenz K, Patrick Favre P, Bernd K, Goran A, Klaus F 2007 Energy hubs for the future *J. IEEE powe. and ener. maga.* **5(1)** 24-30.
[4] Chaudry M, Jenkins N, Strbac G 2008 Multi-time period combined gas and electricity network optimisation *J. Elec. Powe. Syst. Rese.* **78** 1265-79
[5] Pierluigi M 2014 MES(multi-energy systems):An overview of concepts and evaluation models *J. Ener.* **65** 1-17
[6] Jianzhong W, Jinyue Y, Hongjie J, Nikos H, Ned D, Hongbin S 2016 Integrated Energy System *J. Appl. Ener.* **167** 155-157
[7] Thanhtung H, Yongjun Z, V.V.THANG2, Jianang H 2017 Energy hub modeling to minimize residential energy costs considering solar energy and BESS *J. Jour. of Mode. Powe. Syst. and Clea. Ener.* **5(3)** 389-399.
[8] A.T.D. P, Vahid M. N, Dasaraden M, Jean-Louis S 2017 Electrical hubs: An effective way to integrate non-dispatchable renewable energy sources with minimum impact to the grid *J.Appl. Ener.* **190** 232-248
[9] Dimitrios P, Danny P, Goran S 2014 Decentralized coordination of microgrids with flexible demand and energy storage *J. IEEE Tran. on Sust. Ener* **5(4)** 1406-14
[10] Jinghua L, Jiakun F, Qing Z, Zhe C 2016 Optimal operation of the integrated electrical and heating systems to accommodate the intermittent renewable sources *J. Appl. Ener.* **167** 244-254
[11] Xuezhi L, Jianzhong W, NICK J, Audrius B 2016 Combined analysis of electricity and heat networks *J. Appl. Ener.* **162** 1238-50
[12] Yingrui W, Bo Z, Jing G, Jiaqi S, Jianhua Z 2016 Multi-Energy flow calculation method for Integrated Energy System containing electricity,heat and gas *J. Power System Technology* **40(10)** 2942-50
[13] Yue C, Wei W, Feng L, Shengwei M 2017 A multi-lateral trading model for coupled gas-heat-power energy networks *J. Appl. Ener.* **200** 180-191.
[14] Hongjie J, Dan W, Xiandong X, Xiaodan Y 2015 Research on some key problems related to integrated energy systems *J. Auto. of Elec. Powe. Syst.* **9(7)** 198-207
[15] Sneum Daniel M, Sandberg E, Soysal Emilie R, Skytte K, Olsen Ole J 2016 Framework conditions for flexibility in the district heating-electricity interface
[16] Baorong Z, Tingcheng H, Yongjun Z 2017 Reliability Analysis on Microgrid Considering Incentive Demand Response *J. Auto. of Elec. Powe. Syst.* **41(13)** 70-78.

IOP Conf. Series: Earth and Environmental Science **83** (2017) 012003 doi:10.1088/1755-1315/83/1/012003

Reconfiguration of Smart Distribution Network in the Presence of Renewable DG's Using GWO Algorithm

M. Siavash[1], **C. Pfeifer**[1], **A. Rahiminejad**[2], **B. Vahidi**[3]

1 Department of Material Science and Process Engineering, University of Natural Resources and Life Sciences, Vienna, Austria
2 Department of Electrical and Computer Science, Esfarayen University of Technology, Esfarayen, North Khorasan, Iran
3 Department of Electrical Engineering, Amirkabir University of Technology, Tehran, Iran
E-mail: majid.siavash@students.boku.ac.at

Abstract. In this paper, the optimal reconfiguration of smart distribution system is performed with the aim of active power loss reduction and voltage stability improvement. The distribution network is considered equipped with wind turbines and solar cells as Renewable DG's (RDG's). Because of the presence of smart metering devices, the network state is known accurately at any moment. Based on the network conditions (the amount of load and generation of RDG's), the optimal configuration of the network is obtained. The optimization problem is solved using a recently introduced method known as Grey Wolf Optimizer (GWO). The proposed approach is applied on 69-bus radial test system and the results of the GWO are compared to those of Particle Swarm Optimization (PSO), and Genetic Algorithm (GA). The results show the effectiveness of the proposed approach and the selected optimization method.

1. Introduction

Considering three major crises of human societies including financial, energy, and environmental issues, the presence of distributed generation especially renewable ones are increased dramatically in distribution networks. These kinds of resources could enhance the performance of the distribution networks from different aspects such as loss reduction, voltage profile and stability improvement, and reliability increase [1]. However, to achieve these improvements, optimal placement and scheduling of distributed generation are required [2].

Since the output power of Renewable Distributed Generations (RDG) such as solar cells and wind turbines depend on ambient conditions, their production could not be dispatched [3]. In other words, the RDGs output generation fluctuates which causes the operation point of distribution network moving away from the optimal point.

Reconfiguration of distribution networks is an effective and low cost method for enhancing the distribution system performance. The problem of optimal reconfiguration is a discrete optimization problem by which the configuration of the distribution network is changed optimally by altering the state of open/closed switches [4]

The two important issues related to this problem are determining the objective function and the solving algorithm. There are different target functions for the problem of reconfiguration including power loss minimization [5] voltage profile improvement [6], reliability increase [7], and voltage stability enhancement [8]. In this paper, the problem is solved with the aim of power loss reduction and voltage stability improvement. There are different methods proposed for solving the reconfiguration problem in the literature such as genetic algorithm [9], Big Bang–Big Crunch algorithm [10], firefly algorithm [11], cuckoo search algorithm [6], plant growth simulation algorithm [12], Teaching–Learning-Based Optimization algorithm [13], bacterial foraging optimization algorithm [5], and Artificial immune algorithm [14]. Different methods have advantages from different aspects such as final results, time of convergence, convergence behaviour, and robustness.

In this paper, a recently introduced method known as Grey Wolf Optimizer (GWO) [15] is proposed for solving the distribution system reconfiguration problem. The suggested approach is applied on a distribution network equipped with renewable energy resources consists of solar cells and wind turbine. The reconfiguration problem is solved hourly in a day considering the hourly variation of load demand, wind speed, solar irradiance, and temperature. The distribution system is considered equipped by smart metering devices which provide accurate information of network conditions in any moment. The problem is solved with the aim of power loss reduction and voltage stability enhancement.

2. Problem formulation

Distribution networks consist of two types of switches including sectionalizing and tie switches which the former are normally close and the later are normally open. Changing the status of these switches changes the configuration of the system and the power flow path which may result in loss reduction, voltage profile improvement, voltage stability enhancement and etc. Depending on a target, the configuration of the system could be obtained optimally; thus, reconfiguration could be considered as an optimization problem as follow:

$$\begin{aligned} \min \quad & f(x) \\ s.t. \quad & h(x) = 0 \\ & g(x) \leq 0 \end{aligned} \tag{1}$$

where x is decision variables (status of switches), f is objective function, h and g are equality and inequality constraints, respectively.

2.1. Objective function

As mentioned before, the problem of reconfiguration in this paper is solved for two objectives as loss reduction and voltage stability enhancement. Herein, these two objective functions are explained.

2.1.1. Loss reduction

The first objective function is active power loss reduction which is the wasted power through feeders. The total power loss can be calculated as follow:

$$P_{loss} = \sum_{k=1}^{N_b} R_k . I_k^2 \tag{2}$$

$$I_k = \frac{V_j - V_i}{R_k + jX_k} \tag{3}$$

where P_{loss} is the total power loss of the network, N_b is the total number of branches, I_k, R_k and X_k are current, resistance and inductance of branch k, and V_i and V_j are the voltages at sending and receiving end of branch k.

2.1.2. Voltage stability enhancement

Voltage stability is a factor which determines the weak buses which are sensitive to load changes and may cause voltage collapse. There are different indices introduced for voltage stability determination such as Shin criterion [16], Aman criterion [17], Kayal criterion [18]. In this paper, the Kayal criterion is used as the Voltage Stability Factor (VSF) which is as follow:

$$VSF_{total} = \sum_{m=1}^{N-1} (2 \times V_{m+1} - V_m) \tag{4}$$

where N is the total number of buses, V is voltage, VSF_{total} is the total VSF of the system. A higher value of VSF means a more stable network.

In order to add the two mentioned objectives for simultaneous optimization of both targets, the objective functions are normalized as follow:

$$f = f_1 + f_2$$
$$f_1 = \frac{P_{loss}}{P^0_{loss}} \tag{5}$$
$$f_2 = \frac{VSF^0_{total}}{VSF_{total}}$$

where P^0_{loss} and VSF^0_{total} are the values of loss and voltage stability factor in the base case of the network.

2.1.3. Constraints

During the optimization procedure, there are some constraints which must be satisfied. These constraints are as follow:

Power balance: for any proposed configuration of distribution network, the power balance equation must be checked using power flow program.

Voltage magnitude: the magnitude of bus voltages must be remained in a predefined range

$$V_i^{\min} < V_i < V_i^{\max} \tag{6}$$

Thermal limit of feeders: the power flows through the feeders must be lower than their thermal limit

$$S_i < Limit_i \tag{7}$$

where S_i and $Limit_i$ are power flow and thermal limit of feeder *i*.

Radial structure: for any suggested configuration, the structure of the network must be remained radial.

3. Grey Wolf Optimizer (GWO)

Grey Wolf Optimizer (GWO) algorithm is a population based algorithm which was introduced by Mirjalili in 2014 [15]. This method simulates the hunting procedure of a wolf pack. The leadership hierarchy of the wolf pack is modeled by defining the wolves α, β, δ, and ω. The first level of leadership is the wolf α which is responsible for lots of decisions such as hunting, sleeping place, and time of wake. Mathematically, the best solution of the populations is considered as the wolf α. The second level of hierarchy is the wolf β which is the second best solution of the population. The third best solution is δ and the other wolfs are considered as lowest ranking wolves ω. The wolves ω are the last group of wolves which are allowed to eat. These wolves are always ready to submit to higher ranking level wolves.

The main phases of hunting procedure are as follows:

1- Tracking, chasing, and approaching the prey

2- Pursuing, encircling, and harassing the prey until it stops moving

3- Attack towards the prey

Grey wolves encircle the pray before attacking to it. The encircling procedure could be mathematically modelled using the following equations.

$$\vec{D} = \left|\vec{C}.\vec{X}_P(t) - \vec{X}(t)\right| \tag{8}$$

$$\vec{X}(t+1) = \vec{X}_P(t) - \vec{A}\vec{D} \tag{9}$$

where t is iteration number, $\vec{A}$ and $\vec{C}$ are coefficient vectors, $\vec{X}_P$ is prey position, and $\vec{X}$ is the position of grey wolf. The coefficient vector can be calculated as follow:

$$\vec{A} = 2\vec{a}\vec{r}_1 - \vec{a} \tag{10}$$

$$\vec{C} = 2\vec{r}_2 \tag{11}$$

where $\vec{r}_1$ and $\vec{r}_2$ are random vectors in the range of [0, 1], and $\vec{a}$ is vector which reduces from 2 to 0 during the iterations.

As mentioned before, the first three high ranking wolves (α, β, and δ) have a better knowledge of the prey location. Thus, these three wolves are remained unchanged in the current iteration and the other

IOP Conf. Series: Earth and Environmental Science **83** (2017) 012003 doi:10.1088/1755-1315/83/1/012003

wolves position updated based on the position of three first wolves. The updated procedure of wolves ω is as follows:

$$\vec{D}_\alpha = \left|\vec{C}_1 \vec{X}_\alpha - \vec{X}\right|, \vec{D}_\beta = \left|\vec{C}_2 \vec{X}_\beta - \vec{X}\right|, \quad \vec{D}_\delta = \left|\vec{C}_3 \vec{X}_\delta - \vec{X}\right| \tag{12}$$

$$\vec{X}_1 = \vec{X}_\alpha - \vec{a}_1 \vec{D}_\alpha, \vec{X}_2 = \vec{X}_\beta - \vec{a}_2 \vec{D}_\beta, \quad \vec{X}_3 = \vec{X}_\delta - \vec{a}_3 \vec{D}_\delta \tag{13}$$

$$\vec{X}(t+1) = \frac{\vec{X}_1 + \vec{X}_2 + \vec{X}_3}{3} \tag{14}$$

The final position would be in a random place within a circle which is defined by the positions of alpha, beta, and delta in the search space. In other words alpha, beta, and delta estimate the position of the prey, and other wolves updates their positions randomly around the prey.

4. Simulation and results

In this section the proposed approach is applied on 69-bus radial test system which the single line diagram of the network is depicted in Figure 1. The system was used in many previous researches [19]-[22] and the line and bus data of this system can be found in [20].

This system has a total load of 3802.2 kW and 2694.6 kVAR which is considered as the peak load situation. In this situation the total active power loss is 224.98 kW and the minimum voltage is 0.9092 (in p.u.). In addition, the total VSF for this network in the peak situation is 66.04. The daily load profile of this network is depicted in Figure 2 which is a percentage of peak load demand.

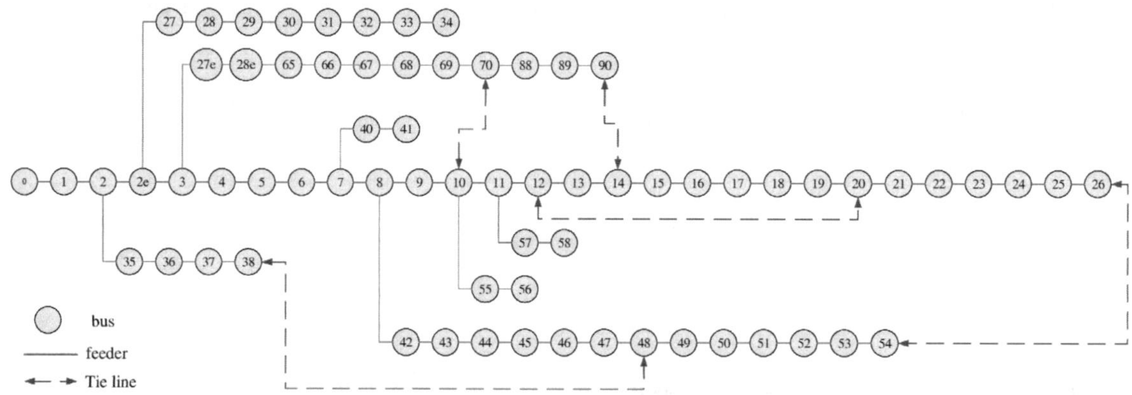

Figure 1. Single line diagram of 69 bus test system

4.1. Presence of RDG's

In order to investigate the effect of RDG's on network performance a 500 kW set of PV array and a 800 kW wind turbine are added to the buses 17 and 50 respectively. The location of RDG's are the optimal location based on [23]. The hourly variation of wind power generation and PV power output are depicted in Figure 3. In the presence of these resources the active power loss, minimum voltage, and total VSF at the peak hour are, respectively, 176.5, 0.9198, and 66.4. Thus, it can be concluded that the presence of RDG' improves the performance of the network. However, because of output fluctuation of RDG's, the performance of the network may move away from the best operation point. Thus, optimal reconfiguration is performed.

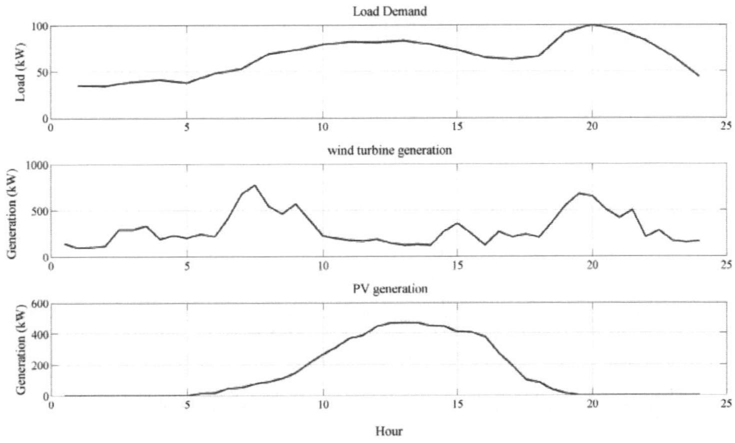

Figure 2. Hourly load profile and generation profile of wind turbine and PV array

4.2. Optimal reconfiguration

In order to enhance the network performance in the presence of RDG's, optimal network reconfiguration is performed hourly using GWO. As mentioned before, the objective functions of optimization problem are loss reduction and voltage stability enhancement.

The hourly loss of the system in three casas (i.e., before RDG's, after RDG's and no reconfiguration, and reconfiguration in the presence of RDG's) are depicted in Figure 3. As it is obvious, by the placing the RDG's, loss of the system is decreased. It also can be concluded that the reconfiguration of distribution network enhances the system performance dramatically from loss reduction point of view. Figures 4, and 5 depict the minimum voltage and total VSF in different hours in different cases. It also can be concluded that the proposed approach moves the operation point of the system to the optimal point.

The percentage of loss reduction is illustrated in Table 1. As it is obvious, the presence of RDG's results in loss reduction. However, using network reconfiguration, the network performance is highly improved and optimized. Thus, it can be concluded that using the proposed approach, not only the advantages of renewable resources are achieved, but also the network operation point is optimized in different ambient conditions.

In order to demonstrate the better performance of the selected optimization algorithm, the network reconfiguration results in peak load situation using GWO are compared to those of GA, and PSO. The convergence behaviours of these three algorithms are compared in Figure 6. As it is obvious, the proposed method not only reaches the better results, but also finds the best answer in lower iterations. In other words, the selected method outperforms the GA.

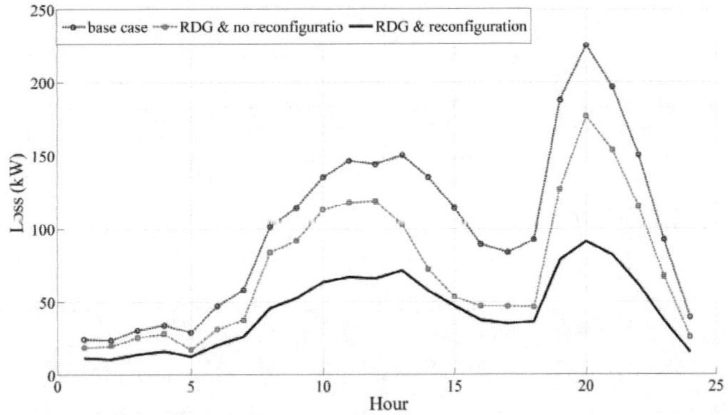

Figure 3. Hourly loss of the system in different cases

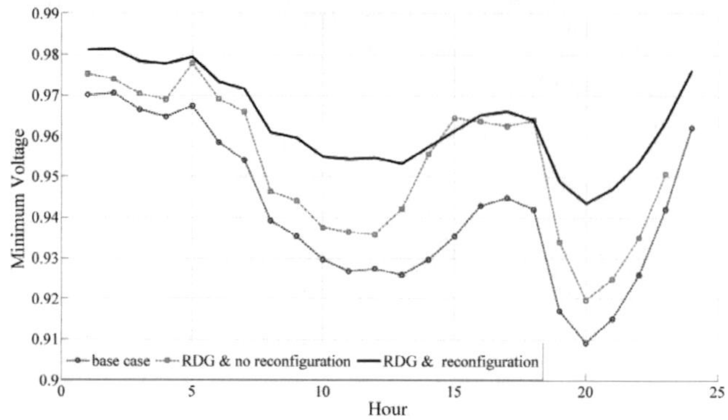

Figure 4. Hourly minimum voltage in different cases

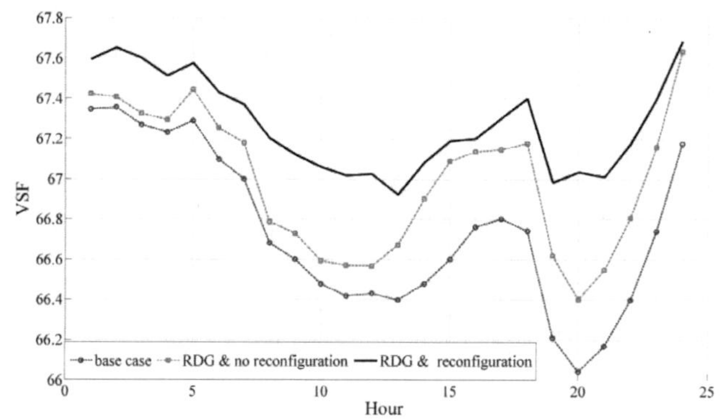

Figure 5. Hourly total VSF in different cases

Table 1. Style and Font Size

hour	PLR* Case 1	PLR Case 2	Hour	PLR Case 1	PLR Case 2
1	24.1	52.6	13	31.5	52.5
2	17.3	55	14	46.5	57.4
3	16.5	54.7	15	53.3	58.6
4	17.6	52.7	16	47	58
5	40.2	57	17	43.7	58
6	34	55.9	18	49.4	60.5
7	34.8	54.6	19	32.2	58.2
8	17.4	54.7	20	21.5	59.5
9	19.7	53.8	21	21.8	58.3
10	16.6	53	22	23.5	59
11	19.8	54.3	23	27.1	59.7
12	17.8	54.2	24	33.9	60.6

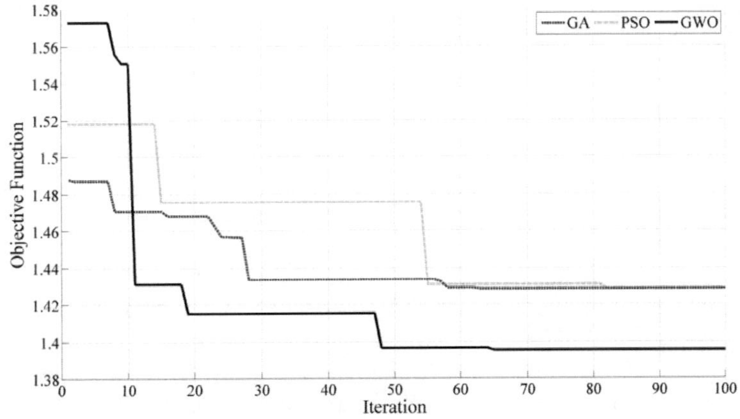

Figure 6. convergence behaviour of different methods

5. Conclusion

In this paper, the reconfiguration of distribution network in the presence of renewable distributed generation (RDG) is performed. It is obvious that the RDG's improve the network performance; however, since the output generation of these kinds of resources are fluctuated by the ambient conditions variation, the network operation point may move away from the optimal point. Thus, reconfiguration of distribution network as a low cost tool is employed to enhance the network performance in different conditions. In this paper, the distribution network is considered to be equipped with smart metering devices; thus, the network conditions (i.e., the load demand, wind speed, and solar irradiance) are known in any moment. The hourly optimal reconfiguration of network is implemented using a recently introduced method known as Grey Wolf Optimizer (GWO). The results show that by the proposed approach, the network performance is highly improved. In addition, by comparison of the results of the GWO with those of some other methods, it is concluded that the GWO outperforms other methods from final results and convergence behaviour points of view.
The future studies can be performed on the integration and optimal management of storage systems considering the variable output of RDGs. Moreover, the other objective functions such as reliability, voltage profile improvement, and energy cost minimization could be taken into account. The novel optimization methods can also be employed for possible better answers.

6. References

[1]. Georgilakis PS, Hatziargyriou ND. Optimal distributed generation placement in power distribution networks: models, methods, and future research. 2013 *IEEE Trans Power Syst.*;28(3):3420-8.
[2]. Shaaban MF, El-Saadany EF. Optimal allocation of renewable DG for reliability improvement and losses reduction. In: *Power and Energy Society General Meeting, 2012 IEEE*. IEEE; 2012:1-8.
[3]. Rahiminejad A, Faramarzi D, Hosseinian SH, Vahidi B. An effective approach for optimal placement of non-dispatchable renewable distributed generation. 2017 *J Renew Sustain Energy.*;9(1):15303.
[4]. Rao RS, Ravindra K, Satish K, Narasimham SVL. Power loss minimization in distribution system using network reconfiguration in the presence of distributed generation. 2013 *IEEE Trans power Syst.*;28(1):317-25.
[5]. Naveen S, Kumar KS, Rajalakshmi K. Distribution system reconfiguration for loss minimization using modified bacterial foraging optimization algorithm. 2015 *Int J Electr Power Energy Syst.*;69:90-7.

[6]. Nguyen TT, Truong AV. Distribution network reconfiguration for power loss minimization and voltage profile improvement using cuckoo search algorithm. 2015 *Int J Electr Power Energy Syst.*;68:233-42.
[7]. López JC, Lavorato M, Rider MJ. Optimal reconfiguration of electrical distribution systems considering reliability indices improvement. 2016 *Int J Electr Power Energy Syst.*;78:837-45.
[8]. Sahoo NC, Prasad K. A fuzzy genetic approach for network reconfiguration to enhance voltage stability in radial distribution systems. 2006 *Energy Convers Manag.*;47(18):3288-306.
[9]. Nara K, Shiose A, Kitagawa M, Ishihara T. Implementation of genetic algorithm for distribution systems loss minimum re-configuration. 1992 *IEEE Trans Power Syst.*;7(3):1044-51.
[10]. Sedighizadeh M, Bakhtiary R. Optimal multi-objective reconfiguration and capacitor placement of distribution systems with the Hybrid Big Bang–Big Crunch algorithm in the fuzzy framework. 2016 *Ain Shams Eng J.*;7(1):113-29.
[11]. Shareef H, Ibrahim AA, Salman N, Mohamed A, Ai WL. Power quality and reliability enhancement in distribution systems via optimum network reconfiguration by using quantum firefly algorithm. 2014 *Int J Electr Power Energy Syst.*;58:160-9.
[12]. Rajaram R, Kumar KS, Rajasekar N. Power system reconfiguration in a radial distribution network for reducing losses and to improve voltage profile using modified plant growth simulation algorithm with distributed generation (dg). 2015 *Energy Reports.*;1:116-22.
[13]. Lotfipour A, Afrakhte H. A discrete Teaching–Learning-Based Optimization algorithm to solve distribution system reconfiguration in presence of distributed generation. 2016 *Int J Electr Power Energy Syst.*;82:264-73.
[14]. Souza SSF, Romero R, Pereira J, Saraiva JT. Artificial immune algorithm applied to distribution system reconfiguration with variable demand. 2016 *Int J Electr Power Energy Syst.*;82:561-8.
[15]. Mirjalili S, Mirjalili SM, Lewis A. Grey wolf optimizer. 2014 *Adv Eng Softw.*;69:46-61.
[16]. Shin J-R, Kim B-S, Park J-B, Lee KY. A new optimal routing algorithm for loss minimization and voltage stability improvement in radial power systems. 2007 *IEEE Trans Power Syst.*;22(2):648-57.
[17]. Aman MM, Jasmon GB, Mokhlis H, Bakar AHA. Optimal placement and sizing of a DG based on a new power stability index and line losses. 2012 *Int J Electr Power Energy Syst.*;43(1):1296-304.
[18]. Kayal P, Chanda CK. Placement of wind and solar based DGs in distribution system for power loss minimization and voltage stability improvement. 2013 *Int J Electr Power Energy Syst.*;53:795-809.
[19]. Huang Y-C, Yang H-T, Huang C-L. Solving the capacitor placement problem in a radial distribution system using tabu search approach. 1996 *IEEE Trans power Syst.*;11(4):1868-73.
[20]. Baran ME, Wu FF. Optimal capacitor placement on radial distribution systems. 1989 *IEEE Trans power Deliv.*;4(1):725-34.
[21]. Injeti SK, Kumar NP. Optimal planning of distributed generation for improved voltage stability and loss reduction. 2011 *Int J Comput Appl.*;15(1):40-6.
[22]. Abdelaziz AY, Mohamed FM, Mekhamer SF, Badr MAL. Distribution system reconfiguration using a modified Tabu Search algorithm. 2010 *Electr Power Syst Res.*;80(8):943-953.
[23]. Rahiminejad A, Hosseinian SH, Vahidi B, Shahrooyan S. Simultaneous Distributed Generation Placement, Capacitor Placement, and Reconfiguration using a Modified Teaching-Learning-based Optimization Algorithm. 2016 *Electr Power Components Syst.*;44(14).

Techno-economical Analysis of Rooftop Grid-connected PV Dairy Farms; *Case Study of Urmia University Dairy Farm*

A M Nikbakht[1], N Aste[2], H J Sarnavi[1] and F Leonforte[2]
1 Department of Mechanical Engineering of Biosystems, Urmia University, Iran.
2 Department of Architecture, Built Environment and Construction Engineering, Politecnico di Milano University, Italy.
E-mail: a.nikbakht@urmia.ac.ir

Abstract. The global trends indicate a growing commitment to renewable energy development because of declining fossil fuels and environmental threats. Moreover, the global demographic growth coupled with rising demands for food has escalated the rate of energy consumption in food section. This study aims to investigate the techno-economic impacts of a grid-connected rooftop PV plan applied for a educational dairy farm in Urmia university, with total estimated annual electrical energy consumption of 18,283 kWh, located at the north west part of Iran. Based on the current feed-in tariff and tremendously low electricity price in agriculture section in Iran, the plants with size ranged from 14.4 to 19.7 kWp (initial investment ranged from 26,000 to 36,000 USD) would be satisfied economically.

1.Introduction

Distributed photovoltaic electricity generation can provide electricity for using up locally, as well as, feeding surplus amount in local distribution grid. Distributed renewable energy systems can serve as a complement to centralized energy generation systems, or even as a substitute [1]. The global demographic growth coupled with rising demands for food has escalated the rate of energy consumption in food section. At the same time, the tremendous threat of climate change associated with the use of declining fossil fuels makes supplying this energy source increasingly unpleasant from the standpoint of environmental impacts. Hence, the agricultural sector faces a great balancing act on the choice of its energy sources. Increasing attention is being focused on the renewable energy sources employment in the agricultural sector in several countries of the world purposely to contribute to global reduction in greenhouse gas emissions and sustainable food production. Furthermore, it might seem intuitively reasonable to explain the emergence of new energy policies driving transition to a sustainable energy systems even in agriculture sectors. Decentralized small scale renewable energy plants at the agricultural farms may be a good energy policy [2]. The assessment of small scale rooftop PV plants regarding their techno-economic feasibility is of utmost importance. Farmers wish to know what options offer them the most beneficial investment and potential for reducing operating costs and energy use, while providing the greatest return on investment. This paper presents an investigation of techno-economic impacts of a grid-connected rooftop PV plan applied for a educational dairy farm. The site selected for the study is an educational dairy farm located in Urmia university with large space available on rooftop area (roughly 1000 m2). The geographic and climatic conditions in Urmia are very favorable for renewable solar energy (Figure 1). The country exposes by an outstanding direct normal radiation of up to 5.5 kWh/m^2/day and an average of 300 sunny days per year. The geographic site location and meteorological properties of Urmia city was depicted in Figure 1. Moreover, the herd size up to 50 lactating cow is expected to be kept in this farm which is demonstrated in Figure 2.

37°65′ N - 44°98′ E - 1332 m

Temperature : -7°C to 32°C (min -13°C -max 35°C)
Precipitation falls: 390 mm
The length of the day : 9:34 h at21Dec - 14:46 h at 20Jun
Cloud cover : 16% (mostly clear) to 62% (partly cloudy)
Relative Humidity: 26% to 94% (min 16% max 100%)
Wind speeds 0 - 7 m/s
Precipitation : 15% average chance (30May-19Sep)
Global Solar Radiation : 1800-1900 kWh/m2 annually

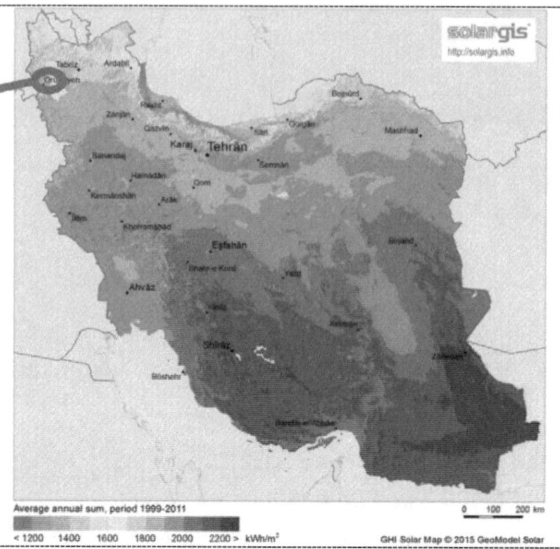

Figure 1. Case Study ; Site location and meteorological properties [3], Iran solar resources map [4].

2. Material and methods

Software simulation is a commonly used approach to design, examine and make decision about application of PV systems configurations. All grid connected systems can meet the load demand at any time making the technical feasibility assured for all grid connected configurations. The assessment structure in the present study is based on the simultaneously interaction between farm load demand profile and PV system configuration as the electricity flow between the system and the grid. TRNSYS software was adopted to develop dairy farm GRPV plant model. MATLAB and EXCEL were also used to post process the TRNSYS results, specifically the economic analysis. The technical performance parameters are also needed to define an overall performance evaluation of RGPV systems with respect to the energy production under the operating conditions. Two important indicators were implemented to provide judgment of plant performance in technical point of view; Capacity Factor (CF) defined by equation (1) and Capacity Value (CV) defined by equation (2). CV refers to the contribution of a power plant to reliably meet demand. Solar plants can be designed and operated to increase their capacity value either capacity factor.

$$CF = \frac{Final\ a.c\ Output\ Electricity\ (kWh)}{Nominal\ d.c\ Power\ (kW)} \tag{1}$$

$$CV = \frac{Met\ Demand\ Load\ by\ PV\ (kWh)}{Demand\ Load\ (kWh)} \tag{2}$$

Cash flows were calculated at the end of each year. Net Present Value (NPV) was calculated as equation (3) at the end life of project (20 years), the same period of guaranteed power purchase agreement. However the usual life time of PV plats is 25 years, this under estimation of life time would cause to better confidence level of modeling results. Where " Se " is annual PV electricity selling income , " Sa " is annual saving in electricity bill, " OM " is annual operation and maintenance cost, and " i " is the year index. Moreover, Internal rate of return (IRR) is defined as a discount rate that makes the NPV of all cash flows equal to zero. IRR calculations rely on the same formula as NPV does.

$$NPV_{20Y} = \sum_{i=1}^{20} \frac{Se_i + Sa_i - OM_i}{(1 + DR)^i} - II \quad (3)$$

2.1. Load profile modelling

Dairy farm energy consumption modelling is the first step in exploring the possible demand response and load compensation opportunities under the grid connected rooftop PV initiative. Generally, dairy farms use between 800 and 1200 kWh/cow-yr.

a) Farm Barns view

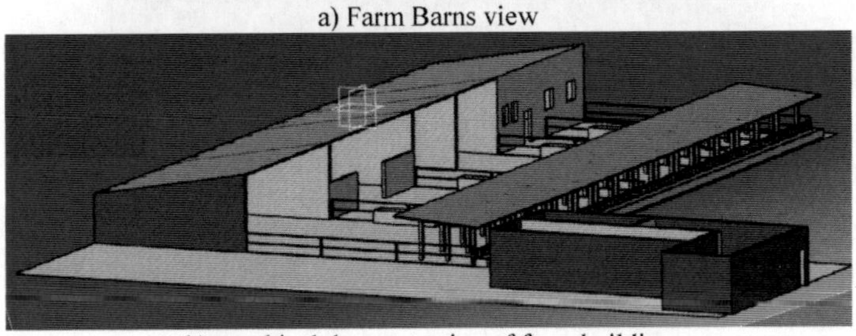

b) graphical demonstration of farm building

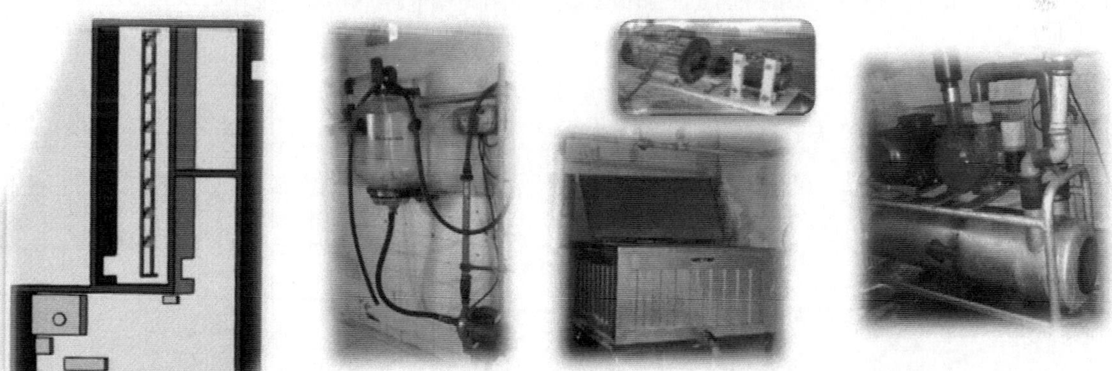

c) 8 stall Milking Parlour, Milk Pump, Milk cooling tank and its agitator, Vacuum Pump

Figure 2. Case studied dairy farm with electrical appliances

Moreover, Kythreotou et al. reviewed several researches which have been conducted in different countries and reported the annual energy consumption of dairy farms put in the wide range of 281-2900 kWh/cow/year [5]. Electrical energy consumption in dairy farms varies largely because of machinery, production systems, working habits and maintenance [6]. Other factors such as herd size, milk production capacity, management practices, and ambient conditions, can also have significant effects on overall system performance. This study is aimed to establish a mathematical model to represent and model the load profile of dairy farm based on the farm herd size, farm milking capacity and farm location. Farm electricity load profiling includes the details on the electrical appliances, its energy requirement, and consumption pattern. The model is generated based on the activity time scheduling depicted in Figure 3. Regarding the technical specification of farm equipments, presented in Table 1, the TRNSYS transient system simulation program was used to develop a model to overcome these limitations.

Table 1. Technical specification of electrical appliances demonstrated in Figure 2-c

Farm Activity	Electrical Appliance	Commercial Brand	Technical Specification
Milking	Vacuum Pump	CAMAK-AGM112M	4 kW - 1425 rpm
Milk Cooling	Compressor & Fan	Paykan	4.75 kW, Scroll Compressor
	Agitator	SYN MOTOR+Gearbox	0.75 kW - 1440 rpm
Pumping	Milk Pump	LOWARA-CEAM70/5/A	0.55 kW - 30~80 lit/min - 28.8~20.2 m
Artificial Lighting	Barn	Fluorescent	25× (2× *32 W) - supplying about 100 lux *
	Milking Parlor	Incandescent light bulb	8 × 100 W - supplying more than 500 lux
	Office	Incandescent light bulb	2 × 100 W - supplying about 500 lux

* Baased on the regulation [7] , not the current lighting system

TRNSYS modules were developed for all major milking center equipment components, including the milking unit, refrigeration unit, artificial lighting and pumping. Models were verified by comparing predicted results to monitored data. Simulation is performed by hourly solar resource and meteorological parameters profiles based on the Typical Meteorological Year (TMY) data set of Urmia. The simulated load profile is successfully validated against the EUI reported by literature.

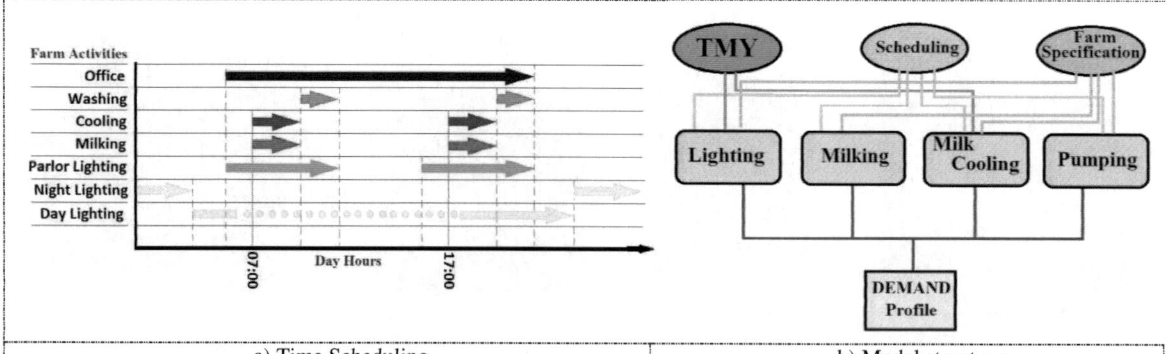

a) Time Scheduling b) Model structure

Figure 3. Dairy farm demand load profile modelling

2.2. Grid-connected rooftop PV system

Rooftops provide a large expanse of untapped area for solar energy generation, and onsite distributed generation could potentially reduce the costs and losses associated with the transmission and distribution of electricity [8]. The size of the RPV installation can vary dramatically and is dependent on the plant design purpose, which can be based on either of the following; attainable roof area, electricity demand, available funding, and the grid operator feed-in capacity. The variety of solar PV technologies are available running on a scale of efficiency, price, durability and flexibility, depending upon the need of projects. For most applications, a mono or polycrystalline solar PV solution is usually the best option, as these established technologies generally provide the right balance of price, efficiency and reliability.

Table 2. Techno-economic properties of the PV panels

Type	Model	Size (mm*mm*mm)	Efficiency (%)	Power (W_p)	Initial Cost ($/$W_p$)	BoS Cost ($/$W_p$)	O&M Cost ($/$W_p$)	Life Time (year)
Poly Crystalline cell	JAP6-240	1650×991×40 (1.64 m2)	16	240	0.68	1.14	0.0053	25

PV modules market is still relatively small and non maturated in Iran compared to global market. Most PV panels range in efficiencies of 13 to 16%, though some high-end model modules can reach percentages as high as 20%. Most of inverters receive benefit of the maximum power point tracker (MPPT) technology, making sure the solar PV modules generated DC power at their best power output at any given time during sunshine hours. The inverter efficiency value almost remained at the range of 0.8-0.9. According to the information gathered from Iran market and quote from informed opinion by screening the experience of local experts, the basic technical specification of PV module and inverter was selected to run simulation as presented in Table 2. The overall DC/AC de-rate factor of 0.83 was selected based on existing literature on the optimal sizing of inverters to minimize the cost of PV-generated electricity [9] [10] [11]. The long axis of the dairy barn is set in the north-south direction to have the maximum benefit of the sun as illustrated in Figure 2-a and Figure 2-b, imposing an azimuth angle of 95 degrees for PV panels. Buildings should be placed so that direct sunlight can reach the platforms, gutters and mangers in the cattle shed. Moreover, based on the pitch angle of 15 degrees for barn roof, it has been selected for tilt angle of PV panels.

2.3. *Local grid tariffs*

The electricity bill tariffs of the agriculture and livestock section in Iran are levelized with respect to the consumption seasons and day time. Two seasons of Cold (23September-19March) and Warm (20March-22September) are considered during the year, and three levels of Low, Mean and Peak Consumption were considered during the day as well, respectively with the tariffs of 0.391, 0.782 and 1.515 USD cent/kWh[1] for small scale consumers (less than 30 kW). For warm season low, mean and peak levels put into 24:00-7:00, 8:00-19:00 and 20:00-23:00, respectively. For cold season the low, mean and peak levels put into 22:00-5:00, 6:00-17:00 and 17:00-21:00, respectively. Moreover, in summer months (21 June to 21 September) these prices will increase by 20%. Most recently guaranteed electricity purchase FiT allocated to the end user consumers and limited to the connection capacity are 22.332 USD cents and 26.067 USD cents per each purchased kWh respectively for PV plants size up to 100kW and up to 20 kW for guaranteed period of 20 years.

3. Results and discussion

Cash flow and energy flow diagrams of the dairy farm is illustrated in Figure 4. January and August represent the cold and warm periods of the year, respectively. In top diagrams of Figure 4, income and bill as a function of day hours are monitored.

[1] With exchange rate of 1 USD=37,480 IRR in free markets on April 2017

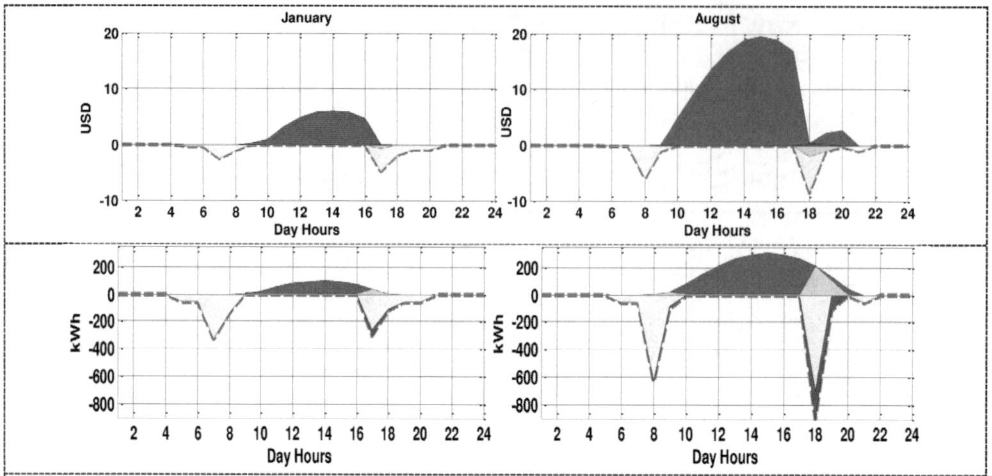

Figure 4. Monthly electrical energy and cash flows. Top diagrams: cash flows, Low diagrams: Electrical energy flows. Right side: August, Left side: January.

Regarding high tariffs and incentives for PV electricity in Iran, incomes surplus the bills wherein, regardless of RGPV, in January and August, 8 and 11 percent of the bills are covered by PV, respectively. This may be because of subsidies for the electricity in the country. However, the CV values for January and August are 8 % and 21 %, respectively. The different amounts of CV and bill coverage originated from the dependency of electricity price on time of use in Iran. On the other hand, income as a result of guaranteed PV buying, is calculated to be 584 $ and 2317 $ for January and August, respectively. Therefore, a farmer in Iran would prefer to sell the produced PV electricity rather than simply consuming it which endangers the policy of renewable energy strategy toward self sufficiency plans. Three major economical features, IRR, NPV and PP are analyzed in Figure 5. To get a more comprehensive perception of the problem, horizontal axis is segmented and scaled on both plant size and capital cost. The influence of incentive policy is now highlighted in the sense that the trends of economical features in plant sizes of less than 20 kWp are quite distinct from that of larger sizes.

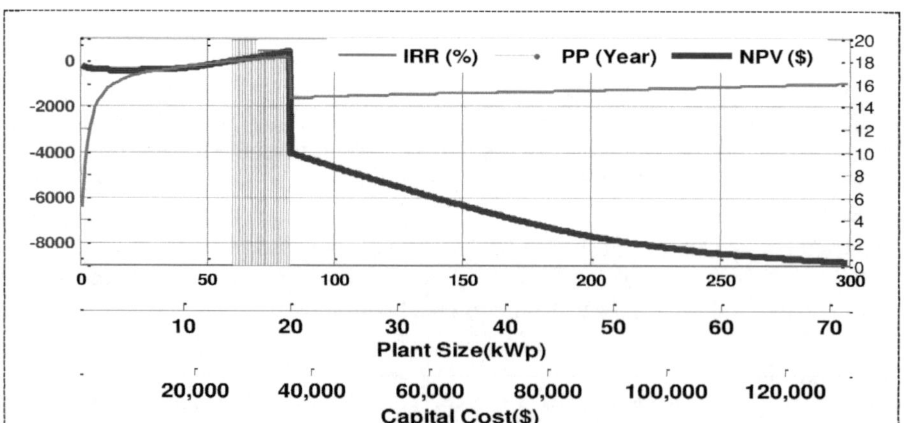

Figure 5. Economical features of GRPV plants respect to the number of PV panels, as well as, the plant size and capital cost in scaled horizontal axis. (Right vertical axe: NPV in USD, Left vertical Axe: IRR in % and PP in Year)

Base year is 2017 and calculations are based on 20-year life-time of the technology. Regarding 18% discount rate, in plant sizes of the range 14.4 kWp to 20 kWp, the utilization of PV would meet economic justification on the basis of IRR (Figure 5, red line). As shown in the figure, NPV is in accordance with IRR and corroborates the fact that up to 20 kWp plant size, the utilization of PV would be rational but as the plant size grows, there is no reason to invest on solar PVs in a dairy farm. The main point to be noticed in the results is that payback period (PP) is not attractive for an investor unless more incentive are employed.

4. Conclusion

The current study analyses the techno-economic impacts of grid connected rooftop PV system on the balance of a educational dairy farm located in cold climate north west of Iran. By means of TRNSYS simulation software, it has been shown that the GRPV plant with 89 PV modules of the nominal plant size of 20 kWp is the economically optimal solution. Up to 20 kWp plant sizes, economical feature including IRR, NPV and PP prove the reasonability of implementation of PV in dairy farms. Different size of PV plants with capacity factor of 1321 kWh/kWp, corresponding to the farmer initial investment was adopted to find the economical characteristics of net present value and payback period of investment after 20 years. Economical evaluation was carried out based on the year 2017, considering 18% discount rate and recent incentive renewable energy policies. Due to approved incentives already employed in Iran, justification of the renewable energy installation in dairy farms require meticulous and further investigations on policies.

5. References

[1] UNDP, "Energy Access Situation in Developing Countries: A Review Focusing on the Least Developed Countries and Sub-Saharan Africa,". (UNITED NATIONS DEVELOPMENT PROGRAMME web site, Access date;may 2017) 2013.

[2] M. Bey, A. Hamidat, B. Benyoucef, and T. Nacer, "Viability study of the use of grid connected photovoltaic system in agriculture : Case of Algerian dairy farms," *Renew. Sustain. Energy Rev.*, vol. 63, pp. 333–345, 2016.

[3] Spark, [Online]. Available:https://weatherspark.com, Access date April 2017.

[4] SOLARGIS, "Solar GIS," *2017*, 2017. [Online]. Available: http://solargis.com.

[5] N. Kythreotou, G. Florides, and S. A. Tassou, "A proposed methodology for the calculation of direct consumption of fossil fuels and electricity for livestock breeding , and its application to Cyprus," *Energy*, vol. 40, no. 1, pp. 226–235, 2012.

[6] M. Rajaniemi, M. Turunen, and J. Ahokas, "Direct energy consumption and saving possibilities in milk production," *Agron. Res.*, vol. 13, no. 1, pp. 261–268, 2015.

[7] INSO, "Dairy Cow Milking Parlor, Iranian National Standardization Organization 16027," , in Farsi, 2013.

[8] P. Gagnon, R. Margolis, J. Melius, C. Phillips, P. Gagnon, R. Margolis, J. Melius, and C. Phillips, "Rooftop Solar Photovoltaic Technical Potential in the United States : A Detailed Assessment Rooftop Solar Photovoltaic Technical Potential in the United States : A Detailed Assessment," 2016.

[9] P. Denholm, R. M. Margolis, S. Ong, and B. Roberts, "Break-Even Cost for Residential Photovoltaics in the United States : Key Drivers and Sensitivities Break-Even Cost for Residential Photovoltaics in the United States: Key Drivers and Sensitivities," *NREL Tech. Rep.*, no. December, 2009.

[10] A. Sharma, C. R. Chen, and N. Vu Lan, "Solar-energy drying systems: A review," *Renew. Sustain. Energy Rev.*, vol. 13, no. 6–7, pp. 1185–1210, 2009.

[11] S. Ravi Kumar, "A techno-economic analysis of a residential solar Photovoltaic system installed in 2010: A comparative case study between California and Germany." 2012.

IOP Conf. Series: Earth and Environmental Science **83** (2017) 012005 doi:10.1088/1755-1315/83/1/012005

An Intelligent Approach to Strengthening of the Rural Electrical Power Supply Using Renewable Energy Resources

F C Robert[1], G S Sisodia[2] and S Gopalan[1]

1 Department of Electronics and Communication Engineering, Amrita School of Engineering, Amritapuri, Amrita Vishwa Vidyapeetham, Amrita University, India.
2 College of Business Administration, Ajman University, Fujairah Campus, UAE
Emails: chidanand@am.amrita.edu, sundar@am.amrita.edu, g.sisodia@ajman.ac.ae

Abstract. The healthy growth of economy lies in the balance between rural and urban development. Several developing countries have achieved a successful growth of urban areas, yet rural infrastructure has been neglected until recently. The rural electrical grids are weak with heavy losses and low capacity. Renewable energy represents an efficient way to generate electricity locally. However, the renewable energy generation may be limited by the low grid capacity. The current solutions focus on grid reinforcement only. This article presents a model for improving renewable energy integration in rural grids with the intelligent combination of three strategies: 1) grid reinforcement, 2) use of storage and 3) renewable energy curtailments. Such approach provides a solution to integrate a maximum of renewable energy generation on low capacity grids while minimising project cost and increasing the percentage of utilisation of assets. The test cases show that a grid connection agreement and a main inverter sized at 60 kW (resp. 80 kW) can accommodate a 100 kWp solar park (resp. 100 kW wind turbine) with minimal storage.

1. Introduction

One of the main differences between developed and developing countries is the quality of electrical supply in rural areas. There are close to 1.2 billion people with no access to electricity. More than half are situated in Africa and about 40% in developing Asia [1]. When available, the quality of supply in rural areas is very low (voltage stability, reliability) [2], [3]. This is one of the causes for low economic activity and unemployment in rural areas and has led to the migration of rural population towards cities [4]. As the population migrates away from villages, the economic activity concentrates more and more in cities, giving less incentive for utilities to invest in rural distribution grids. In the context, if rural areas are provided with basic necessities, the economic growth of the developing country can be healthier and more balanced.

Local and distributed generation is expected to be well suited for powering rural areas. However, in many cases, renewable energy generation require government incentive to be economically viable. There are several ways to improve the economics of renewable energy projects. Besides the optimal location of the project, two complimentary strategies can be used: 1) optimising the design of power equipment by accepting a small percentage of the electricity curtailment, and 2) adding local storage. These combined methodologies lead to better utilisation of assets and a better integration of renewable energy project on the grid (See figure 1).

IOP Conf. Series: Earth and Environmental Science **83** (2017) 012005 doi:10.1088/1755-1315/83/1/012005

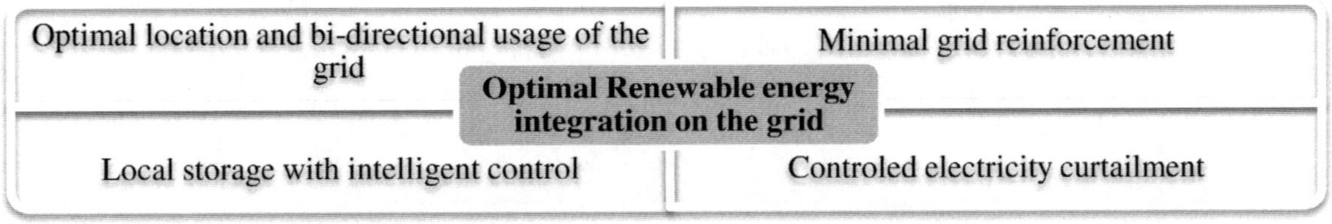

Figure 1. Optimisation of renewable energy integration

The optimal project design and grid integration solution depends on the renewable energy pattern, the cost of storage, the cost of power equipments, the feed-in tariff structure, and the grid reinforcement cost. The present methodology is valid in all situations but is especially beneficial when the grid extension is either costly or time consuming, which is the case in many developing countries where rural grids are often already used near their rated capacity and the grid reinforcement procedure slow.

2. Optimal renewable energy integration

2.1. Literature review

The reduction of losses is a very cost effective way to satisfy more demand without the need for additional generation capacity. It can be achieved by grid reinforcement, distribution automation, by improving the real time data availability on the distribution grid [5], by the use of smart transformers to control the voltage and power flow on the limiting distribution line [6] and by installing local generation closer to the load. With the help of local electricity generation, the transmission and distribution losses can be reduced from 30% [7] to 10% or less (losses in Italy are close to 6%). Such a gain in overall efficiency increases the value of local generation of the same percentage. Considering a low capacity distribution grid in a village, local generation has thus an increased value that can vary between 10% and 30% (losses in rural grid can be as high as 40% or more in developing countries).
Distributed generation can improve reliability and decrease the losses of a distribution system [8]. The preferred trade off between distributed generation investment and the gain in reliability and in reduction of losses provides the right incentives for the utility to invest. In [8] however, the cost of generation equipment is taken in a bulk manner and the size of inverter is not optimised. In addition, the optimal size of generation does not considers low capacity grid limitations which can be a constraint especially in developing countries. Adding local renewable energy on the distribution grid can also be used to improve the reliability and voltage profile [9]. Some of the recent work has focused on the optimal economical design of microgrids by comparing isolated generation with grid extension cost [10]. However the size of the grid connection capacity is rarely discussed and hybrid options have not been considered. In [11], on PV microgrid design cost optimisation, a simple sensitivity analysis shows that the cost of electricity reduces as the connection capacity with the main grid increases from 20% of the microgrid generation capacity up to 40%. However there is no simulation for a grid connection capacity close to 100%, the size of the inverter is also not discussed. The optimal level of connectivity to the main grid and the optimal size of inverter have to be established. The present study adresses these gaps.

2.2. Methodology

Existing methodologies focus on grid reinforcement and the optimal positioning of the new generation. In the scenario described in figure 2a, the wind turbine provides electricity to local loads, reducing the power flow and thus the losses in the main grid. Considering the base load at the end of the line, minor grid reinforcement (for 30 kW to 50 kW) towards the nearest transformer (or nearest greater capacity node) is sufficient to install a 65 kW wind turbine.

IOP Conf. Series: Earth and Environmental Science **83** (2017) 012005 doi:10.1088/1755-1315/83/1/012005

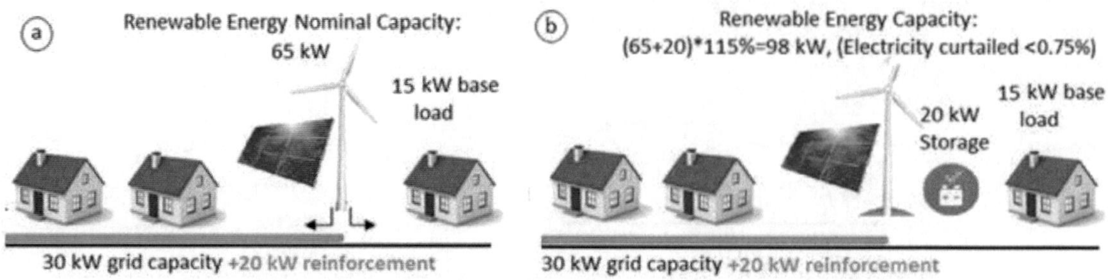

Figure 2. Model for increasing renewable energy integration on low capacity grids

The low capacity grid and the long delays in grid reinforcement may be a limiting factor for the fast development of renewable energy projects in rural areas (famous cases are in Tamil Nadu, India). However, wind and solar farms produce at their nominal value a small portion of the time. Indeed, solar panels produce power above 90% of their nominal value less than 4 hours per day at best. A wind turbine with an average power curve (nominal power at 12.5 m/s and 75% of nominal power at 10 m/s) generates above 75% of its nominal output around 15% of the time only (see table 1). In consequence, a static grid limiting factor may be too restrictive in many configurations. When allowing a small percentage of electricity curtailment (~0.75%), a wind farm of 100 kW can be connected to a grid of 85 kW capacity. 0.75% corresponds to a maximum instantaneous loss of 15% due to 15kW in excess at nominal generation, which is happening less than 15% of the time (see test scenario Section IV for more precise calculations). Thus, increasing grid export capacity above 85 kW may not be financially profitable.

Table 1. Frequency of occurrence of wind generation above 75% of the nominal power

Type of Wind distribution	Weibull K=2.08, Avrg. speed 7.4 m/s	Weibull K=1.74, Avrg. speed 7.7m/s
Percentage of the time where wind generation is above 75% of the nominal power	16.6%	12.1%

In addition to allowing electricity curtailment, local storage can be added to store the peak production and restore it during peak consumption. Storage can thus allow higher renewable energy generation to be connected to low capacity grids while at the same time reducing the nominal sizing of the main inverter. This type of utilisation of storage can be combined with other utilisations such as improving the grid stability or making profit from selling electricity during peak time, through multi-objective control algorithms.

A storage capacity of few hours at 20 kW can enable a wind turbine of 100kW to be connected to a grid of 50kW capacity as illustrated in figure 2b. In this configuration a small percentage of electricity may be curtailed (see Section IV for test case calculations). The storage size can be designed to ensure maximal cost efficiency; the saving in grid reinforcement, in inverter sizing and in energy sells are compared to the capital and maintenance cost of storage and the energy losses (see table 2). In addition storage can be used to improve the reliability of a microgrid in case of failure from the main grid. Most of storage can be made flexible and reused at other locations once the grid is reinforced [12].

Table 2. Benefits Vs cost of storage

Benefits of storage				Cost of storage	
Decreases the grid connection requirement	Decreases the required size for the inverter	Revenue from Peak time energy sells	Improved microgrid reliability	capital and maintenance cost	Round trip efficient: Losses

Solar parks and wind farm require an inverter to be connected to the grid, inverter may cost up to 20% of the total project cost. When reducing the required size of the grid connection with the method mentioned, the size of inverter can be decreased as well in the same proportions.

2.3. Calculations

The percentage of electricity curtailment for different cases of grid (and inverter) power limitations can be established based on the frequency distribution of the solar radiation (resp. wind speed), the power curve of the solar panel (resp. of the wind turbine) and the internal losses of the solar park (resp. wind farm). Two tests cases are presented in the following section, one with solar PV and one with wind turbines. In each case the frequency distribution of the renewable energy resource is taken for one location. The power curve of solar panel and wind turbine are then used to calculate the frequency distribution of the renewable electricity generation power. Two curves are drawn for each case, an ideal one, and one that considers different derating factors. The derating factors for solar PV are shown in table 3 and for wind turbines in table 4. The losses are integrated into the calculations using two types: 1) the losses of availability which impact the frequency of occurrence of any generation power by a uniform percentage, and 2) the other losses, which reduce the output power by the specified percentage.

Table 3. Derating of solar installation

Derating factors	Losses
Temperature derating	-0.5%/°C, above 25°C
Aging losses	1% to 2%/yr
DC wiring losses	2%
Dusting losses	5%
MPPT errors	1%
System availability	2%

Table 4. Derating of Wind Turbine Output

Cause of derating	Derating
Environment effect (dirt, insects)	0,5%
Wind turbulences	1%
Components misalignments	1%
Drive train degradation	1% /yr
Internal Cabling	1%
System availability	3%

A new frequency distribution of power output including the derating factors is then calculated. In this study, the calculations are performed for the first year only. From the frequency distribution of the generated power, it is possible to calculate the amount of electricity produced when the power output at the point of connection and the main inverter is limited (see equations (1) and (2)).

$$E_p = \int_0^{P_M} P_i * f_i \tag{1}$$

$$E_{limit} = \int_0^{P_{limit}} P_i * f_i + \int_{P_{limit}}^{P_M} P_{limit} * f_i \tag{2}$$

Where E_p is the total electricity produced, P_M is the nominal power of the generation equipment, P_{limit} is the maximal power that can be exported through the grid (and the inverter) and f_i the frequency of occurrence of each P_i.

The calculation for the amount of electricity curtailed is straight forward (see equation (3)).

$$L_{\%} = \frac{E_p - E_{limit}}{E_p} \tag{3}$$

Where $L_{\%}$ is the loss due to curtailment in percentage of electricity produced. The amount of electricity curtailment for different grid limitations and inverter power provides represent a valuable input for investors. They are thus able to conduct a financial analysis that compares the cost of grid reinforcement and of the inverter, with the decreased profit due to electricity curtailment. The following test cases demonstrate that significant gains can be achieved when following this approach.

3. Case study: Solar output frequency distribution

3.1. Solar PV in Tunis

The frequency distribution of the solar radiation of Tunis is taken for example [13]. Figure 3 illustrates the frequency distribution of the solar irradiance. The power output from a solar panel is almost proportional to the solar radiation. Assuming that PV panel produce at their nominal power for sun radiations at 1000W/m^2 and that an MPPT ensures the optimal output throughout the day, the frequency distribution of an ideal solar PV output is represented in figure 4.

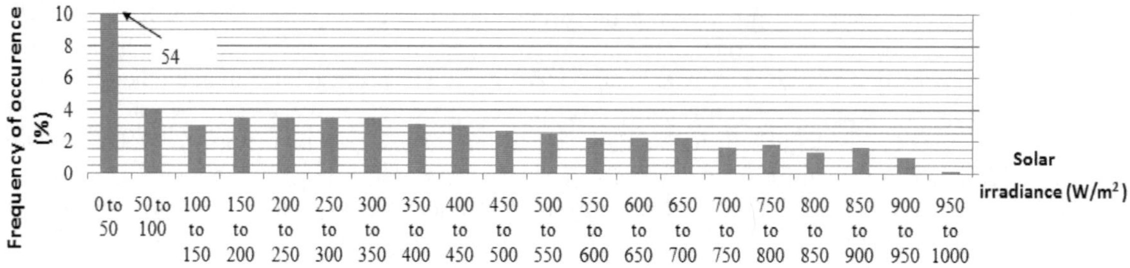

Figure 3. Frequency distribution of solar radiation in Tunis

A more realistic curve is also shown (see figure 4) when considering the derating factors mentioned in table 3. In sub tropical countries the temperature factor plays a significant role. In this study the cell temperature is taken to increase linearly from 25°C in absence of sun, up to 60°C when the sun reaches its peak at 1000 W/m^2.

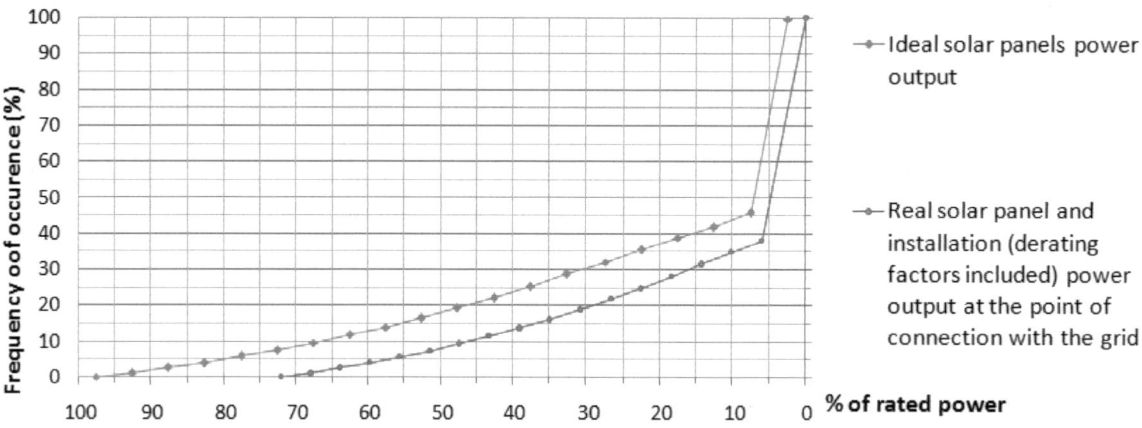

Figure 4. Frequency distribution of solar PV output above X% of its rated capacity

For the first year, the results show that the output of the solar installation almost never reaches above 70% of the rated power capacity of the solar panels. It reaches above 62% for only 2.7% of the time. After a few years, with a yearly derating of 1 to 2%, the output will rarely reach 60%. Thus, in this test case, the grid connection agreement as well as the main inverter can be designed at a rated power between 60% and 70% of the rated power of the solar panels, even without any storage. It is to be noted that the load factor of this project is 0.125, which is within the norm. Considering that the cost of inverter and grid connection represent above 20% of the total cost the project; these results validate that such approach to solar design can lead to significant gain for the developers and thus for the final customer.

3.2. Wind turbine in Tiruchirapalli

The Wind distribution at 10m height of Tiruchirapalli, India is taken for this second test case. It is very close to a Weilbull distribution k=1.83 with an average wind speed of 4.97 m/s [14]. The power curve of the 100 kW Nothern Power NPS100C-21 wind turbine is taken for this study (see figure 5). This wind turbine is installed at a standard hub height of 29 m. As wind speed increases non-linearly with height, the wind data are extrapolated. The wind distribution at 29 m height is calculated using the logarithmic rule with a roughness index of 0.01 corresponding to rough pasture (see figure 6). Figure 7 illustrates the frequency distribution of the power output above certain percentage of the rated power, in the ideal case and considering the derating factors as well.

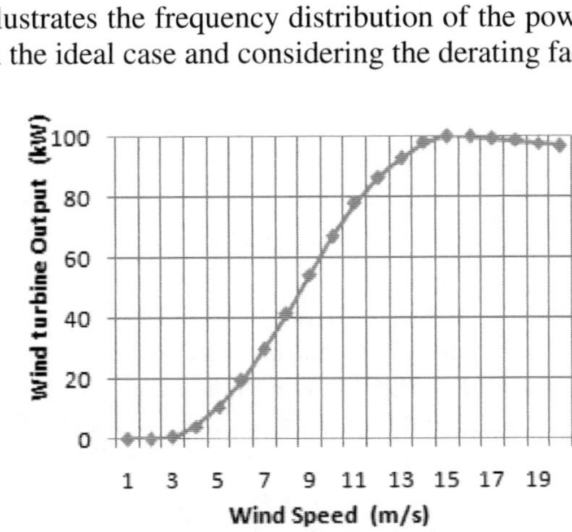

Figure 5. Power Curve of the 100 kW Nothern Power wind turbine

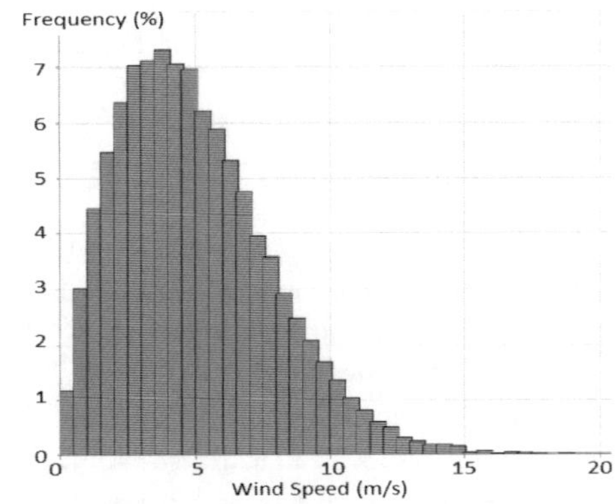

Figure 6. Wind Speed Distribution at Tiruchirapalli at 29 m height

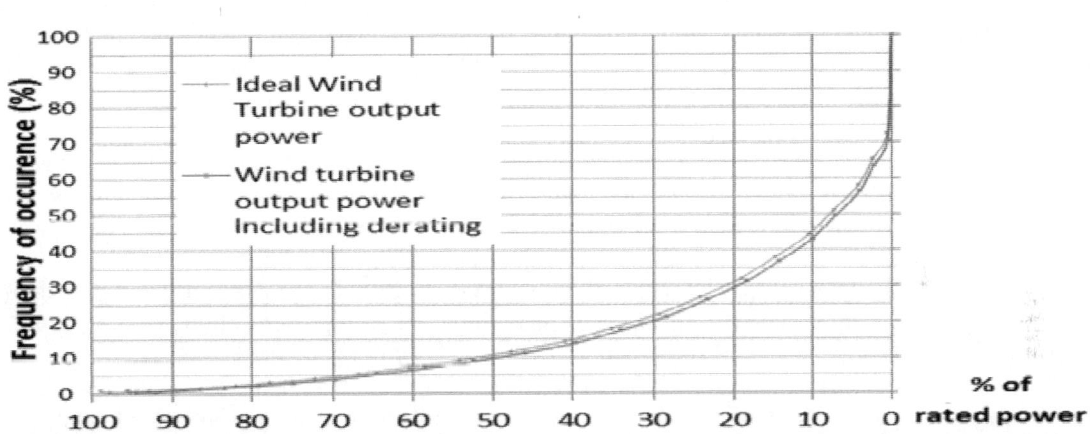

Figure 7. Frequency distribution of the wind turbine output power above X% of its rated power

The results show that the output power reaches beyond 90% only 1.2% of the time. It reaches beyond 80%, 2.7% of the time, and beyond 70%, 4.7% of the time. There is less potential gain when compared to solar due to the smaller derating factors. The following section provides an analysis of this data to quantifying the possible reduction in grid connection and inverter sizing for both the configurations.

3.3. Results and analysis

In both test scenarios it appears clearly that dimensioning the main inverter and the grid connection agreement with the utility, at the rated power of the generation equipment (either solar panels or wind turbines) may not lead to the financial optimal design. Indeed, even the first year of the solar operation, the solar output at the point of connection with utility rarely reaches above 70% of their rated capacity. This is mainly due to the temperature derating in a hot climate. The output is between 60% and 70% only 5% of the time. It is between 50 and 60%, 10% of the time. For the wind turbine test scenario, the derating has much less impact on the output power. The wind distribution is the main parameter to consider. In the case of Tiruchirapalli, the wind generation reaches beyond 90%, only 0,9% of the time; it is between 80% and 90%, 1.3% of the time, and between 70% and 80% 2.7% of the time.

The percentage of electricity curtailed (relative to the electricity produced) is calculated in both scenarios for various rated power of the inverter and the grid in table 5. In the solar (resp. wind) test case, dimensioning the grid connection agreement and inverter for 60% (resp. 80%) of the rated power of the panel (resp. wind turbine), will only lead to an average loss of 1,2% (resp.1%) due to curtailment for the first year.

Table 5. Percentage of electricity curtailment for various grid and inverter power limitations

	P_{limit} 90%	P_{limit} 80%	P_{limit} 70%	P_{limit} 60%	P_{limt} 50%	P_{limit} 40%
Solar Tunis	0%	0%	0,1%	1,2%	4,1%	8,8%
Wind Tiruchirapalli	0.2%	1%	2,5%	4,4%	6,3%	8,1%

To go one step further, a local storage with a maximal charge current at 10% of the rated capacity of the panels can be installed to absorb the peak production and lower further the inverter sizing requirement and grid connection sizing at 50% in the case of solar (resp. 70% in the wind energy case). This storage will be used only 5% (resp. 2% for wind) of the time (the power output is between 50% and 60% of the rated capacity only 5% of the time for solar). Such solution is cost effective in case the storage is considered as a shared resource to achieve other objectives such as supporting the grid during peak time, or provide back up for the microgrid and improve the reliability. In this case a multi-objective algorithm can be designed. As the derating increases with years, storage may not be needed anymore for this optimisation, making it available to achieve other objectives or for being used at other locations.

4. Conclusion

The development of many developing countries has been focused on urban areas. However the current weaknesses of the power system can be worked around. This article has illustrated that wind and solar energy generation can be dimensioned beyond the limiting grid capacity with minimal amount of electricity curtailment. The concept of using the available storage to limit the peak production has also been presented to improve further the grid integration of wind and solar projects. In hot climate such as in India, the benefits of this approach are magnified in case of PV technology due to the important derating of the solar generation related to the negative temperature coefficient.

A test scenario has shown that solar PV projects rarely produce at their rated capacity and thus do not require a grid connection at their rated capacity. When the maximum power is limited to 60% of the rated power of the solar panels, the losses represent only 1.2% of the generated electricity for the first year of operation. Considering a derating of at least 1%/yr (recent studies show an average degradation of 2%/yr in India), the required grid capacity agreement is thus close to 60% of the rated capacity of the solar panels. The test case with wind energy generation, shows that when the maximum power is limited to 90% of the rated power of the turbine, the curtailed electricity represents only 0.2% of the electricity produced. When storage is added and controlled by an intelligent algorithm, more solar or wind turbine can be installed for the same maximum output limitation. Both these methodologies render the solar and wind energy projects more appropriate and cost effective in rural areas.

In the course of this study it appeared very clearly that such methodology can be applied to any solar or wind project in order to optimise the cost effectiveness of its design. Considering that grid connection can account for 10% of the project cost and that inverter can account for 20% of the project cost, there is much possibility for optimisation of the design using this method. This article also presented the concept of using storage as a shared resource to achieve multi-objectives. More research is necessary to assess the potential benefits such approach.

Overall, this work provides a valuable input for researchers and renewable energy developers to optimize their design and the integration of renewable energy generation on the grid.

5. References

[1] 2016 *World Energy Outlook Electricity access database* OECD/IEA.

[2] Banerjee S G and Barnes D 2014 *Power for all: Electricity access challenge in India* (New York U.S.: World Bank Publications)

[3] Oyuke A, Penar P H and Howard B 2016 *off-grid or 'off-on': Lack of access, unreliable electricity supply still plague majority of Africans* Afrobarometer

[4] Tacoli C, Mc Granahan G and Satterthwaite D 2015 *World migration report* International Organization for Migration
[5] Nithin S, Sivraj P, Sasi K. K, and Lagerstom R 2014 *Power and Energy Systems: towards Sustainable Energy* (India: IEEE)
[6] Jishnu Sankar C V C, Haritha G and Nair M G 2016 *Int. Conf. on Energy Efficient Technologies for Sustainability* (India: IEEE)
[7] World Energy Council India and Power Grid Corporation of India Limited 2013 *Transmission & Distribution in India* Transmission and distribution in India a report (India: WEC-IMC and power grid corporation of India limited
[8] Moradi M, Khandani A 2014 *Evaluation economic and reliability issues for an autonomous independent network of distributed energy resources* International Journal of Electrical Power & Energy Systems **56** pp75-82
[9] Priya T, Sanjana V, Gohila B, Lavanya R, Anbazhagan A, Veerasundaram M et al. 2015 *Procedia Technology* **21** pp139-146
[10] Amutha W and Rajini V 2016 *Cost benefit and technical analysis of rural electrification alternatives in southern India using HOMER* Renewable and Sustainable Energy Reviews **62** pp236-246.
[11] Mao M, Jin P, Chang L and Xu H 2014 *Economic Analysis and Optimal Design on Microgrids With SS-PVs for Industries* IEEE Transactions on Sustainable Energy 2014 **5** (4) pp1328-36.
[12] Schreurs M, De Boer P, and Hooiveld R 2010 *2010 China Int. Conf. on Electricity Distribution* (ICED)
[13] El Ouderni A, Maatallah T, El Alimi S, Ben Nassrallah S 2013 *Experimental assessment of the solar energy potential in the gulf of Tunis, Tunisia* Renewable and Sustainable Energy Reviews **20** pp 155-68
[14] Sarkar A, Gugliani G and Deep S. 2017 *Weibull model for wind speed data analysis of different locations in India* KSCE Journal of Civil Engineering pp1-13

IOP Conf. Series: Earth and Environmental Science **83** (2017) 012006 doi:10.1088/1755-1315/83/1/012006

A Comparison of Prediction Methods for Design of Pump as Turbine for Small Hydro Plant: Implemented Plant

Hossein Naeimi[1], Mina Nayebi Shahabi[1], Sohrab Mohammadi[1]

1Water and Wastewater Co., Urmia, Iran
E-mail: hosseinnaimi@alumni.ut.ac.ir

Abstract. In developing countries, small and micro hydropower plants are very effective source for electricity generation with energy pay-back time (EPBT) less than other conventional electricity generation systems. Using pump as turbine (PAT) is an attractive, significant and cost-effective alternative. Pump manufacturers do not normally provide the characteristic curves of their pumps working as turbines. Therefore, choosing an appropriate Pump to work as a turbine is essential in implementing the small-hydro plants. In this paper, in order to find the best fitting method to choose a PAT, the results of a small-hydro plant implemented on the by-pass of a Pressure Reducing Valve (PRV) in Urmia city in Iran are presented. Some of the prediction methods of Best Efficiency Point of PATs are derived. Then, the results of implemented project have been compared to the prediction methods results and the deviation of from measured data were considered and discussed and the best method that predicts the specifications of PAT more accurately determined. Finally, the energy pay-back time for the plant is calculated.

1. Introduction

About 81% of the total primary energy supply in the world is obtained through fossil fuels. In 2014 about 32,381 million tons of CO_2 are emitted through the world. Iran relies mainly on fossil fuels based energy production due to high reserves of oil and natural gas in the country. Around 16.4% of the electricity generation in world is through hydro power [1]. At the moment In Iran, just about 0.34% of total primary energy supply is obtained through hydro power [2]. Conventional production with fossil fuels presents problems associated with the high cost, rapid depletion and detrimental environmental effects of these fuels. Renewable energy is probably the best solution. Due to rapid increase in energy consumption, the requirement of such alternatives for electricity generation has been increased.

It has been estimated that RE technologies can generate sufficient energy to fulfil all electricity demand in Iran by the year 2030 at a price level of 40.3– 45.3 €/MWh, depending on the sectoral integration [3]

Among all renewable resources, small hydropower (SHP) is one of the most promising sources of energy generation. In developing countries, small and micro hydropower plants are very effective source for electricity generation. The energy pay-back time (EPBT) and greenhouse gas (GHG) emissions for SHP generation system are less than other conventional electricity generation system [4]. So, encouragement of small hydropower schemes can solve the problem of energy crises of the country. Different countries are following different criteria to classify hydro power plants. Although definitions vary, The United States Department of Energy (DOE) defines A Classification of hydro power plants, as follows in Table 1 [5]:

IOP Conf. Series: Earth and Environmental Science **83** (2017) 012006 doi:10.1088/1755-1315/83/1/012006

Q	discharge, m^3/s	*Greek Symbols*	
H	head, m	φ	discharge number
N_s	specific speed, (m, m^3/s)	π	power number
D	impeller diameter, m	η	efficiency
BEP	best efficiency point	ψ	head number
g	gravitational acceleration, m/s^2	Subscripts	
n	rotational speed, rps	p	pump
N	rotational speed, rpm	t	turbine
PAT	pump as turbine		
P	power, kW		
GHG	Green House Gas		
O&M	Operation and Maintenance		

Table 1. Classification of hydro power plants

Type	Capacity
Large Hydro-Power	have a capacity of more than 30 megawatts
Small Hydro-Power	generate 10 MW or less of power
Micro Hydro-Power	have a capacity of up to 100 kilowatts

Using centrifugal pumps in reverse is one of the efficient alternatives for recovering Electricity through small and micro hydro powers. The concept of electricity generation through reverse running centrifugal pump is not new. Around 80 years ago, the research on this field had been started [6].
In this paper, some of the prediction methods of BEP of PATs are derived. Besides, the measured data of a small-hydro site implemented on the by-pass of a Pressure Reducing Valve (PRV) in Urmia city in Iran are presented. The aim is evaluation of deviations between the prediction methods and the measured data in order to identify the more accurate methods. The results of implemented project have been compared to the prediction methods results available in the literature and the best method that predicts the specifications of PAT more accurately determined. Finally, the energy pay-back time for the implemented small-hydro plant is calculated.

2. Typical turbines

Hydropower systems generate electrical energy by converting the energy of falling water to mechanical energy with a turbine and from mechanical to electrical energy by generators coupled to turbines.
The turbine is situated after the pipeline and can be either classified as low or high head. With high head systems normally using turbines such as Pelton wheels or Turgo runners, according to Western North Carolina Renewable Energy Initiative [7]. Low head systems typically use Francis, Kaplan or crossflow turbines to turn the generator. A rough guide to turbine choice is given in Fig.1 (a).

3. Pump as turbines

Using pump as turbine (PAT) is an attractive and significant alternative [8]. In such a system a pump is operated in reverse so that is functions as a turbine. This is especially popular in areas where the availability of turbines is limited as pumps are typically easier to get hold of [9].
Pumps are relatively simple and easy to maintain. They also have a competitive maximum efficiency when compared to conventional turbines [8]. The mass production of pumps means that they are comparatively much more cost-effective than conventional turbines [10].

3.1. Advantages and limitations of PAT

Pumps are relatively simple machines and available in a wide range of duties. They are easily installed, operated, maintained and repaired. Besides they are available at lower cost and cheap from economical point of view. And the Capital payback period of PATs in the range of 5–500 kW is two years or less [8] [11].
The main disadvantage of PAT Is that the characteristics curves in turbine mode are not usually supplied with the pump [12]. Besides they are not as well documented as turbines. Manufacturers do not provide the characteristic of the pump running in reverse. Pump operates in turbine mode with higher head and discharge at the same rotational speed [13].

4. PAT field applications and overview on prediction methods

Due to inadequate experimental data for pumps working as turbines, the field applications of these machines are not yet well defined [8].There are many different types of pumps that can be used as a turbine. Fig. 1 (b) gives the rough guide to aid the choice [14]. Many correlations based on theoretical approaches are available to predict the performance of a PAT. Several researchers have presented correlations for predicting the performance of a pump-as-turbine [20].

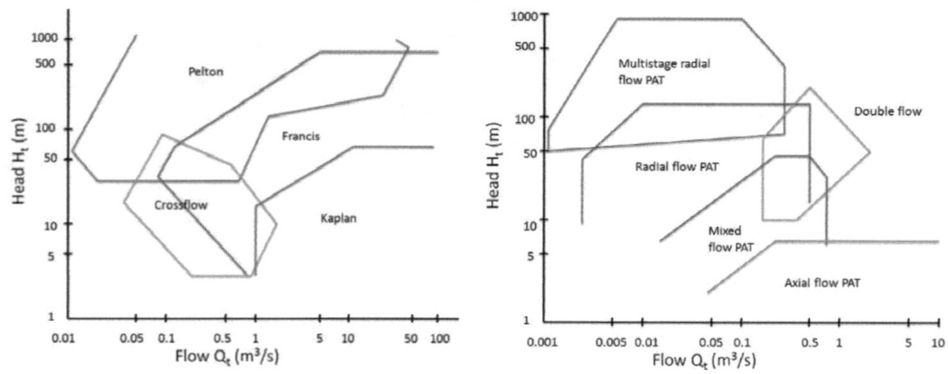

Figure 1. (a).Rough guide to turbine type operating ranges; (b) Choice of pumps for PAT [14])

In the following equations different methods of prediction pump reverse characteristics are presented. They are based on theoretical or experimental analyses. This equations calculate the head, flow rate at the BEP in reverse mode using the efficiency, head and flow rate value at the BEP in direct mode. Correlations are presented by following ($N_{st} = N_s.\eta_p$): Stepanoff: Equation (1) [15], Alatorre-Frenk: Equation (2) [19], Sharma: Equation (3) [17], Schmiedl: Equation (4) [21], Grover: Equation (5) [22], Hergt: Equation (6) [18] and Childs: Equation (7) [16].

$$\frac{H_t}{H_p} = \frac{1}{\eta_p};\ \frac{Q_t}{Q_p} = \frac{1}{\sqrt{\eta_p}};\ \eta_t = \eta_p;\ N_{st} = N_s \eta_p; \quad (1)$$

$$\frac{H_t}{H_p} = \frac{1}{0.85\eta_p{}^5 + 0.385};\ \frac{Q_t}{Q_p} = \frac{0.85\eta_p{}^5 + 0.385}{2\eta_p{}^{9.5} + 0.205};\ \eta_t = \eta_p - 0.03; \quad (2)$$

$$\frac{H_t}{H_p} = \frac{1}{\eta_p{}^{1.2}};\ \frac{Q_t}{Q_p} = \frac{1}{\eta_p{}^{0.8}};\ \eta_t = \eta_p;\ P_t = P_p\ ; \quad (3)$$

$$\frac{H_t}{H_p} = -1.4 + \frac{2.5}{\eta_p};\ \frac{Q_t}{Q_p} = -1.5 + \frac{2.4}{\eta_p{}^2};\ \frac{\eta_t}{\eta_p} = 1.158 - 0.265 N_{st}; \quad (4)$$

$$\frac{H_t}{H_p} = 2.693 - 0.0229 N_{st};\ \frac{Q_t}{Q_p} = 2.379 - 0.0264 N_{st};\ \frac{\eta_t}{\eta_p} = 0.893 - 0.0466 N_{st}; \quad (5)$$

$$\frac{H_t}{H_p} = 1.3 - \frac{6}{N_{st} - 3};\ \frac{Q_t}{Q_p} = 1.3 - \frac{1.6}{N_{st} - 5}; \quad (6)$$

$$\frac{H_t}{H_p} = \frac{1}{\eta_p};\ \frac{Q_t}{Q_p} = \frac{1}{\eta_p};\ \eta_t = \eta_p; \quad (7)$$

5. Field implementation Case study-Urmia Small Hydro Plant

The small-hydro site has been implemented on the by-pass line of a pressure reducing valve located on a pipeline which supply water to a water treatment plant in Urmia city, Iran Fig. 2 (a). A schematic is shown in Fig. 3. The flow rate and available head at the input of PRV were about 0.47 m^3/s and 85 m. Based on Fig. 1 (b), a double suction pump (TVP-300.500) with impeller diameter of 490 mm and specific speed 29.4 (m, m^3/s) was selected to install and operate as PAT. To measure the pressure a sensor ranging 0-25 bar was used and the flow rate was measured using an ultrasonic flowmeter at upstream of PAT. A 4-pole squirrel cage motor of 315 kW was coupled to the PAT with a synchronous speed of 1500 rpm. By means of a plc system, all the data are logged and valves are remotely controlled with electronic actuators and all the operation of the PAT can manually or automatically controlled by a homemade software program as shown in Fig. 2 (b).

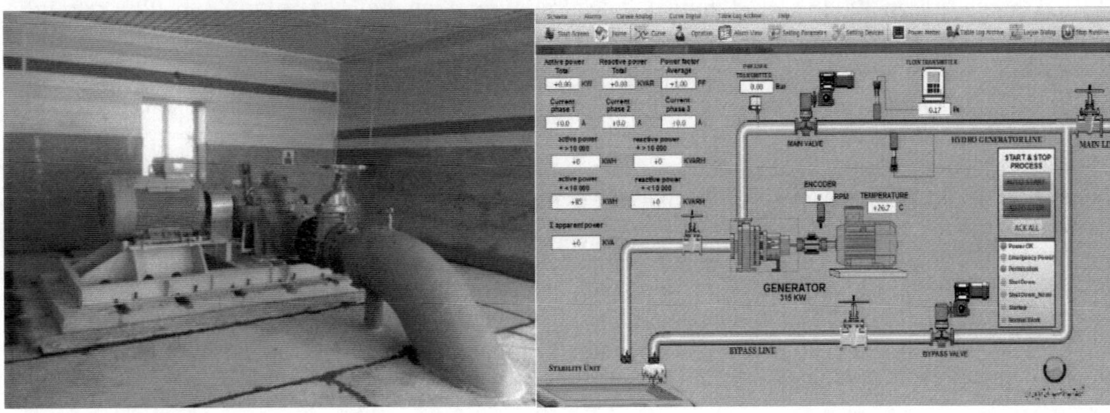

Figure 2. (a) PAT installed in Small-Hydro site, Urmia; (b) Software developed for monitoring and control of system

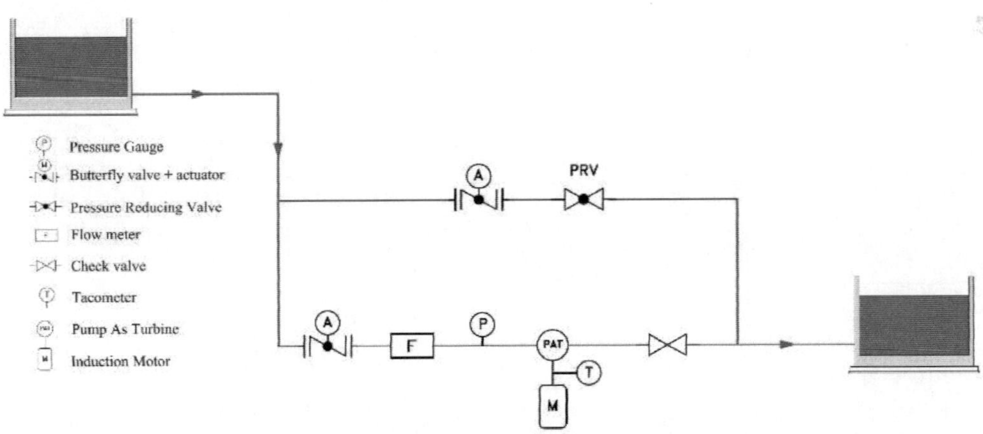

Figure 3. Schematic layout of implemented plant

6. Results

Before installation of PAT, a performance test was carried by the manufacturer in pump mode. Based on test sheet, the following results were derived at BEP of pump. Q_{pBEP}= 1406 m3/h, H_{pBEP}= 62.55 m, P_{pBEP}=334 kW and η_{pBEP}= 71.7 %. The results and comparisons at the dimensionless scales are shown in Fig. 4. As compared to pump operation, the pump operates at higher head and discharge values in turbine mode. The non-dimensional parameters are expressed as follows:

$$\psi = \frac{g.H}{n^2.D^2}, \phi = \frac{Q}{n.D^3}, \pi = \frac{P}{\rho.n^3.D^5} \quad (8)$$

7. Comparison of prediction methods

The results of implemented Small- hydro site have been compared to the results of prediction methods available in the literature as listed in Table 2. In Table 3 the deviations between the predictions of the methods and the measured data are presented.
It is easy to observe that large deviations are shown in methods of Alatorre-Frank and Grover for both head and flow rate values. The flow rate calculated using the methods of Stepanoff and Hergt are very close to field measurement. Also, the head value calculated using the methods of Stepanoff, Sharma, Schmiedl and Childs are close to those of measured by a margin of ±5%. The method of Hergt underestimates the head value 33.3% and the method of Schmiedl overestimates the flow rate value 37.5%. A prediction method is valid only if it accurately predicts both head ratio and flow rate ratio simultaneously for a given range of specific speeds. So, in this case, method of Stepanoff acceptably estimates both head and flow rate ratios for this centrifugal pump with N_s= 29.4 (m, m^3/s).

8. Payback time

In table 4 payback time Analysis is presented. The total cost for plant included installation, operation and maintenance are 76000$. So Energy Production Costs is calculated 0.038 ($/kWh). With regard to Purchase tariff for Small-Hydro Installation on the pipelines in Iran year 2016 [23] equal to 1500 IRRs/kWh and calculation of the total income, the Capital payback period of this PAT is less than one year. In addition, the total GHG emission savings, with average minimum of 775 g CO_2-equiv./kWh GHG Emissions (from Specific Energy Sources like Lignite, Hard coal, Oil, Industrial and Natural gas) [24] is calculated 1560 tons per year.

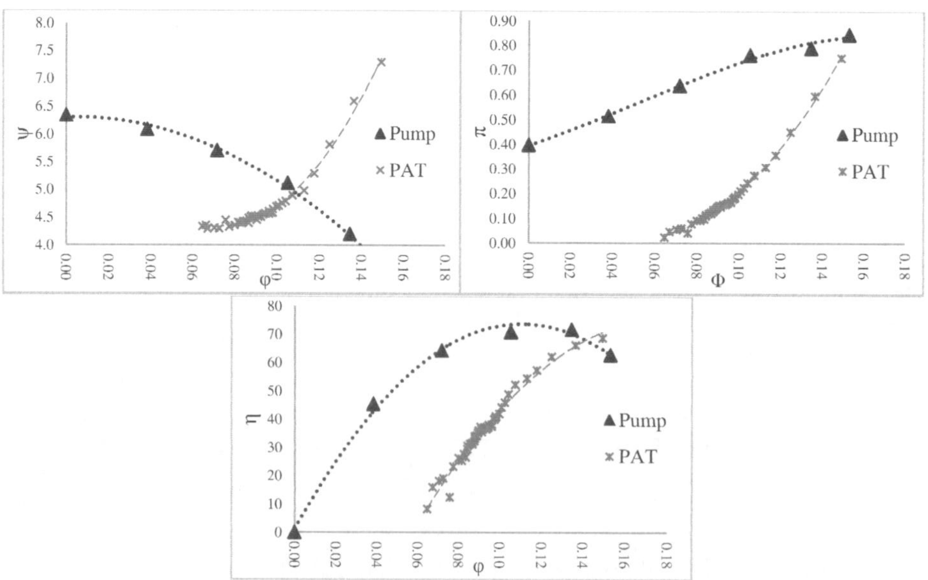

Figure 4. Dimensionless Head, Power and Efficiency curves in pump and turbine modes.

Table 2. Comparison between results of prediction methods and Measured Values in Field

METHODS	H_t (*m*)	Q_t(*l/s*)	η_t (%)
Measured in Field	90.7	463	68
Stepanoff	87.2	461	71.7
Alatorre-Frank	114.5	736	68.7
Sharma	93.2	510	71.7
Schmiedl	93.6	636	80.4
Grover	138.2	712	64.5
Hergt	60.6	469	-
Childs	87.2	545	71.7

Table 3. Prediction methods comparison with Measured Values in Field. Head, Capacity and Efficiency Percentage Errors

METHODS	Error-H (%)	Error-Q (%)	Error-η (%)
Stepanoff	-3.9%	-0.3%	5.4%
Alatorre-Frank	26.2%	59.1%	1.0%
Sharma	2.8%	10.2%	5.4%
Schmiedl	3.1%	37.5%	18.3%
Grover	52.4%	53.9%	-5.2%
Hergt	-33.3%	1.3%	-
Childs	-3.9%	17.7%	5.4%

Table 4. Energy Payback Time Analysis

Installation Cost *($)*	O&M Costs *($/y)*	Energy produced *(MWh/y)*	Energy Production Costs *($/kwh)*	Energy purchase tariff *($/kwh)*	Total Income *($/y)*	Payback Time *(Year)*	Emission Savings *(tons)*
72000	4000	2000	0.038	0.043	86000	≤1 years	1560

9. Conclusions

A centrifugal pump of specific speed 29.4 (m, m3/s) was installed in a SHP. As compared to pump operation, the pump operates at higher head and discharge values in turbine mode. The best efficiency in turbine mode was found 2.7% lower than best efficiency in pump mode. As a comparison in prediction methods in this study, method of Stepanoff acceptably estimates both head and flow rate values. Also, the payback time of this PAT is less than one year. However some uncertainties are still remains in prediction of turbine mode characteristics using pump operation data, future works and more experimental data can improve all methods.

10. References

[1] Key world energy statistics, International Energy Agency (IEA), 2016.
[2] Iran Energy Statistics, Iranian Minister of Energy, 2013.

[3] *The Role of a 100% Renewable Energy System for the Future of Iran: Integrating Solar PV, Wind Energy, Hydropower and Storage,* Lappeenranta University of Technology.

[4] I. K. Varun Bhat and Ravi Prakash, "Life cycle analysis of run-of river small hydro power plants in India," *The Open Renewable Energy Journal,* no. 1, pp. 11-16, 2008.

5] "Types of Hydropower Plants," The United States Department of Energy (DOE), [Online]. Available: https://energy.gov/eere/water/types-hydropower-plants.

[6] A. Tamm, A. Braten, B. Stoffel and G. Ludwig, "Analysis of a standard pump in reverse operation using CFD," in *20th International Association for Hydro-Environment Engineering (IAHR)*, Charlotte North Carolina USA, 2000.

[7] "Fact Sheet: Microhydro," Western North Carolina Renewable Energy Initiative, Appalachian State University, 2007.

[8] S. Derakhshan and A. Nourbakhsh, "Theoretical, numerical and experimental investigation of centrifugal pumps in reverse operation," *Experimental Thermal and Fluid Science,* no. 32, p. 1620–1627, 2008.

[9] B. h. Teuteberg, Design of a Pump-As-Turbine Microhydro System for an Abalone Farm, Department of Mechanical and Mechatronic Engineering, Stellenbosch University, 2010.

[10] S. Baumgarten and Guder, W, "Pump as Turbines," *Techno digest, KSB Aktiengesellschaft,* no. 11, 2005.

[11] A. S. Aidhen and Pratibha H. Gaikwad, "Pump As Turbine With Induction Generators In Pico Hydro For Electrification of High Terrain Areas: A Review," *International Journal of Current Engineering and Technology,* vol. 6, no. 5, 2016.

[12] A. Williams, "Pumps as turbines for low cost micro hydro power," in *Renewable Energy Conference*, Denver, USA, 1996.

[13] O. Paish, "Micro-hydro power: status and prospects," *Journal of Power and Energy, Part A,* vol. 216, pp. 31-40, 2002.

[14] J. M. Chapallaz, P. Eichenberger and G. Fischer, Manual on Pumps Used as Turbines. Volume 11, Deutsches Zentrum fur Entwicklungstechnologien GATE, 1992.

[15] A. Stepanoff, Centrifugal and Axial Flow Pumps, New York, USA: John Wiley, 1957.

[16] S. Childs, "Convert pumps to turbines and recover HP. Hydro Carbon Process. Pet. Refin," no. 41, p. 173–174, 1962.

[17] K. Sharma, "Small Hydroelectric Project-Use of Centrifugal Pumps as Turbines: Technical Report," Kirloskar Electric Co., Bangalore, India, 1985.

[18] T. Wang, F. Kong, K. Chen, X. Duan and Q. Gou, "Experiment and analysis of effects of rotational speed on performance of pump as turbine.," *Trans. Chin. Soc. Agric. Eng,* no. 32, pp. 67-74, 2016.

[19] Alatorre-Frenk, Cost Minimization in Micro Hydro Systems Using Pumps-Asturbines. Ph.D. Thesis,, Coventry, UK: University of Warwick, 1994.

[20] H. Nautiyal and A. K. Varun, "Reverse running pumps analytical, experimental and computational study: A review.," *Renew. Sustain. Energy Rev,* no. 14, pp. 2059-2067, 2010.

[21] E. Schmiedl, "Serien-Kreiselpumpen im Turbinenbetrieb; Pumpentagung," Karlsruhe, Germany, 1988.

[22] K. M. Grover, "Conversion of Pumps to Turbines," GSA Inter Corp., Katonah, NY, USA, 1980.

[23] "Guaranteed elec. Purchase tariff for Small-Hydro Installation on the pipelines,Guaranteed Feed in Tariffs(FiTs)," Renewable Energy Organization of Iran, [Online]. Available: http://www.suna.org.ir/suna_content/media/image/2016/09/4815_orig.pdf.

[24] R. Dones, C. Bauer, R. Bolliger, B. Burger, M. Faist, R. Frischknecht, T. Heck, N. Jungbluth and A. Röder, Life Cycle Inventories of Energy Systems: Results for Current Systems in Switzerland and other UCTE Countries, Duebendorf, Switzerland: Paul Scherrer Institut, Villigen and Swiss Centre for Life Cycle Inventories, 2003.

IOP Conf. Series: Earth and Environmental Science **83** (2017) 012007 doi:10.1088/1755-1315/83/1/012007

Investigation of waste heat recovery of binary geothermal plants using single component refrigerants

M. Unverdi [1]

1 Department of Mechanical Engineering, Sakarya University, Sakarya, Turkey
E-mail: muratunverdi@sakarya.edu.tr

Abstract. In this study, the availability of waste heat in a power generating capacity of 47.4 MW in Germencik Geothermal Power Plant has been investigated via binary geothermal power plant. Refrigerant fluids of 7 different single components such as R-134a, R-152a, R-227ea, R-236fa, R-600, R-143m and R-161 have been selected. The binary cycle has been modeled using the waste heat equaling to mass flow rate of 100 kg/s geothermal fluid. While the inlet temperature of the geothermal fluid into the counter flow heat exchanger has been accepted as 110°C, the outlet temperature has been accepted as 70°C. The inlet conditions have been determined for the refrigerants to be used in the binary cycle. Finally, the mass flow rate of refrigerant fluid and of cooling water and pump power consumption and power generated in the turbine have been calculated for each inlet condition of the refrigerant. Additionally, in the binary cycle, energy and exergy efficiencies have been calculated for 7 refrigerants in the availability of waste heat. In the binary geothermal cycle, it has been found out that the highest exergy destruction for all refrigerants occurs in the heat exchanger. And the highest and lowest first and second law efficiencies has been obtained for R-600 and R-161 refrigerants, respectively.

1. Introduction

Renewable and eco-friendly energy sources are becoming more and more important every day. Especially with decreasing amount of fossil fuels, this importance also affects energy conversion systems using renewable energy sources (geothermal, wind, solar) which are less harmful to the environment than fossil fuels (petroleum, coal, natural gas). Recovery of waste heat generated by energy conversion systems using renewable energy sources also has an impact on system performance. The performance of an energy conversion system can be improved by the addition of equipment or of systems with different energy conversion techniques. Due to low temperature of waste heat in such systems, working fluid of choice is also very important for energy conversion. Due to low temperature of waste heat, working fluid should provide the necessary conditions at lower temperature than that of waste heat. In the literature, refrigerants are preferred in the recovery of waste heat at low temperature. We can summarize some studies using these fluids as follows [1]-[3]. Satanphol et al. interested heat recovery alternative for low grade heat. The types of fluid, the composition and the operating conditions that achieved the maximum net work output were determined through flowsheet modeling and optimization in Aspen Plus v.8.4 simulation software. They used single component refrigerants in Organic Rankine Cycle (ORC). Among the group of pure working fluids in their study, the ORC using R-227ea provided the best performance in terms of net work output [4]. Zeyghami studied performance of the combined flash-binary geothermal power cycle for geofluid temperatures between 150 and 250°C. A thermodynamic model is developed, and the suitable binary working fluids for different geofluid temperatures are identified from a list of thirty working fluid candidates, consisting environmental friendly refrigerants and hydrocarbons. Thirty environmental friendly fluids, including refrigerants and hydrocarbons are selected and evaluated to identify suitable working fluids, which yield highest system efficiency. The results show that for low-temperature heat sources using refrigerants as binary working fluids result in higher overall cycle efficiency and for medium and high-temperature resources, hydrocarbons are more suitable. For combined flash-binary cycle, secondary working fluids; R-152a, Butane and Cis-butane show the best performances at geofluid temperatures 150, 200 and 250°C respectively [5]. Astolfi et al. investigated the potential of ORC (organic Rankine cycles) for the exploitation of low-medium enthalpy geothermal brines. A Matlab code was created in order to define the optimal combination of fluid, cycle configuration and cycle parameters. An extensive thermodynamic analysis is performed considering geothermal sources in the temperature range of 120-180°C. All the assumptions for calculating the plant components performance are set on the basis of data from literature and real power plants data sheets. Thermodynamic optimization results, shown in terms of reduced variables, allow defining some general rules for the selection of the optimal combination of working fluid and cycle configuration [6]. Brown et al. investigated working fluids for organic Rankine cycle (ORC) applications with a goal of identifying "ideal" working fluids for five renewable/alternative energy sources. A wide range of "theoretical" working fluids are investigated with the goals to identify potential alternative working fluids and to guide future research and development efforts of working fluids. . The study suggests a working fluid's critical temperature and its critical ideal gas molar heat capacity have the largest impact on the cycle efficiency and volumetric work output, with "ideal" working fluids for high efficiency possessing critical temperatures on the order of 100%-150% of the source temperature and possessing intermediate values of critical ideal gas molar heat capacity [7]. Tchanche et al. presented various Rankine cycle architectures for single fluids and other improved versions operating with ammonia/water mixture. Waste heat resources and their potential for driving organic Rankine cycles have been outlined. The nature state and temperature of the heat source significantly influences the choice of the type of organic Rankine cycle machine. Potential of these sources and temperature have been revealed, and led to enormous potential for recovery using organic Rankine cycles and other technologies. Characteristics of a module were recorded: heat source temperature, power output, thermal efficiency, etc. The maximum thermal efficiency of ORC is found close to 25%. The selection of an ORC will be primarily based upon application, heat source temperature and desired power output [8]. Wang et al. examined the relationships between the critical parameters of working fluids and the Rankine cycle performances under defined condensing pressure. The results revealed that the relationship between critical temperature and thermal efficiency for pure fluid would depend on the initial operating conditions. For the binary mixtures, the trend of critical pressure with the quality was basically opposite to that of evaporator irreversibility rate. Therefore, the critical properties could be considered as predictive index for selecting working fluids to achieve a preferred efficiency for pure fluids and reduce the evaporator irreversibility rate for mixture fluids [9]. This study modeled a binary geothermal power plant for waste heat recovery in Germencik Geothermal Power Plant and investigated system performances for 7 different single component refrigerants. The terms energy and exergy were used in the availability of the waste heat for the designed model, and energy and exergy efficiencies were obtained for the refrigerants. In addition, exergy destructions and net power productions for the equipment which is effective in the system performance were obtained from the model.

2. Aydın Germencik geothermal area

Aydin Germencik geothermal area is a high-temperature field located in the west of Menderes Graben and within the boundaries of Allangullu-Omerbeyli. The reservoir temperatures of the field are between 200°C and 215°C. The highest reservoir temperature in the field is 232°C. The total surface area of the field is 50 km^2. A power generating plant with a net capacity of 47.4 MWe has been established by Gurmat Electricity Company since March 2009. Due to its characteristics, it is a steam-dominant field and a double-flash steam power cycle is used in power generation due to high reservoir temperature. There are 6 generation wells and 7 reinjection wells in the field. Geothermal fluid at an average of 210°C and 650 kg/s flow rate is used in the injection wells for power generation at the power plant. A water-cooled condenser is used for the condensation of the gases which cannot be condensed at the power plant. The geothermal fluid at 110°C, which completes its cycle in the power plant, is transferred to the reinjection wells. More detailed information on the power plant is given in Reference [10].

3. Waste heat recovery and binary geothermal power plant model

This study investigated the availability of the waste heat for recovery purposes before the geothermal fluid, which completes its cycle in Germencik Geothermal Power Plant, is transferred to the reinjection wells. A simple binary cycle was used for the availability of the waste heat. 7 different R-134a, R-152a, R-227ea, R-236fa, R-600, R-143 and R-161 single-component refrigerants were used in the cycle model. Binary geothermal power plants consist mainly of pumps, heat exchangers (evaporators), turbines and condensers. Figure 1 shows the binary geothermal power plant cycle used in this study. The operating conditions for the binary geothermal power plant proposed for waste heat recovery at Germencik Geothermal Power Plant were determined under certain assumptions.

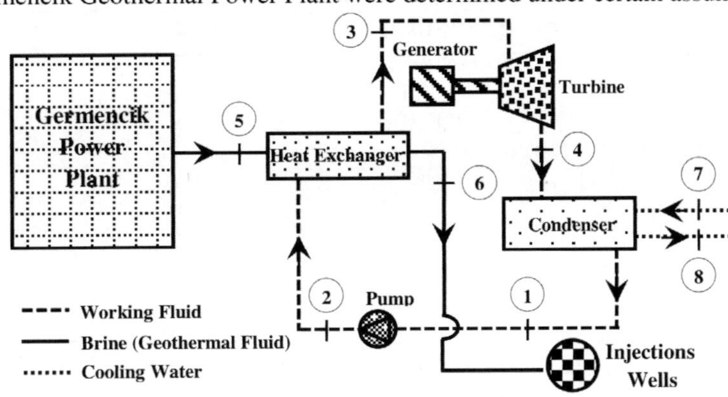

Figure 1. Schematic diagram of Organic Rankine cycle.

The actual operating conditions of the power plant were also taken into consideration for the determination of the operating conditions for the proposed cycle. In this way, the cycle was investigated in similar situations and under similar thermodynamic conditions applied to real binary power cycles. In the proposed binary geothermal power plant, geothermal fluid used as waste heat is completely re-ejected. Geothermal fluid is a low-temperature reservoir and in the liquid phase during the power cycle. Geothermal fluid (waste heat) is assumed to enter the evaporator at 100 kg/s flow rate, T=110°C (state number: 5) and under P=160 kPa and to be 70°C (state number: 6) by the time it exits the evaporator. In addition, the thermal efficiency of the heat exchanger is assumed to be 90%. The circulation of refrigerants (working fluids) is based on Organic Rankine Cycle (ORC), which is a closed cycle. The working fluid enters the pump in the saturated liquid phase at T=20°C (state number: 1) and is assumed to be sent to the evaporator under P=900 kPa (state number: 2).The isentropic efficiency of the pump is assumed to be 75%. The working fluid exists the evaporator at T=85°C and under 900 kPa (state number: 3) as superheated steam. Having 90% isentropic efficiency and being in the superheated vapor phase, the working fluid is sent to the water cooled condenser after it generates power in the turbine. The cooling water is assumed to enter the water-cooled condenser at T=10°C (state number: 8) and under P=500 kPa and exit the condenser at T=18°C (state number: 7). Under these conditions, such properties required for 7 different refrigerants as the mass flow rate of the working fluid, amount of cooling water, net power generated and cycle efficiencies were calculated.

3.1. Selected working fluid

Refrigerants are very suitable secondary fluids for binary geothermal cycles. Selection of a suitable working fluid plays an important role in determining system performance. In the evaporator and condenser in a binary cycle, refrigerants transfer the heat from a place to another. Selection of appropriate refrigerants largely depends on the initial investment and system operating costs as well as fulfilling heat transfer expectations. On the other hand, refrigerants' climatic damages to the ozone layer and environment should also be taken into consideration. Another important environmental factor is global warming. The fact that refrigerants generate greenhouse gases should also be taken into account to make sure that refrigerants of choice contribute to the global warming as little as possible. Since refrigerants are chemical substances, they can also adversely affect human health. The main harmful characteristics of refrigerants are that they are toxicity and flammable. In applications, the toxicity and flammability characteristics of refrigerants are very important. Although low toxicity and low flammability are the desired properties in refrigerants, flammability is less important if there is no ignition in a given system.

Table 1. Physical properties of 7 single working fluids and their some thermodynamic properties.

Type	Semi-Empirical ODP	GWP (100 year)	Critical Temperature (K)	Critical Pressure (kPa)
R-134a	0	1,430	374.2	4059
R-152a	0	124	386.4	4520
R-227ea	0	3,220	376	2990
R-236fa	0	9,810	398.1	3200
R-143m	0	-	377.9	3635
R-161	0	12	375.3	5010
R-600	0	4	425.1	3796

To briefly summarize the refrigerants used in this study; R-134a, R-152a, R-227ea, R-236fa, R-143m and R-161 are hydrofluorocarbon (HFC) refrigerants while R-600 is a hydrocarbon (HC) refrigerant. HFC refrigerants are commonly used in refrigeration systems. HFC refrigerants do not harm the ozone layer, however, potentially contribute to global warming. HFCs are extremely safe in terms of flammability and toxicity. HCs, natural and non-toxic refrigerants, have low global warming potential and do not harm the ozone layer. Having high efficiency, HC refrigerants are eco-friendly, however, the only problem they have is that they are flammable. Table 1. shows some thermophysical properties of the refrigerants used in this study [11]-[16].

4. Analysis

4.1. Balance equations of mass, energy and exergy

Mass, energy and exergy analyses were carried out in the binary geothermal power plant examined for the availability of waste heat in Germencik Geothermal Power Plant. During the analyses, energy and exergy values for each point in the flow diagram (Figure 1.) were calculated using the formulas defined in thermodynamics [17-20]. The assumptions made in the calculations are as follows. (a)The power plant is at the steady-state conditions. (b) Potential and kinetic energy exchange has been neglected. (c) Heat transfer from the confines of the system (adiabatic) has been neglected. (d) Reference temperature and pressure in the model are 278 K and 101.3 kPa respectively. (e) Friction and pressure drop losses have been neglected. (f) Kinetic exergy, potential exergy and chemical exergy changes have been neglected. (g) The Engineering Equation Solver (EES) program was used for the calculations. (h) The thermophysical properties of geothermal water have been taken as steam_nbs, and therefore the chemical properties of non-condensable gases and other geothermal waters have been neglected.

The overall mass balance equation at continuous flow condition is as follows.

$$\sum_{i=1}^{i=n} \dot{m}_i = \sum_{i=1}^{i=n} \dot{m}_e \tag{1}$$

Where $\dot{m}$ represents the mass flow rate, and subscript i and e represent input and output respectively.

Energy balance is represented by Equation (2).

$$0 = \dot{Q}_{cv} - \dot{W}_{cv} + \dot{m}_i h_i - \dot{m}_e h_e \tag{2}$$

Energy balance of the power plant at continuous flow condition is given in Equation (2) and subscript cv, i and e represent control volume, input and output conditions, respectively. Q, W and h are enthalpy terms for net heat transfer, net work flow and unit volume, respectively. The first law of thermal systems (energy) efficiency is

$$\eta_I = \dot{W}_{net}/\sum \dot{E}_i \tag{3}$$

as described in Equation (3) and $\dot{E}_i = \dot{m}_i(h_i - h_0)$ obtained. Exergy is expressed as in

$$\dot{X} = \dot{X}_{PH} + \dot{X}_{KH} + \dot{X}_{PT} + \dot{X}_{CH} \tag{4}$$

Equation (4) where in subscripts PH, KN, PT, CH refer to physical exergy, kinetic exergy and potential exergy and chemical exergy, respectively. Specific exergy ψ and exergy X expression for the plant is as follows,

$$\dot{X} = \dot{X}_{PH} = \dot{m}\psi = \dot{m}[(h - h_0) - T_0(s - s_0)] \tag{5}$$

are defined as. Exergy destruction of system can be defined as follow:

$$\dot{X}_{dest} = \sum \dot{X}_i - \sum \dot{X}_o \tag{6}$$

The second law (exergy) efficiency and exergy losses of organic Rankine power plant are expressed as

$$\eta_{II-plant} = \dot{W}_{net}/\sum \dot{X}_i \tag{7}$$

$$\dot{I} = \sum \dot{X}_i - \dot{W}_{net} \tag{8}$$

Equations (7)-(8).

5. Results and discussion

This study investigated the availability of waste heat in Germencik Geothermal Power Plant and analyzed the performance of 7 different single-component refrigerants in the simple binary geothermal power plant. Mass, energy and exergy balance equations were solved under the same inlet conditions to verify calculations for 7 different refrigerants. Selection of an appropriate working fluid has a great effect on the performance of binary geothermal power plants. For this reason, single-component refrigerants were preferred for the power plant given in Figure 1. The refrigerants analyzed in our study are R-134a, R-152a, R-227ea, R-236fa, R-600, R-143 and R-161. These refrigerants were selected and analyzed as they were suitable for the operating conditions of the power plant. The critical temperature and pressure of the selected refrigerants were taken into account and the inlet

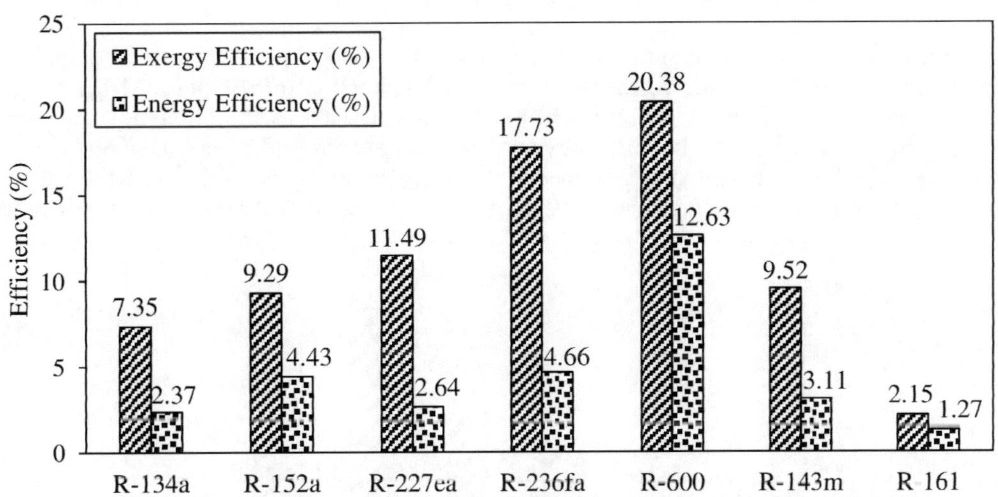

Figure 2. Energy and exergy efficiencies for the refrigerants investigated.

temperature and pressure of the working fluid in the power plant were selected as 85°C and 900 kPa, respectively. In addition, the freezing point of the refrigerants is below the ambient temperature, which

IOP Conf. Series: Earth and Environmental Science **83** (2017) 012007 doi:10.1088/1755-1315/83/1/012007

may cause solidification during the operation of the system. As also stated in section 3, the selected refrigerants work under the conditions given in the binary cycle. Owing to these reasons, this study investigated the performance of 7 different refrigerants with different characteristics. In this study, the dead state temperature and pressure are 278 K and 101.325 kPa, respectively. EES software program was used to calculate the properties of the geothermal water, working fluids (refrigerants) and cooling water. Energy and exergy efficiency values of the cycle for the binary geothermal power plant were calculated in Figure 2, which shows that R-600 obtained the best energy and exergy efficiency. R-236fa also showed similar results to R-600 in terms of exergy and energy. R-161 refrigerant, on the other hand, showed the worst energy and exergy performance. Although R-600 ($\dot{m}$= 24.51 kg/s) and R-161($\dot{m}$= 25.63 kg/s) refrigerants circulated in the cycle at similar mass flow rates, R-600 performed better than R-161. This is due to the thermophysical properties of R-161 refrigerant under the selected operating conditions for the cycle.

Figure 3 shows the exergy destructions for various components in the power plant.

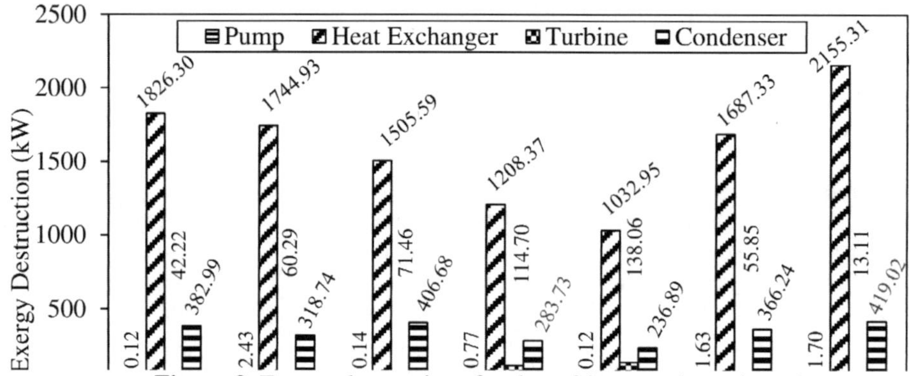

Figure 3. Exergy destructions for the refrigerants investigated.

The results indicate that the highest exergy loss occurred in the heat exchangers and that, in the cycle, the other exergy losses were lower than that in the heat exchangers, which shows that heat exchangers play an important role in determining the performance of a power plant. Net work output is an important parameter affecting the performance of binary geothermal power plants. High net work output improves the power plant performance. Net work is defined as the difference between the generated and consumed work. In this study, the net work is the difference between the work generated in the turbine and the work consumed in the pump. Figure 4 shows the net work output for the refrigerants studied. According to the results of the analysis, R-600 had the highest net work

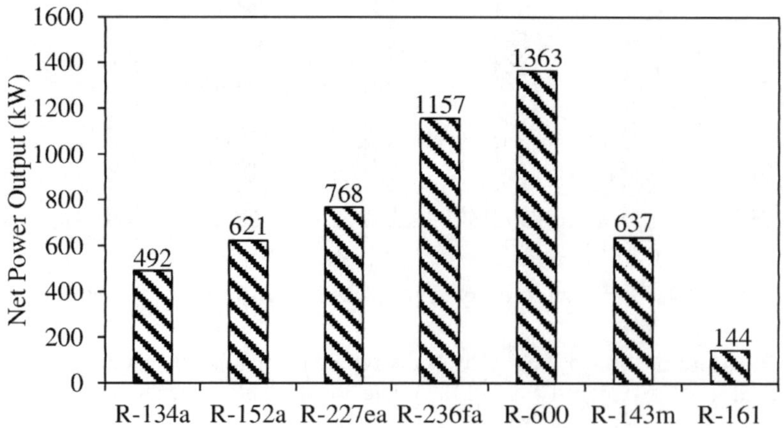

Figure 4. Net power output for the refrigerants investigated.

followed by R-236fa, R-227ea, R-143m, R-152a, R-134a and R-161 refrigerants.

6. Conclusions

It is a known fact that working fluids used in binary geothermal power plants, inlet temperature and pressure of waste heat geothermal fluid, re-injection temperature of geothermal fluid and environmental conditions have an effect on power plant performance. The analyses indicate that the heat exchanger (evaporator) has a great impact on the performance of the power plant. Increased exergy losses in the heat exchanger reduced the energy and exergy efficiency. It was observed that R-161 refrigerant had low energy and exergy efficiency and the lowest net work output, while it caused the maximum exergy destruction in the heat exchanger. In the exact opposite situation is obtained for R600 refrigerant. It is, therefore, important to take exergy destructions in the heat exchanger into account when selecting a suitable working fluid. The selected working conditions for the power plant in question also significantly affect net power generation and the losses in the power plant equipment. According to the results of analysis, the working fluid mass flow rates obtained for R-161 and R-600 refrigerants are very close to each other. While there was maximum exergy loss in R-161 refrigerant in the heat exchanger, the net power generation was the lowest. This is due to the small difference in enthalpy between the turbine inlet and outlet, which is one of the operating conditions of the designed power plant. In the exact opposite situation, the net power generation increased as the difference in enthalpy between the turbine inlet and outlet of R-600 refrigerant was higher than the operating conditions. In geothermal binary power plants, energy efficiencies range from 1 to 14% while exergy efficiencies range from 8 to 55%. The exergy efficiency was higher than the energy efficiency in the geothermal binary power plant in question which uses low-temperature geothermal resources as energy source. These results are consistent with the literature [21, 22]. Due to their good performance, single-component refrigerants are preferred in ORC power plants worldwide in general. Irem [23] and Dora-I,II [24], [25] binary geothermal power plants in Aydin, Turkey, operate according to an ORC and use n-pentane fluid as a secondary working fluid. N-pentane was not selected for this study as it did not meet the operating conditions of the power plant in question. In this study, R-600 refrigerant showed the best performance depending on the operating conditions of the power plant. Therefore, R-600 can be the refrigerant of choice in similar ORC power plants. However, air and environmental pollution, global warming and ozone depletion should also be taken into consideration when selecting a suitable working fluid. In addition, the flammability and toxicity values of an appropriate working fluid should be as low as possible. Selection of a suitable working fluid for geothermal power plants should be based on the optimum thermodynamic, thermophysical and environmental properties. Considering all the factors; high energy and exergy efficiencies, reasonable operating conditions, low ozone depletion rate, global warming potential, non-toxicity and non-flammability, it is not easy to find an ideal working fluid.

7. References

[1] Ratlamwala TAH, Dincer I, Gadalla MA. Performance analysis of a novel integrated geothermal-based system for multi-generation applications. *Applied Thermal Engineering* 2012; 40:71-9.

[2] Wang, X., Liu, X., Zhang, C., Parametric optimization and range analysis of Organic Rankine Cycle for binary-cycle geothermal plant. *Energy Conversion and Management* 80 (2014) 256–265.

[3] Hsieh, J.C.,Fu, B.R.,Wang, T-W., Cheng, Y., Lee, Y.R., Chang, Y.C., Design and preliminary results of a 20-kW transcritical organic Rankine cycle with a screw expander for low-grade waste heat recovery. *Applied Thermal Engineering* 110 (2017) 1120–1127.

[4] Satanphol, K., Pridasawas, W., Suphanit, B., A study on optimal composition of zeotropic working fluid in an Organic Rankine Cycle (ORC) for low grade heat recovery. *Energy* 123 (2017) 326-339.

[5] Zeyghami, M., Performance analysis and binary working fluid selection of combined flash-binary geothermal cycle. *Energy* 88 (2015) 765-774.

IOP Conf. Series: Earth and Environmental Science **83** (2017) 012007 doi:10.1088/1755-1315/83/1/012007

[6] Astolfi, M., Romano, M. C., Bombarda, P., Macchi, E., Binary ORC (organic Rankine cycles) power plants for the exploitation of mediumelow temperature geothermal sources e Part A: Thermodynamic optimization. *Energy* 66 (2014) 423-434.

[7] Brown, J.S., Brignoli, R., Quine, T., Parametric investigation of working fluids for organic Rankine cycle applications. *Applied Thermal Engineering* 90 (2015) 64-74.

[8] Tchanche, B.F., Pétrissans, M., Papadakis, G., Heat resources and organic Rankine cycle machines. *Renewable and Sustainable Energy Reviews* 39 (2014) 1185–1199.

[9] Wanga, T., Zhanga, Y., Wanga, P., Pengb, Z., Shua, G., Prediction of the roles of critical properties for pure and binary mixture working fluid in Rankine cycle performances. *Energy Procedia* 66 (2015) 57 – 60.

[10] Unverdi, M., Cerci, Y., Performance analysis of Germencik Geothermal Power Plant. *Energy* 52 (2013) 192-200.

[11] *Montreal Protocol on Substances that Deplete the Ozone Layer: 2006 Report of the Rigid and Flexible Foams Technical Options Committee (FTOC); 2006 Assessment* (PDF), Nairobi, Kenya: United Nations Environment Programme (UNEP) Ozone Secretariat, March 2007, ISBN 978-92-807-2826-2, retrieved 2010-12-16.

[12] P. Forster; V. Ramaswamy; P. Artaxo; T. Berntsen; R. Betts; D.W. Fahey; J. Haywood; J. Lean; D.C. Lowe; G. Myhre; J. Nganga; R. Prinn; G. Raga; M. Schulz; R. Van Dorland (2007). *"Chapter 2: Changes in Atmospheric Constituents and in Radiative Forcing"*.

[13] Global *Warming Potentials of ODS Substitutes*. Science - Ozone Layer Protection. US EPA. 2007. Archived from the original on 2010-10-16. Retrieved 2010-12-16.

[14] https://en.wikipedia.org/wiki/List_of_refrigerants (access date: 04.01.2017)

[15] Solomon, S., Wuebbles, D., Isaksen, I., Kiehl, J., Lal, M., Simon, P., Sze, N.-D., Ozone Depletion Potentials, Global Warming Potentials, and Future Chlorine/Bromine Loading, *Scientific Assessment of Ozone Depletion*: Chapter 13, 1994.

[16] Daniel, J.S., Velders, G.J.M., Douglass, A.R.,Forster, P.M.D., Hauglustaine, D.A., Isaksen, I.S.A., Kuijpers, L.J.M., McCulloch, A., Wallington, T.J., Halocarbon Scenarios, Ozone Depletion Potentials, and Global Warming Potentials. *Scientific Assessment of Ozone Depletion*: Chapter 8, 2006.

[17] Soundararajan K, Ho H. K, Su B, Sankey diagram framework for energy and exergy flows. *Applied Energy* 2014; 136:1035–1042.

[18] Kotas TJ. *The Exergy Method of Thermal Plant Analysis*. Krieger Publishing Company Malabar, Florida 1995.

[19] Cengel YA, Boles MA. *Thermodynamics, An Engineering Approach 5th edition*. McGraw-Hill College, Boston, MA, 2006.

[20] Bejan A, Tsatsaronis G, Moran M. *Thermal Design and Optimization.* John Wiley & Sons, Inc. United States of America, 1996.

[21] DiPippo R. *Geothermal Power Plants, Principles, Applications, Case Studies and Environmental Impact, Third Edition*. Butterworth-Heinemann Elsevier; 3 editions, 2012.

[22] DiPippo R. Second Law assessment of binary plants generating power from low-temperature geothermal fluids. *Geothermics* 2004; 33:565–586.

[23] Unverdi, M., Cerci, Y., Thermodynamic analysis and performance improvement of Irem Geothermal Power Plant in Turkey: A Case Study of Organic Rankine Cycle. *Environmental Progress & Sustainable Energy*. (Decision State).

[24] http://www.megeelektrik.com.tr/tr/mege/sirket-profili (access date: 04.01.2017)

[25] Ganjehsarabia, H., Gungor, A., Dincer, I., Exergetic performance analysis of Dora II geothermal power plant in Turkey. *Energy* 101-108 (46), 2012.

IOP Conf. Series: Earth and Environmental Science **83** (2017) 012008 doi:10.1088/1755-1315/83/1/012008

Study of a Combined Power and Ejector Refrigeration Cycle with Low-temperature Heat Sources by Applying Various Working Fluids

S Jafarmadar[1] and A Habibzadeh

1 Mechanical Engineering Department, Urmia University, Urmia, Iran.
E-mail: s.jafarmadar@urmia.ac.ir, a.habibzadeh@urmia.ac.ir

Abstract. A power and cooling cycle which combines the organic Rankine cycle and the ejector refrigeration cycle supplied by waste heat energy sources is discussed in this paper. Thirteen working fluids including one wet, eight dry and four isentropic fluids are studied in order to find their performances on the combined cycle. First and second law analysis has been performed by using a computer program in order to investigate various operating conditions' effects on the proposed cycle by fixing power/refrigeration ratio and varying waste heat source and evaporator temperature. According to the results, in general, dry and isentropic ORC fluids have better performance compared with wet fluids. The increase in evaporator temperature leads to the decrease in exergy efficiency. On the other hand, exergy efficiency rises with the turbine inlet temperature decrease and an increase of heat source temperature. Rising expansion ratio and inlet temperature of the turbine causes an increase in the thermal efficiency of the cycle.

1. Introduction

In recent years, scientist and engineers have tried to find more efficient power systems to reduce environmental problems such as atmospheric pollution, acid precipitation, ozone depletion and global warming. Low-temperature heat sources, such as waste heat and renewable energies (geothermal energy and solar energy) exist in the considerable quantities. Due to these reasons, exploring combined power and refrigeration cycles which use such low-grade heat sources, has attracted more and more attention.

A novel combined power and ejector refrigeration cycle was proposed by Dai et al. [1]. The cycle combined the Rankine cycle and the ejector refrigeration cycle by adding a turbine between the boiler and the ejector. The vapor from the boiler could be expanded through the turbine to generate power, and the turbine exhaust can drive the ejector.

Wang et al. [2] studied this cycle with R123. According to the results, the biggest exergy destruction occurs in the heat recovery vapor generator; it can be reduced by increasing the area of heat transfer and the coefficient of heat transfer in the HRVG. For more performed investigation related to ORCs and combined power and ejector, refrigeration cycles see [3]-[7].

In the present study, a combined power and refrigeration cycle is proposed to produce both power and refrigeration by utilizing different working fluids and a low-grade heat source. This cycle combines the ORC and the ejector refrigeration cycle. First and second law analysis is conducted to compare different working fluids and different working conditions.

2. Cycle operation and assumptions

In this study, the waste heat is used as the heat source to simulate the combined power and ejector refrigeration cycle shown in Figure 1. To simplify the modeling of the combined cycle, the following assumptions are made:

- The system runs in a steady state.
- The kinetic and potential energies, as well as friction losses, are neglected.
- Vapor generator, evaporator, turbine, ejector, and condenser are assumed adiabatic.
- The expansion valve process is at constant enthalpy (isenthalpic).
- The working fluid at the evaporator outlet is saturated vapor.
- The outlet state from the condenser is saturated liquid.
- A temperature difference of 10 K is assumed between state 2 and state 9.
- A temperature difference of 10 K is assumed between state 4 and state 16.
- A temperature difference of 10 K is assumed between state 13 and state 14.

The base case conditions for the simulation of the combined cycle are summarized in Table 1.

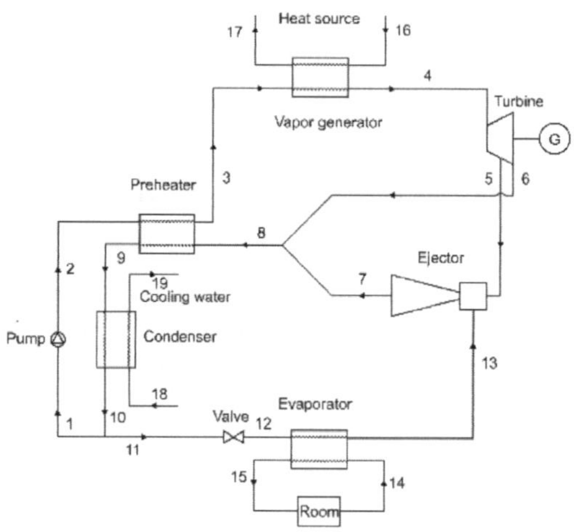

Figure 1. Schematic diagram of the combined power and ejector refrigeration cycle.

Table 1. Physical, safety and environmental data for analyzed fluids

Environment temperature (K)	298.15
Environment pressure (kPa)	101.325
Turbine inlet pressure (Mpa)	0.6
Turbine inlet temperature (K)	373.15
Turbine extraction pressure (MPa)	0.2
Extraction ratio	0.35
Turbine isentropic efficiency (%)	85
Pump inlet temperature (K)	293.15
Pump isentropic efficiency (%)	80
Evaporator temperature (K)	263.15
Heat source mass rate (kg/s)	75
Power refrigeration ratio	2.5
Cooling water mass rate (kg/s)	20
Cooling water inlet temperature (K)	288.15

3. Choice of working fluids

One of the main concerns for choosing a working fluid is its environmental effects. Ozone depletion potential (ODP), global warming potential (GWP) and the atmospheric lifetime (ALT) are three important factors that should be regarded. Fortunately, most working fluids used in the ORC cycle can be used in the ejector refrigeration cycle.

4. Thermodynamic analysis

4.1. Energy analysis

$$\dot{Q}_{con} = \dot{m}_9(h_9 - h_{10})$$

$$\dot{Q}_{eva} = \dot{m}_{12}(h_{13} - h_{12})$$

$$\dot{Q}_{ph} = \dot{m}_1(h_3 - h_2)$$

$$-\dot{W}_p = \dot{m}_1(h_2 - h_1)$$

$$\eta_p = \frac{h_{2s} - h_1}{h_2 - h_1}$$

$$\dot{W}_t = \dot{m}_4(h_4 - h_5) + (\dot{m}_4 - \dot{m}_5)(h_5 - h_6)$$

$$\dot{m}_5 h_5 + \dot{m}_{13} h_{13} = \dot{m}_7 h_7$$

$$\eta_t = \frac{h_4 - h_6}{h_4 - h_{6s}}$$

$$R_{extr} = \frac{\dot{m}_5}{\dot{m}_4}$$

$$\beta = \frac{P_4}{P_5}$$

$$\dot{Q}_{vg} = \dot{m}_3(h_4 - h_3) \qquad \dot{W}_{net} = \dot{W}_t + \dot{W}_p$$

4.2. Exergy analysis

Energy efficiencies provide neither information of how nearly the performance of a system approaches ideality nor the reversibility aspects of the thermodynamic processes. To determine more meaningful efficiencies, a quantity which provides a measure of an approach to an ideal is required. Thus, exergy efficiency must be introduced. Exergy destruction equations for condenser, ejector, evaporator, preheater, pump, expansion valve, turbine, and vapor generator are as follows:

$$\dot{I}_{con} = \dot{E}_9 + \dot{E}_{18} - \dot{E}_{10} - \dot{E}_{19} \qquad \dot{I}_{eje} = \dot{E}_5 + \dot{E}_{13} - \dot{E}_7$$

$$\dot{I}_{eva} = \dot{E}_{12} + \dot{E}_{ref} - \dot{E}_{13} \qquad \dot{I}_{ph} = \dot{E}_2 + \dot{E}_8 - \dot{E}_3 - \dot{E}_9$$

$$\dot{I}_p = -\dot{W}_p + \dot{E}_1 - \dot{E}_2 \qquad \dot{I}_{exv} = \dot{E}_{11} - \dot{E}_{12}$$

$$\dot{I}_t = \dot{E}_4 - \dot{E}_5 - \dot{E}_6 - \dot{W}_t \qquad \dot{I}_{vg} = \dot{E}_3 + \dot{E}_{16} - \dot{E}_4 - \dot{E}_{17}$$

$$\dot{I}_{tot} = \dot{I}_{con} + \dot{I}_{eje} + \dot{I}_{eva} + \dot{I}_{ph} + \dot{I}_p + \dot{I}_{exv} + \dot{I}_t + \dot{I}_{vg}$$

4.3. Efficiency

$$\eta_{th} = \frac{\dot{W}_{net} + \dot{Q}_{eva}}{\dot{Q}_{vg}} \qquad \eta_{exe} = \frac{\dot{W}_{net} + \dot{E}_{ref}}{\dot{E}_{in}}$$

4.4. Entrainment ratio

The performance of an ejector is evaluated by its entrainment ratio, which is defined as the mass flow rate ratio of the secondary fluid to that of the primary fluid:

$$\mu = \frac{\dot{m}_{sf}}{\dot{m}_{pf}}$$

5. Validation

Based on the above analysis, a simulation program using EES software [8] for the combined ORC and ejector refrigeration cycle was developed. Obtained solution is validated with the results of Dai et al. [1] in Table 2 in which R123 was selected as the working fluid which shows a very good agreement.

Table 2. Validation of the numerical model with the previously published data

Parameter	This work	Dai et al.
Generating temperature (K)	413	413
Condensing temperature (K)	293	293
Evaporating temperature (K)	263	263
Heat input (kJ/kg)	1263	1246.96
Entrainment ratio	0.396	0.389
Pump work (kJ/kg)	3.45	3.45
Turbine work (kJ/kg)	115.8	114.14
Net work (kJ/kg)	112.35	110.69
Refrigeration capacity (kJ/kg)	61.61	60.44
Thermal efficiency (%)	13.77	13.72
Exergy efficiency (%)	22.53	22.2

6. Results and discussion

The detailed data of the analyzed cycles for 13 different working fluids are listed in Table 3.

Table 3. Comparison of the combined power and refrigeration cycle with 13 different working fluids

Element	Unit	R123	R124	R134a	R141b	R142b	R152a	R227ea	R236fa	R245fa	R600	R600a	R601a	RC318
μ	-	0.1364	0.1692	0.1549	0.1124	0.1384	0.1219	0.2576	0.1995	0.1574	0.1708	0.191	0.1792	0.2805
$\dot{m}$	kgs^{-1}	2.049	4.075	6.507	1.678	3.133	4.587	6.41	4.04	2.716	3.151	4.05	2.498	5.473
$\dot{I}_{con}$	kW	6.506	10.02	14.5	6.722	9.632	14.66	12.23	9.921	8.698	14.34	15.09	12.65	9.982
$\dot{I}_{eje}$	kW	8.495	14.95	19.37	10.8	12.55	18.8	23.25	5.811	5.372	8.091	10.41	22.07	8.591
$\dot{I}_{eva}$	kW	0.2373	0.3796	0.6818	0.07173	0.4053	0.6525	0.6363	0.9225	0.4603	2.183	0.8372	0.498	0.6151
$\dot{I}_{exv}$	kW	0.2648	0.36	0.4311	0.2262	0.3447	0.4292	0.5734	0.4726	0.3761	0.7698	0.9279	0.7215	0.5796
$\dot{I}_{p}$	kW	0.1821	0.202	0.03708	0.1805	0.2182	0.1087	0.2445	0.272	0.2398	0.5336	0.5406	0.5275	0.3013
$\dot{I}_{ph}$	kW	1.415	9.043	41.92	1.356	5.888	28.64	23.33	8.215	3.874	10.14	18.35	4.992	12.34
$\dot{I}_{t}$	kW	9.857	9.857	11.29	10.34	11.34	14.24	10.79	11.29	11.9	24.3	24.32	24.1	10.67
$\dot{I}_{vg}$	kW	17.57	38.86	94.8	14.43	76.51	97.4	89.11	62.28	37.08	51.75	77.96	46.53	62.88
$\dot{Q}_{eva}$	kW	21.22	22.34	25.59	23.44	25.71	32.27	24.47	25.46	26.97	55.21	55.12	54.63	24.18
$\dot{Q}_{in}$	kW	426.6	701.1	1318	458.7	740.4	1448	855.6	707.9	622.1	1357	1551	1059	694.9
$\dot{W}_{t}$	kW	53.04	55.86	63.98	58.6	64.27	80.69	61.17	64	67.41	138	137.8	136.6	60.44
$\dot{W}_{p}$	kW	0.9105	1.01	0.1854	0.9027	1.091	0.5436	1.223	1.36	1.199	2.668	2.703	2.638	1.507
$\dot{W}_{net}$	kW	52.13	54.85	63.8	57.7	63.18	80.14	59.95	62.64	66.22	135.4	135.1	133.9	58.94
Refrigeration exergy	kW	1.942	2.342	2.374	2.145	2.353	2.954	2.343	2.33	2.468	5.053	5.045	5	2.213
η_{th}	%	17.19	11.01	6.781	17.69	12.01	7.762	9.866	12.46	14.98	14.05	12.26	17.8	11.96
η_{exe}	%	14.23	14.97	17.41	15.75	17.25	21.87	16.37	17.1	18.07	36.95	36.88	36.57	16.09

6.1. Effects of evaporator temperature $(T_{12} = T_{13})$

IOP Conf. Series: Earth and Environmental Science **83** (2017) 012008 doi:10.1088/1755-1315/83/1/012008

Figure 2 shows that the exergy efficiency decreases with the increase in the evaporator temperature. The reduction of the refrigeration output exergy and the entrainment ratio of the ejector are the main reasons for the exergy decrease.

The results shown in Figure 3 indicate that the entrainment ratio of the cycle for all of the working fluids decreases with increasing evaporator temperature.

6.2. Effects of turbine inlet temperature (T_4)

It is found from Figure 4 that the exergy efficiency of the cycle increases with increasing turbine inlet temperature. As the power/refrigeration ratio is kept constant, turbine inlet temperature will affect the ejector entrainment ratio which leads to the increase of the refrigeration output exergy.

As shown in Figure 5 the total exergy destruction in the cycle increases monotonically with the turbine inlet temperature.

According to Figure 6, when the turbine inlet temperature goes up, the thermal efficiency rises, too. As the inlet temperature increases, the turbine power, the net power output and the entrainment ratio rise correspondingly.

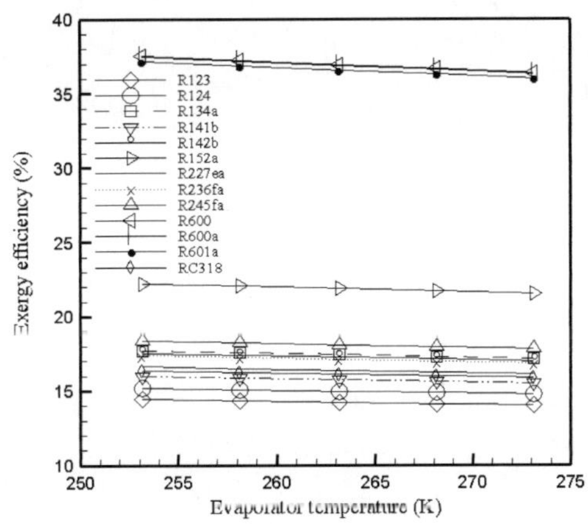

Figure 2. Effect of the evaporator temperature on the exergy efficiency.

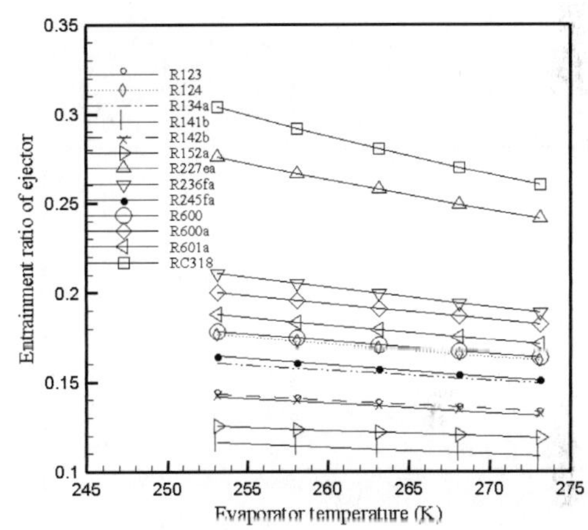

Figure 3. Effect of the evaporator temperature on the entrainment ratio of the ejector.

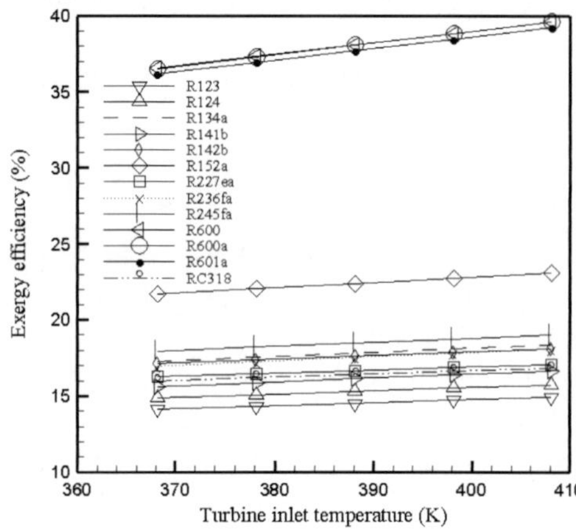

Figure 4. Effect of the turbine inlet temperature on the exergy efficiency.

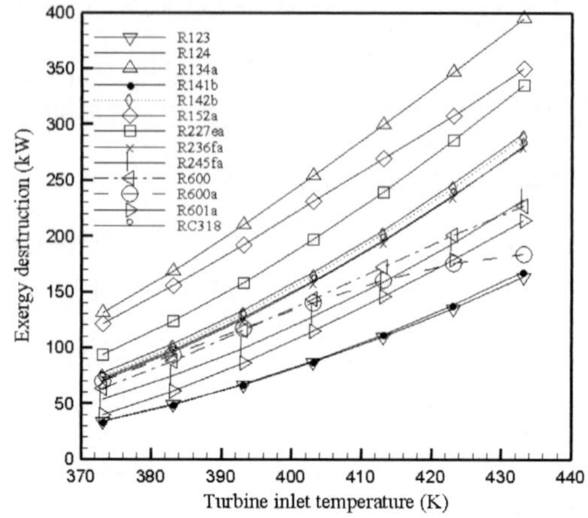

Figure 5. Effect of the turbine inlet temperature on the total exergy destruction.

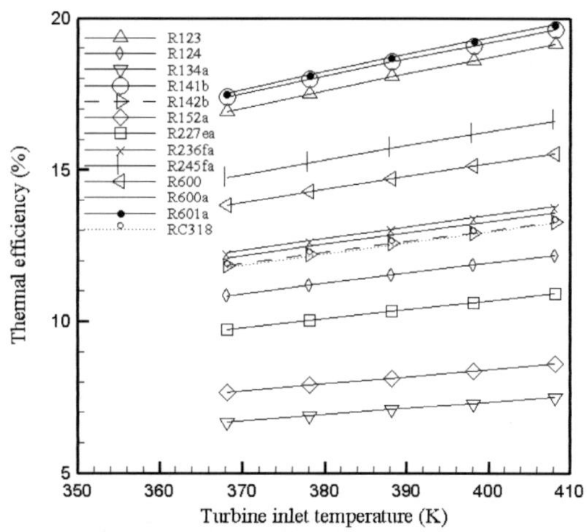

Figure 6. Effect of the turbine inlet temperature on the thermal efficiency.

6.3. Effects of heat source temperature (T_{16})

Figure 7 shows that increase in the heat source fluid temperature leads to the increase in the exergy efficiency. Since the mass flow rate of the turbine and the entrainment ratio of the ejector rise as the vapor generator temperature goes up, the turbine work output and the cooling capacity increase similarly.

6.4. Effects of expansion ratio

From Figure 8 it is apparent that when the expansion ratio of the turbine increases from 2 to 6, the thermal efficiency rises for all of the working fluids. The reason for this is that increasing expansion ratio leads to a decrease in the temperature and pressure of the primary flow entering the ejector.

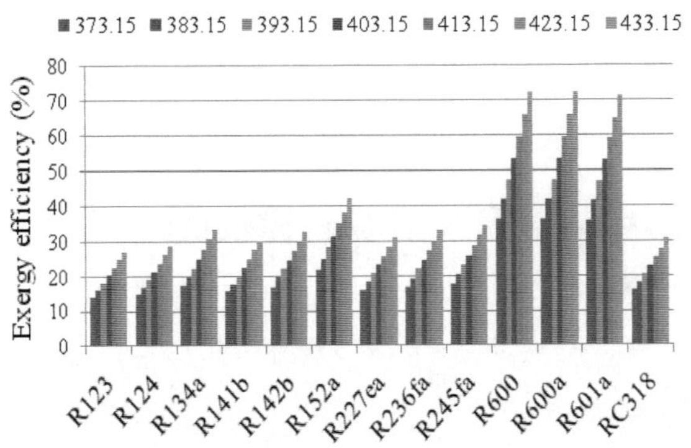

Figure 7. Effect of the heat source temperature on the exergy efficiency.

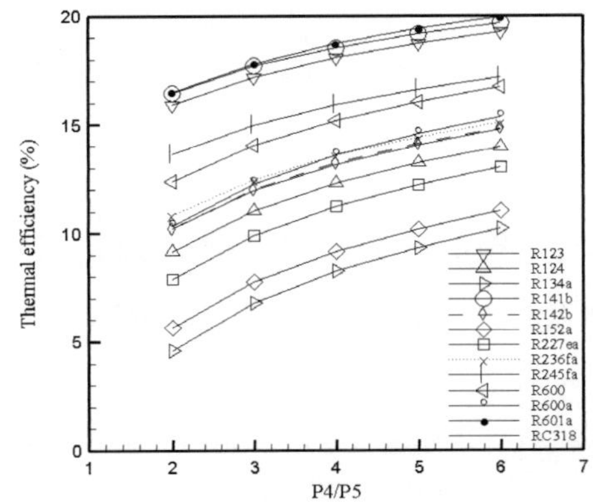

Figure 8. Effect of the expansion ratio on the thermal efficiency.

7. Conclusions

The main conclusions from this study are as follows:

(1) The results confirm the thermodynamic superiority of dry and isentropic ORC fluids over the wet fluids.
(2) Exergy efficiency decreases with increasing evaporator temperature but increases with decreasing turbine inlet temperature and increasing heat source temperature.
(3) Thermal efficiency increases with the increase in the turbine inlet temperature and expansion ratio of the turbine.
(4) Entrainment ratio of the ejector decreases as the evaporator temperature rises.
(5) From the exergy efficiency and environmental friendly point of view, R600 and R 600a are the most suitable working fluids for the proposed combined cycle among the different working fluids studied in this paper.

8. References

[1] Dai Y, Wang J and Gao L 2009 *J. Appl Therm Eng,* **29**, 1983–1990.
[2] Wang J, Dai Y and Sun Z 2009 *Int. J. Ref.* **32**, 1186-1194.
[3] Rashidi MM, Anwar Bég O, Basiri Parsa A and Nazari F 2011 *J Proceedings of the Institution of Mechanical Engineers, Part A: Journal of Power and Energy*. **225** , 701-717.
[4] Sheykhlou H, Jafarmadar S 2016 *Iranian Journal of Science and Technology Transactions of Mechanical Engineering*. **40**, 57-67.
[5] Agrawal BK, Karimi MN 2013 *International Journal of Sustainable Building Technology and Urban Development*. **4**, 170-176.
[6] Cayer E, Galanis N and Nesreddine H 2010 *J Applied Energy*. **87**, 1349–1357.
[7] Tzivanidis C, Bellos E, Antonopoulos KA 2016 *Energy Conversion and Management*. **126**, 421–433.
[8] Klein SA 2004 *Engineering equation solver version 7.171 (*McGraw Hill).

IOP Conf. Series: Earth and Environmental Science **83** (2017) 012009 doi:10.1088/1755-1315/83/1/012009

Flywheel-Based Fast Charging Station – FFCS for Electric Vehicles and Public Transportation

Hossam A.Gabbar[1,2] and Ahmed M. Othman[2,3]

1 Energy Systems and Nuclear Science, University of Ontario Institute of Technology, Canada.
2 Engineering and Applied Science, University of Ontario Institute of Technology, Canada.
3 Faculty of Engineering, Zagazig University, Zagazig, Egypt
E-mail: Hossam.Gabbar@uoit.ca

Abstract. This paper demonstrates novel Flywheel-based Fast Charging Station (FFCS) for high performance and profitable charging infrastructures for public electric buses. The design criteria will be provided for fast charging stations. The station would support the private and open charging framework. Flywheel Energy storage system is utilized to offer advanced energy storage for charging stations to achieve clean public transportation, including electric buses with reducing GHG, including CO2 emission reduction. The integrated modelling and management system in the station is performed by a decision-based control platform that coordinates the power streams between the quick chargers, the flywheel storage framework, photovoltaic cells and the network association. There is a tidy exchange up between the capacity rate of flywheel framework and the power rating of the network association."

1. Introduction

Flywheel kinetic energy storage offers very good features such as power and energy density. Moreover, with some different-range vehicles, this technology can be enough to supply all the energy to the power train. The challenges to be met to integrate such technology in vehicles are the mass, the efficiency and especially the cost. Then, in this project, a techno-economic optimization of a flywheel energy storage system is presented. It is made up of a flywheel, a permanent magnet synchronous machine and a power converter. For each part of the system, physical and economical models are proposed. Finally, an economic optimization is done on a short-range ship profile.
Flywheel energy storage became one of the important energy storage in the world. That is why; flywheels energy storage used a lot in power systems and Microgrid recently as they are flexible, smart, and active. In addition to, they fit more with renewable resources and considered to be friendly to the environment [1]-[4].
Fast charging of modern energy storage has turned into a standard charging innovation because of the operational reserve funds, expanded profitability and wellbeing that this innovation offers. Clients have understood the advantages of fast charging and proceed to understand the advantages at assembling plants and dissemination focuses all through many applications.
This paper proposes a control system for module electric vehicle with Flywheel-Based Fast Charging Station (FFCS). The fundamental part of the FFCS is to trade off the predefined charging profile of PEV battery and incorporated with the arrangement of a hysteresis dynamic power converter supported to the power framework [5], [6].

In that sense, when the dynamic power is not being separated from the network, FFCS gives the power required to maintain the persistent charging procedure of PEV battery. A key trademark of the entire control framework is that it can work without any discrete correspondence between the frameworks tied and FFCS converters.

2. Fast charging vs. other charging approaches

An electric vehicle charging station, additionally called electric reviving point, charging point is a component in a foundation that provisions electrical energy charge for the energizing of electric vehicles, for example, module electric vehicles. Most of the charging platforms are on-road locations associated by electrical service networks and others are situated within retail commercial malls or to be worked through numerous privately owned businesses.
Charging stations are divided categorized into four main groups:

- Private charging stations: An EV connects to when returns home and then auto recharge through night times. A private station ordinarily does not need any client validation, nor metering system; and it is just need simple installation for a devoted circuit. Other convenient chargers will likewise need partition installation as in stations.
- Public stations: Businesses scheme for an expense or without fees that is serviced in organization through the holders of certain parking area. That charging process might as moderate mode that urges vehicles holder to charge the autos during they exploit adjacent shopping or any other stuff. It can incorporate stopping malls and centers or for a business' own particular representatives.
- Fast or quickly charging: for charging demand greater than 40 kW, conveying more than 120 km within timeframe of [10, 25] minutes. They will likewise need routine utilization by suburbanites at metropolitan places, and to be charged during stopped for short and/or long timeframes, see Fig. 1.
- Swaps or changes Process for batteries in less than 15-20 minutes. Predetermined focus on zero- and low-emission vehicle and its range in less than 15 minutes. It can be done with vehicle battery by swaps and also on Hydrogen Fuel Cell vehicles. That plans by coordination the request to refill from customary holders.

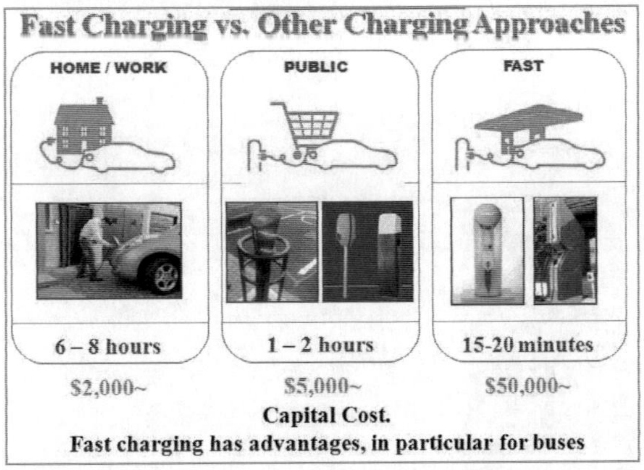

Figure 1. Fast Charging vs. Other Charging Approaches

A critical increment of EVs in develop control markets would animate a develop in power request giving some comfort to utilities working in energy markets as Fig. 2, while likewise helping framework administrators adjust control free market activity by means of vehicle-to-network ventures. The amassing of network associated EV batteries would likewise energy booster development by smoothing supply irregularity and encouraging their incorporation into the power network. As shown, there is high power demand for EV, which confirms the need for Fast Charging Infrastructures [7]-[9].

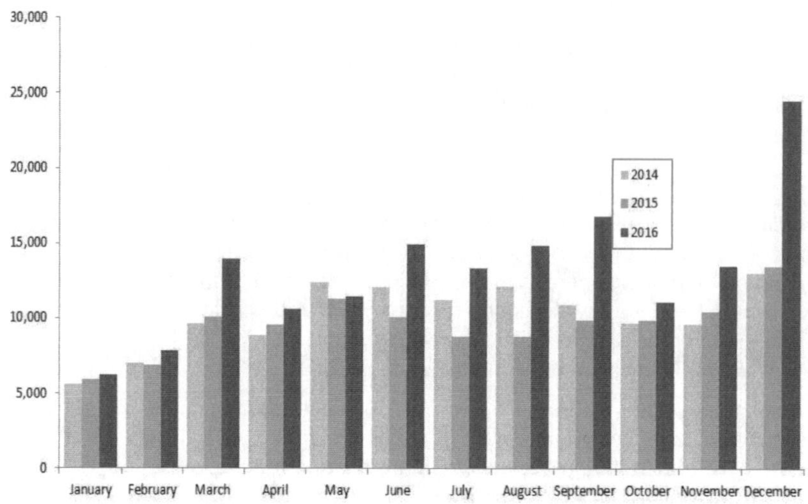

Figure 2. Sales trend of Electric Vehicles till end of 2016, USA

3. Flywheel technology

Flywheel energy storage systems utilize active kinetic rotation that can be stored the rotated mass with low levels of losses in the friction component, see Fig. 3. The input energy as electrical increases the acceleration of the mass to speed by means of an incorporated engine generator. That energy will be stored and depend on the moment of inertia and speed of the rotating shaft, and can be expressed as follows:

Kinetic energy: $E_k = \frac{1}{2} I \omega^2$ (1)

Moment of inertia (I) : $I = \int r^2 dm$ (2)

For a cylinder the moment of inertia: $I = \frac{1}{2} r^4 \pi a \rho$ (3)

Energy is increased if ω increases or if I increases.
The optimization of the energy related to mass can be achieved by spinning flywheel with maximum possible speed.

- Max. Speed: $v_{max} = \sqrt{\frac{2K\sigma_{max}}{\rho}}$ (4)

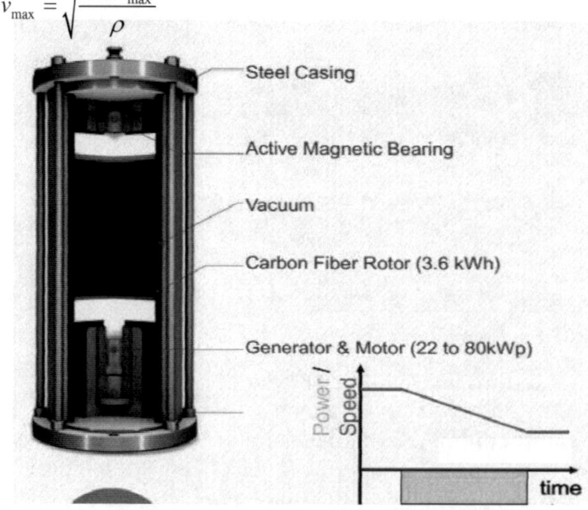

Figure 3. Example Flywheel Technology

The reasons for selecting Flywheel as energy storage system are:

- Higher Power Densities: Especially for EV and HEV applications, flywheel is significant to supply much power density than batty.

- High Reliability and Cycle Life: The power from Flywheels is not degraded, and remarkable by reliability and high cycle life.
- Environmental impact: The environmental impact of flywheels is so friendly where there is no hazardous substance than can be recycled.
- Temperature Sensitivity: Flywheels are for the most part less delicate to the surrounding temperature than batteries. Some essential warm breaking points are set by the temperature in: the windings, to abstain from softening; the magnets, to maintain a strategic distance from demagnetization; and the composite material, to abstain from blazing it.

4. Flywheel-Based Fast Charging Station – FFCS

4.1 Fast charging stations - design criteria

The design criteria of a fast charging station is to cover both of residential and public charging infrastructure, Flywheel-based Fast Charging Station (FFCS) will be built as shown in Fig. 4, 5.

4.1.1 Fast charging stations – covering factor

Fast Charging Stations can achieve a covering factor up to 95% of total electric charging demand. This can be achieved by three factors: 1. waiting time at the FFCS; 2. number empty charging spots per FFCS per time; and 3. maximum number of required charging spots per FFCS.

EVs can be charged at wherever that offers an electrical attachment. This incorporates open carports, auto parks, workplaces, grocery stores, healing centers, lodgings, homes of companions and colleagues, shopping centers, and neighborhoods. The increase of FFCS installations can replace some of the above electric charging options. This can be achieved by planning optimization and control.

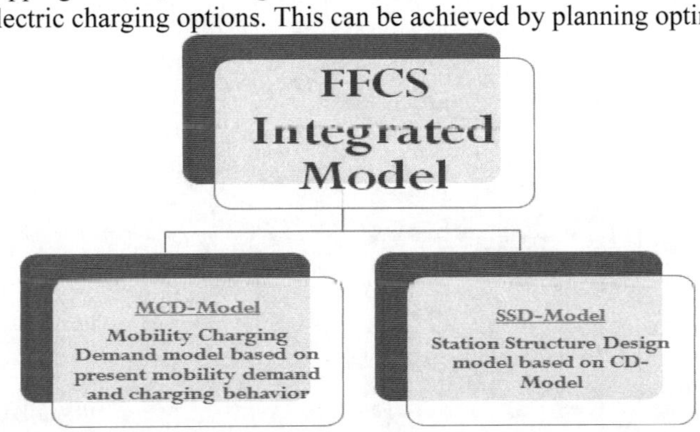

Figure 4. FFCS Integrated Model

4.1.2 Mobility behaviour

In MCD model, the mobility and charging demand of EVs is determined depending on the Mobility Behavior. Designing a fast charging station starts manily by study the profile of the concerned charging load (These include social and technical analysis),

- Social: The popoulation od the assigned region and how frequent they charges their EVs.
- Technical: the limits of the charging infrastructure.

The second step is the modelling that perform simulaion of the real fast charging station within various scenrios.

4.2 Mobility integrated study:

The starting point for the mobility model can be with 50 groups of 1500 vehicles each. Each vehicle is controlled in its charging on the behavior of the onwer:

- The case of the EV as per it is running or stopping (μv);
- The amount of consumed power from the electricity (μm);
- The places where the car usually stands at (μc); where it is house places, or working place or others.
- The vehicle model (μt); the sizing of the EV.

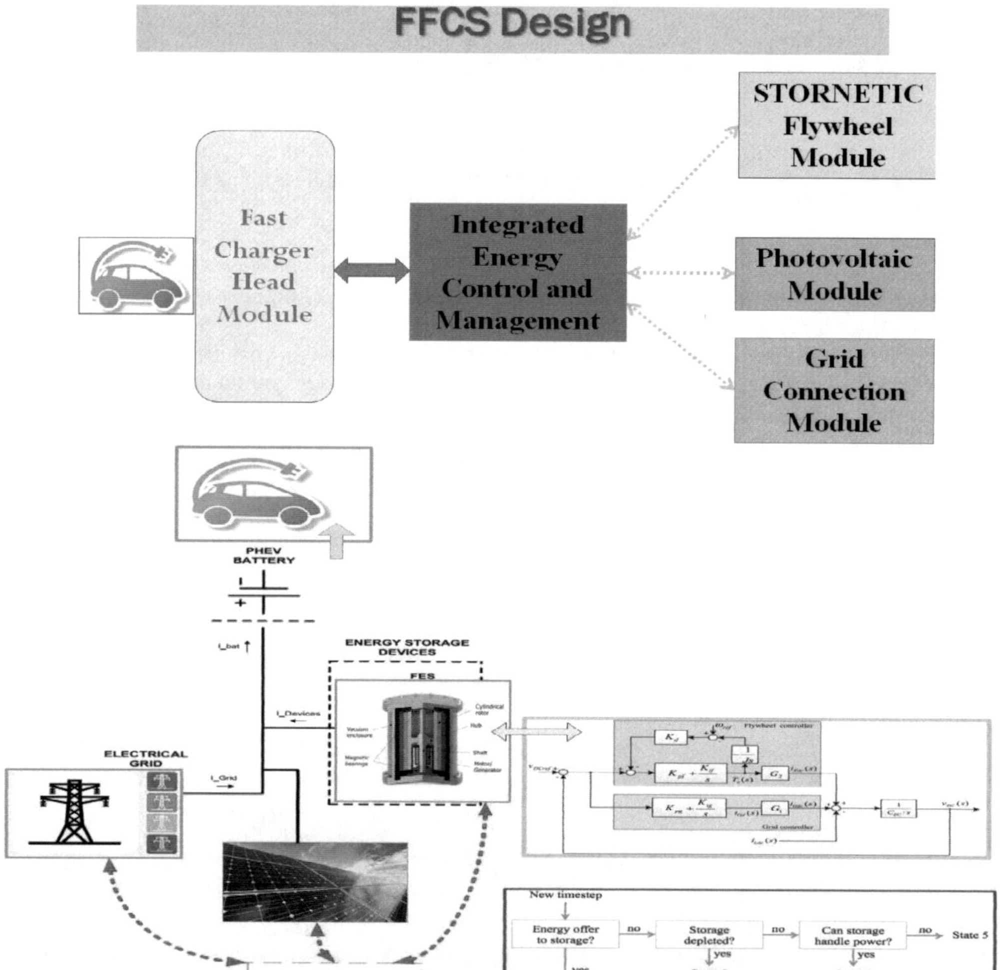

Figure 5. FFCS Design

4.3 FFCS – multi-level circuit design

Multi three phase 2-level AC/DC converter as grid interface and flywheel converters is shown in Figure 6, it depends on Fast DC/DC converter to quick charging where Each component is installated by connection via a common DC bus, see Fig. 6.

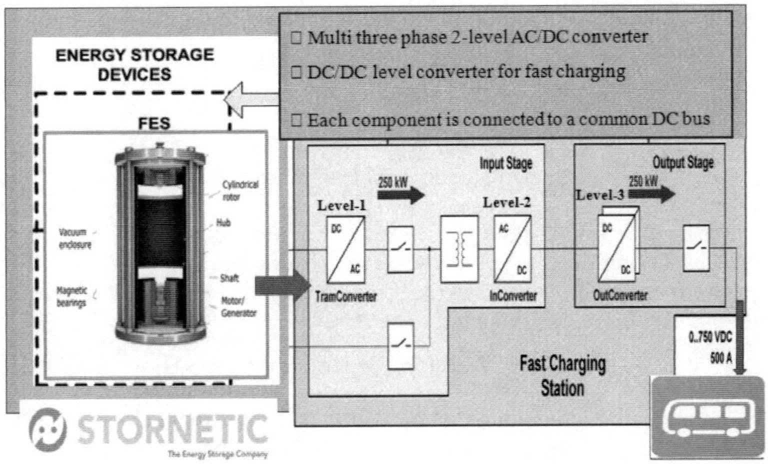

Figure 6. FFCS Multi-level Circuit

4.4 Control of flywheel by hysteresis controller

Points of interest of utilizing hysteresis control, as in Fig. 7, are significant dynamic response and capacity to control the peak demanding and current swell in assigned hysteresis band range. It is likewise extremely successful for physically execution [10]-[13].

Likewise, it diminishes the harrmonics, where it takes a shot at essential by identifying consonant current to compute the measure of the compensated current required for sustaining back to the power framework.

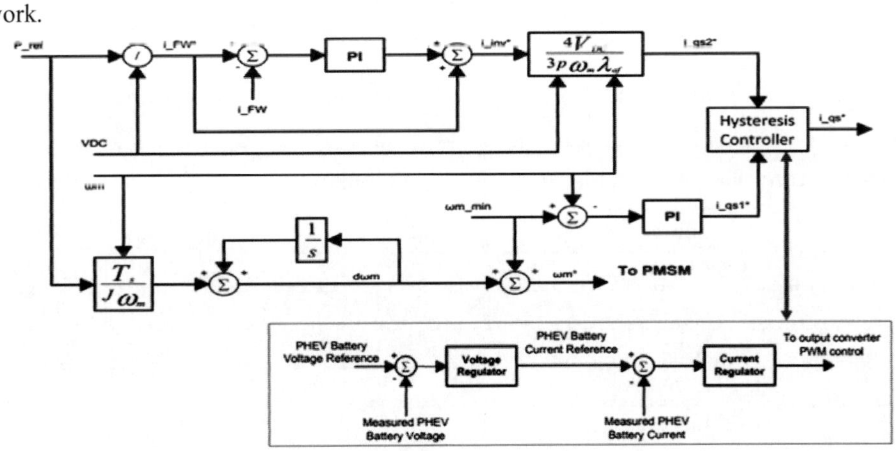

Figure 7. FFCS control

5. Discussion and analysis of fast charging

DC Fast Chargers replace Level 1 and Level 2 charging stations, and are intended to accuse electric vehicles rapidly of an electric yield extending between 50 kW – 120 kW. Most present day completely electric vehicles can be charged with DC fast charge capacity, and there are right now about 2,200 rapid chargers in the Assembled States fit for adding huge range to an EV in very little longer than the time it takes to fill your gas tank.

Fast charging can be achieved using Level-2 (basic) or Ultimate (Level-3)

AC Level 1:117V 16A Max (Normal Charging), AC Level 2:240V 32A or 70A (Basic Fast Charging)

Charger Specifications Level 3 (Ultimate - Direct DC) by FFCS:

- Input 3-phase 200V

- Output:

  Max DC 45-50 kW (can reach 200 kW)

  Max DC Voltage 700V

  Max DC Current 750A

Fast Charging by FFCS has such benefits to demand charge distribution network in big cities which is very congested and peak demand reduction is a topic of interest. These benefits are grid services to provide peak power by turn peak load into base load. In addition to, grid stabilization can be achieved by support for grid expansion and maintain promised levels of supply.
Demonstration and experimentation are done through capabilities of software and hardware in Energy Safety and Control Lab (ESCL), University of Ontario Institute of Technology.

Table 1. Gas vs. Electric Charging

Charging-Vehicle	Charge as per 100 km	Cost per unit	Cost
Electric FFCS	16 kWh	× $0.08/kWh	= $1.28
Gasoline	8.4 L	× $1.35/L	= $11.40
As another comparison, Fuel = $0.13 per km , EV = $0.02 per km			
Electricity is Cheaper 9 times more than Gasoline Electricity saves around $2,000 - $3,000 a year per car, based on 30,000 km / year			

Table 2. Charging Time of different charging categories

considering 100 km, charging time	Voltage	Max. Current	Power
6–8 hours	220 Vac	16-18 A	3.2 kW
3–5 hours	220 Vac	32-36 A	7.4 kW
2–3 hours	400 Vac	16 A	10 kW
1–2 hours	400 Vac	32 A	22 kW
20–30 minutes	400 Vac	63 A	43 kW
20–30 minutes	450–505 Vac	120–135 A	50 kW
10 minute	350–550 Vac	320–380 A	120 kW

Table 1 shows the difference between gas and electric charging while Table 2 shows the charging time of different charging categories.

ROI and Benefits of FFCS

The benefits and ROI ratio between normal charging and FFCS is between: 3-6 times

* With respect to production within same time Frame:

For *X* kwh of charging production, in time *t*

- The amount of normal charging = *X* kwh, The amount of FFCS charging = *3X* kwh

* With considering saving of time reduction of FFCS:

For fixing X kwh of charging production

- The production time of normal charging = $3t$ to $6t$, The production time of FFCS charging = t

6. Conclusions

The proposed FFCS is analysed in view of user and technology requirements, where fast charging is proposed based on flywheel technology. Advantages of fast charging over traditional charging is expressed, in particular when dealing with congested cities where short charging time is critical. The design of FFCS is presented based on flywheel energy storage platform (FESP) which is presented to implement fast charging for transportation infrastructures. Implementation schemes for eBuses and EVs.

7. Acknowledgement

Authors would like to thank STORNETIC for providing data and discussions during this research.

8. References

[1] M. Musio, A. Serpi, C. Musio, and A. Damiano, “Optimal Management Strategy of Energy Storage Systems for RES,” IECON 2015 - Conference of the IEEE Industrial Electronics Society pp. 5044–5049, 2015.

[2] PNNL , Sandia, DR Conover, SR Ferreira, AJ Crawford, DA Schoenwald, J Fuller DM Rosewater, SN Gourisetti, V Viswanathan. Protocol for Uniformly Measuring and Expressing the Performance of Energy Storage Systems. s.l. : PNNL, Sandia, 2016. SAND2016-3078 R, PNNL-22010 Rev.

[3] S. Abapour, M. Abapour, K. Khalkhali, and S. M. Moghaddas-Tafreshi, “Application of data envelopment analysis theorem in plug-in hybrid electric vehicle charging station planning,” IET Gener. Transm. Dis., vol. 9, no. 7, pp. 666–676, 2015.

[4] CEMAC, Donald Chung, Emma Elgquist, Shriram Santhanagopalan. Automotive Lithiumion Cell Manufacturing: Regional Cost Structures and Supply Chain Considerations. s.l. : NREL, April 2016.

[5] P. Fan, B. Sainbayar, and S. Ren, “Operation Analysis of Fast Charging Stations with Energy Demand Control of Electric Vehicles,” IEEE Trans. Smart Grid, vol. 6, no. 4, pp. 1819–1826, 2015

[6] J. Aubry, H. Ben Ahmed, and B. Multon, “Sizing optimization methodology of a surface permanent magnet machine-converter system over a torque-speed operating profile: Application to a wave energy converter,” *IEEE Transactions on Industrial Electronics*, vol. 59, no. 5, pp. 2116–2125, 2012

[7] Z. Xu, W. Su, Z. Hu, Y. Song, and H. Zhang, “A hierarchical framework for coordinated charging of plug-In electric vehicles in China,” IEEE Trans. Smart Grid, vol. 7, no. 1, pp. 428–438, 2016.

[8] T. Franke and J. F. Krems, “What drives range preferences in electric vehicle users?,” Transport Policy, vol. 30, pp. 56–62, 2013.

[9] H. Zhang, Z. Hu, Z. Xu, and Y. Song, “Optimal planning of PEV charging station with single output multiple cables charging spots,” IEEE Trans. Smart Grid, pp. 1–10, 2016.

[10] H. Xu, S. Miao, C. Zhang, and D. Shi, “Optimal placement of charging infrastructures for large-scale integration of pure electric vehicles into grid,” Int. J. Elec. Power., vol. 53, no. 1, pp. 159–165, 2013.

[11] J. Cavadas, “A MIP model for locating slow-charging stations for electric vehicles in urban areas accounting for driver tours,” Transport. Res. E: Log., vol. 75, pp. 188–201, 2015.

[12] H. Zhang, Z. Hu, Z. Xu, and Y. Song, “An integrated planning framework for different types of pev charging facilities in urban area,” IEEE Trans. Smart Grid, vol. 7, no. 5, pp. 2273–2284, 2016.

[13] H. Cai, “Siting public electric vehicle charging stations in Beijing using big-data informed travel patterns of the taxi fleet,” Transport. Res. D: Tr. E., vol. 33, pp. 39–46, 2014.

IOP Conf. Series: Earth and Environmental Science **83** (2017) 012010 doi:10.1088/1755-1315/83/1/012010

Piezoelectric Cylindrical Design for Harvesting Energy in Multi-Directional Vibration Source

M S Nguyen[1, 2], S H Ng[3], P Kim[4] and Y J Yoon[2]

1 Energy Research Institute @ NTU, Interdisciplinary Graduate School, Nanyang Technological University, Singapore 639798, Singapore
2 School of Mechanical and Aerospace Engineering, Nanyang Technological University, Singapore 639798, Singapore
3 A*STAR's Singapore Institute of Manufacturing Technology (SIMTech), 2 Fusionopolis Way, #08-04, Innovis, Singapore 138634.
4 School of Mechanical Design Engineering, College of Engineering, Chonbuk National University, 567 Baekje-daero, Deokjin-gu, Jeonju-si, Jeollabuk-do 54896, South Korea
Email: nguyenmi001@e.ntu.edu.sg (M S Nguyen), yongjiny@ntu.edu.sg (Y J Yoon)

Abstract. Vibration Energy Harvester (VEH) has attracted a great attention recently both in academia and industry. One of the most challenging issues in VEH is the possibility to harvest vibration energy in multiple directions. In fact, Conventional VEH (CVEH) using cantilever beam's structure may possibly become inefficient for the application under multi-directional vibration sources. To overcome this shortcoming of CVEH, this paper proposes a novel design of piezoelectric cylindrical energy harvester (PCEH) which is using patches of piezoelectric material attached to the surface of a cylindrical structure. The Finite Element Method (FEM) analysis using COMSOL Multiphysics software package showed that PCEH has a great potential for the applicability of VEH in the multi-directional vibrating applications such as wearable devices and biomedical devices.

1. Introduction

The development of electronic industry has led to a new generation of the small electronic device which has the micro-scale volume and requires lower power consumption [1]-[3]. However, the lack of scalable energy sources has been a challenge for the implementation of those devices in real-life applications. Batteries, which are conventionally the most common power sources, have not kept pace with the demand for the miniaturization of these devices [4]. In this context, the vibration energy harvesting (VEH), which can scavenge mechanical energy from surrounding and convert it into electricity, have received great interest from academia and industry. Within the vast field of micro-power source, the mechanical vibrations have been proved to be the most suitable energy source for the small scale devices due to its abundance, versatility and power density [5], [6]. With the scalable energy harvester, the electronic devices are expected to operate with a continuous and autonomous power source without any interaction from the surrounding. Among the different types of vibration energy harvester, the piezoelectric energy harvesting (PEH) has been reported to be the most common choice due to its significant advantages in the power density, the simple structure and the ease of application [7]. At the early stages of development, most of the piezoelectric energy harvesters were designed based on cantilever beam structure owing to its large deflection response under the purely bending mode shape. However, for the applications which are excited by a multi-directional vibration source, CVEH may possibly show a poor energy harvesting performance. To overcome this shortcoming of CVEH, several researchers have proposed a variety of multi-directional design for PEHs. Mei et al [8] presented a double-wall cylindrical energy harvester which can harvest vibration energy in various directions. The authors claimed that by using the double wall design, the number of resonant modes is increased, and thus the operating bandwidth of the energy harvester is broadened. Chen et al [9] proposed a dandelion-like multi-directional vibration energy harvester with a good performance in various testing directions. Wei-Jiun et al [10] presented a tri-directional harvester which comprises two piezoelectric cantilevers vibrating along two perpendicular directions and a spring-mass system under magnetic effect. In general, the existing designs of multi-directional PEH commonly require auxiliary structures. These structures increase the complexity and the size of the PEH and therefore may decrease the applicability of the device. In an attempt to treat multi-directional energy harvesting challenge, this paper proposes a simple design of piezoelectric cylindrical energy harvester (PCEH) which is using piezoelectric material attached to the surface of a cylindrical structure as shown in figure 1. The study using finite element method in COMSOL Multiphysics software package revealed that PCEH has a great potential for the VEH in the multi-directionally vibrating application.

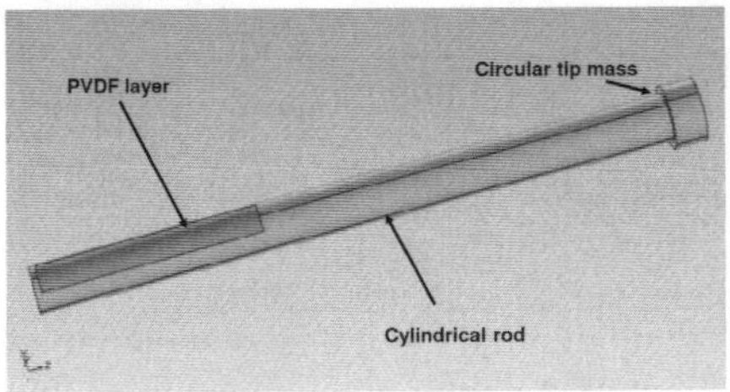

Figure 1. Design of piezoelectric cylindrical energy harvester.

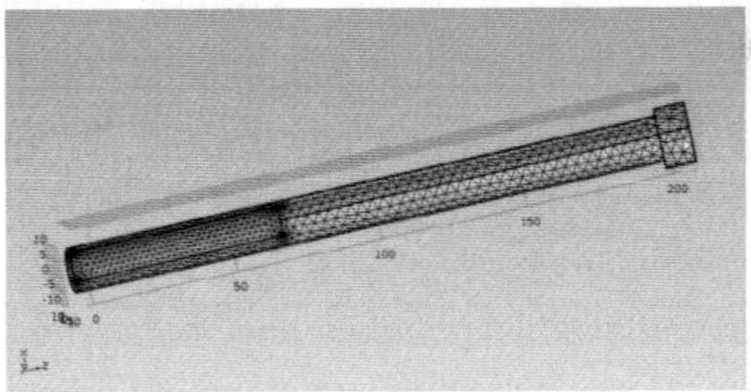

Figure 2. The meshing of PCEH in COMSOL Multiphysics.

2. Theoretical background of piezoelectric materials

Piezoelectric material has a long history of development. These materials specifically generate electric charges under mechanical deformation. Among various types of piezoelectric materials, the piezoelectric polymer such as polyvinylidene fluoride (PVDF) has attracted a great attention owing to its great flexibility and ease of application. In this research, this material will be laminated on the surface of the plastic rod. Under the deformation of the plastic rod, the curvature stress and strain will appear in the piezoelectric layer. The electric charges, which are generated during the bending deformation of the piezoelectric layers, will be collected through the electrodes. These electrodes are then connected to an external electrical circuit. And the output voltage will be measured through a resistor load. In general, a stress-charge form of the piezoelectric constitutive equations is given as follow,

$$\mathbf{T} = \mathbf{c}^{\mathbf{E}}\mathbf{S} - \mathbf{e}^{t}\mathbf{E} , \tag{1}$$

$$\mathbf{D} = \mathbf{e}\,\mathbf{S} + \varepsilon^{\mathbf{S}}\mathbf{E} . \tag{2}$$

where the $\mathbf{T}$, $\mathbf{S}$, $\mathbf{E}$, $\mathbf{D}$ denote the stress component vector, the strain component vector, the electric field component vector, and the electric displacement component vector, respectively; the $\mathbf{c}^{\mathbf{E}}$, $\mathbf{e}$, $\varepsilon^{\mathbf{S}}$ denote the elasticity matrix, piezoelectric coupling matrix, and electric permittivity coefficient matrix, respectively; the superscripts $\mathbf{E}$, $\mathbf{S}$ denote the parameter at constant electric field and at constant strain, respectively; the superscript t stands for the transpose.

The generated charges are calculated through the electric displacement D by,

$$Q = \iint_{\psi} D d\psi , \tag{3}$$

where ψ is the area of the effective electrode.

The voltage V_{output} is measured through the resistor load R_L as $V_{output} = i_{cir} R_L$, with $i_{cir} = \dot{Q}$, where i_{cir} is the current across the circuit and the overdot denotes the derivative with respect to time. All of these mathematical equations are embedded into the COMSOL model.

3. Design and modeling

The design comprises a circular tip mass mounted to the tip of a cylindrical acrylic rod as shown in figure 1. The radius of the tip mass and the cylindrical rod are 7.5 mm and 10 mm, respectively. A piezoelectric polyvinylidene fluoride (PVDF) patch with the size 70x10x0.3 mm^3 is laminated on the surface of the cylindrical rod. The lengths of the cylindrical rod and the circular tip mass are 200 mm and 10 mm, respectively. Under the external excitation, the cylindrical rod will establish a bending mode and accordingly the piezoelectric layer will be compressed and stretched. The system is then connected to an electrical circuit. The voltage output will be measured through the resistive load of 10^6 Ω. A model of PCEH has been built in COMSOL Multiphysics 5.2 Software Package. The detailed material properties and parameters are provided in the Table. 1. Solid Mechanics, Electrostatics, Piezoelectric Effect, and Electrical Circuit Modules were used to model and simulate the design.

Table 1. Material properties of PCEH.

Model Segment	Materials	Property	Value	Unit
Cylindrical Rod	Acrylic Plastic	Density	1190	$kg.m^{-3}$
		Young's Modulus	3.2e9	Pa
		Poisson's Ratio	0.35	
Circular Tip Mass	Neodymium	Density	7500	$kg.m^{-3}$
Piezoelectric layer	PVDF	Density	1780	$kg.m^{-3}$
		Relative permittivity	12	

For the piezoelectric layer, polyvinylidene fluoride (PVDF) is using owing to its high mechanical flexibility compare to others piezoelectric material. The values in the elasticity matrix $\mathbf{c}^{\mathbf{E}}$ and the coupling matrix $\mathbf{e}$ of the PVDF are provided as following: $\mathbf{c}^{\mathbf{E}}$ ={10^{12}/39.5, -10^{12}/10.2, 10^{12}/42, -10^{12}/25.4, -10^{12}/19.5, 10^{12}/99.9, 0, 0, 0, 10^{12}/1.82, 0, 0, 0, 0, 10^{12}/1.69, 0, 0, 0, 0, 0, 10^{12}/1.43} Pa and $\mathbf{e}$ ={0, 0, 0.069, 0, 0, 0.069, 0, 0, -0.099, 0, -0.069, 0, -0.081, 0, 0, 0, 0, 0} C/m^2. Notes that for anisotropic elastic materials the stress-strain matrix is symmetric.

A three-dimensional finite element model in COMSOL is considered for this research. Figure 2 showed the meshing of the piezoelectric layer and the cylindrical rod with the circular tip mass. Tetrahedral elements were chosen for meshing, and a mesh refinement was applied to the region near the piezoelectric layer for a better calculating accuracy. The acrylic rod is fixed at one end and the other end with the attached circular tip mass is set free for vibration. The gravity is chosen in the opposite direction of the Z-direction, which is in the length direction of the rod. This gravity force will be taken into account during the calculation for a more realistic study.

4. Results and Discussion

4.1 Modal analysis

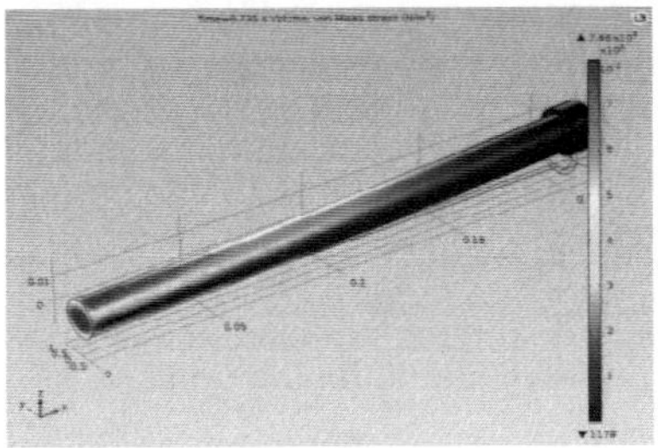

Figure 3. Stress distribution of the PCEH under static force.

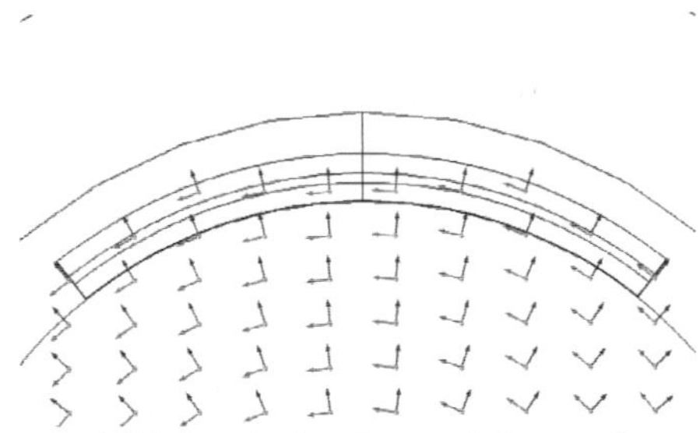

Figure 4. The polarization and the static electric potential of PVDF layer.

A stationary analysis was conducted using Static Study Module in COMSOL. The maximum stress is calculated with Von Mises criteria. The maximum stress was equal to 7.96 [MPa] at the fixed end of the rod and the minimum stress is equal 0.1178 [MPa] at the free end of the rod. The results in figure 3 showed the obvious tendency of the stress and strain distribution with the high stress concentrated at the fixed end and the maximum deflection was found at the free end of the rod. Based on that observation, the PVDF layers were attached near the fixed end of the rod to maximize the energy harvesting efficiency. As the PVDF layers were rolled on the surface of the cylindrical rod, it is essential to use the Rotated Coordinate System in COMSOL to specify the poling direction of the layer. In particular, the PVDF layer was polarized in radius direction as illustrated by the arrows in figure 4. In this figure, the radial blue arrows indicate the poling directions.

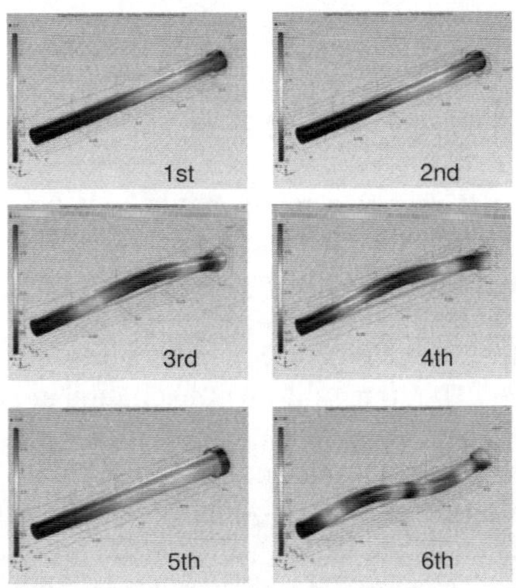

Mode shapes	Resonant frequency (Hz)
1	57.891
2	65.158
3	416.75
4	446.28
5	739.54
6	1207.4
7	1254.9

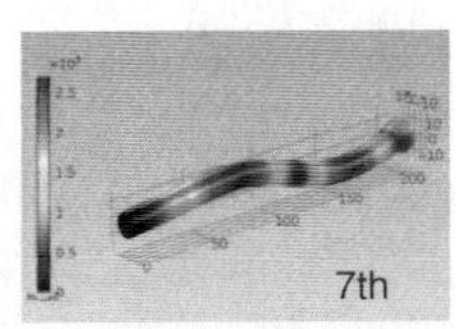

Figure 5. First seven modes of PCEH and the associated natural frequencies.

Eigenfrequency study showed the first seven eigenmodes obtained at various frequencies ranging from 57.891 Hz to 1254.9 Hz. These seven modes and their associated frequencies are shown in figure 4. It should be noted that the 1st and 2nd modes are purely bending modes while the 3rd, 4th, 5th, 6th and 7th modes are the combination of bending mode and torsional mode. The 1st and 2nd modes are under consideration in this research. These modes are more suitable for actual ambient vibration sources due to their low natural frequencies. The frequency response study was also conducted as shown in figure 6. The swept sine frequency test has been utilized with the frequency range from 10 Hz to 1300 Hz and the frequency step of 0.5 Hz. The peak amplitude was detected at around 65 Hz, 446.5 Hz, 739.5 Hz, 1207.5 Hz and 1255 Hz which showed a good agreement with the eigenfrequency study. Furthermore, it is worth noting in figure 6(b) that the output voltage of bending mode (at 65 Hz) and torsional mode (at 446.5 Hz) are comparable.

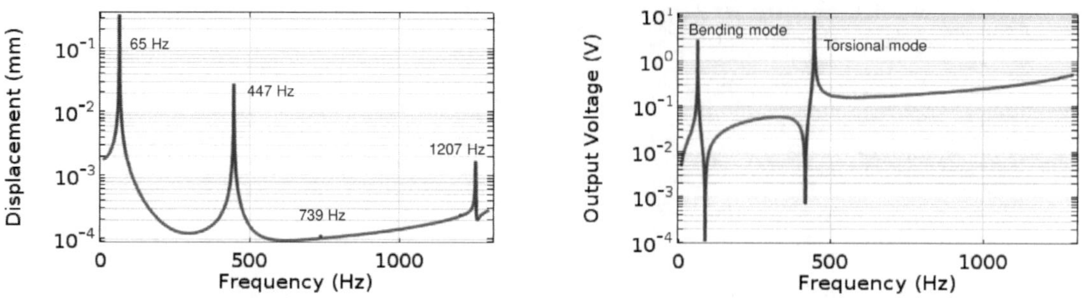

(a) (b)

Figure 6. Frequency response curves for (a) the tip deflection and (b) the output voltage obtained with base excitation in X-direction and the base acceleration of 0.5 m/s^2.

4.2 Response to multi-directional vibration source

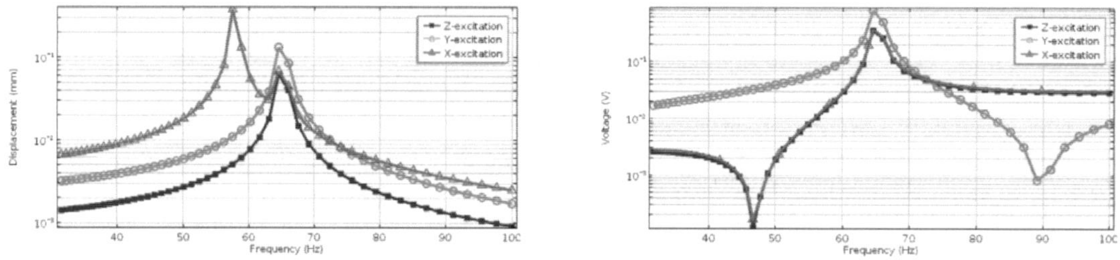

(a) (b)

Figure 7. Frequency response curves for (a) the tip deflection and (b) the output voltage obtaining with the base excitation in X, Y, Z-direction and the base acceleration of 0.5 m/s^2.

As mentioned in the previous subsection, the purely bending mode will be the dominating mode within the frequency bandwidth where the frequency of actual vibration sources is usually found. Therefore, this study will focus on the purely bending mode from now on. Under each direction -X, Y, and Z direction of the base excitation, the frequency response test with the base acceleration of 0.5 m/s^2 and the frequency bandwidth from 30Hz to 100Hz are performed as shown in figure 7. The results showed that the output voltages of PCEH at three different excitation conditions are on similar scales. It should also be noted that there is a small discrepancy in the results for X-direction. Particularly, the peak output amplitudes were found at around 57 Hz and 65 Hz for the case of X-excitation. Meanwhile, in Y- and Z-excitation condition, only the peak value at 65 Hz is observed. This is resulted from the unsymmetrical design of the PCEH - due to the present of the piezoelectric layers. More precisely, the PVDF was only attached to the rod's surface which is perpendicular to the Y-direction in this case. However, the overall results showed that in all three excitation condition, the peak voltage outputs are comparable. In other words, this indicated that PCEH can harvest energy in three different directions for the low base excitation at 0.5 m/s^2.

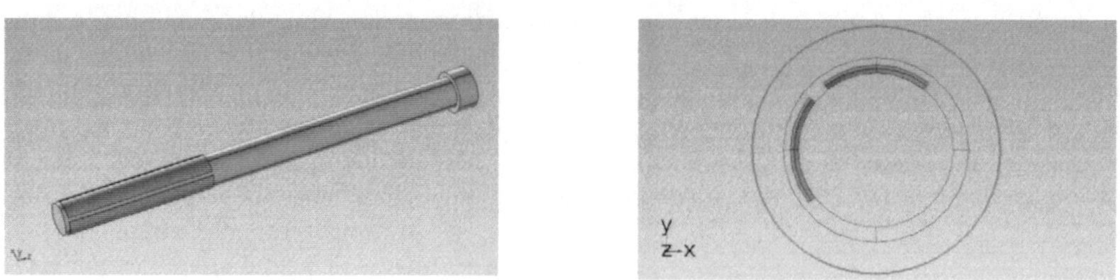

(a) (b)

Figure 8. Design of the PCEH with 2 PVDF layers attached on the rod's surface in X and Y direction: (a) normal view and (b) projected view.

IOP Conf. Series: Earth and Environmental Science **83** (2017) 012010 doi:10.1088/1755-1315/83/1/012010

For a higher base excitation, the effect of unsymmetrical design may possibly become larger. Therefore, to further examine the potential of PCEH in this case, another PVDF patch is attached to the rod's surface which is perpendicular to X-direction as shown in figure 8. By conducting the same simulation, the frequency response results are shown in figure 9. Notes that the excitation acceleration has been increased to 5 m/s^2 in this case.

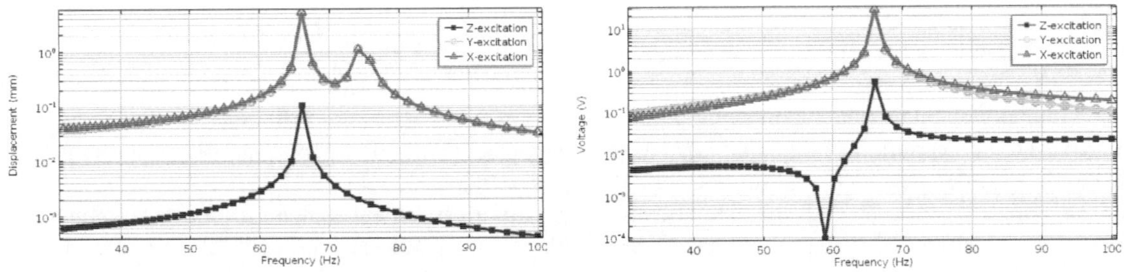

(a) (b)

Figure 9. Frequency response curves for (a) the tip deflection and (b) the output voltage obtaining with the base excitation in X, Y, Z-direction and the base acceleration at 5 m/s^2.

In this simulation, the displacement and the voltage output in the Z-excitation condition are smaller when compared to the ones in X- and Y-excitation condition. However, it is shown that the results in X- and Y-direction are approximately in the same high displacement and voltage values (peak amplitude at 5.5 mm for both cases). This can be explained by the similar structure from the viewing direction in X- and Y-direction.

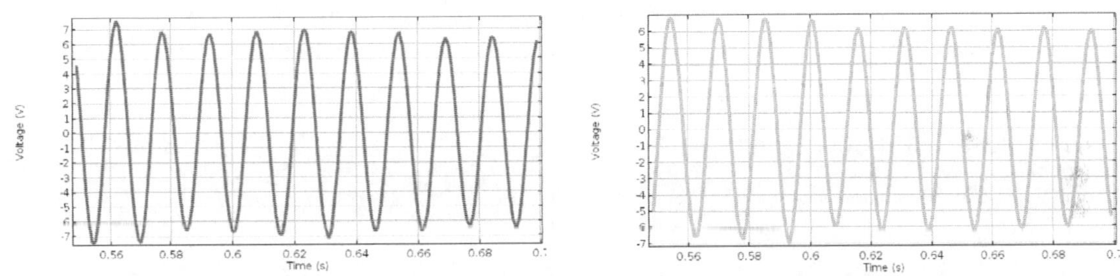

(a) (b)

Figure 10. Time response of output voltage of the PCEH with the excitation in (a) X- and (b) Y-direction, obtained at the excitation frequency of 65 Hz and the base acceleration of 5 m/s^2.

In order to further verify this phenomenon, a time response study has been conducted with the excitation frequency of 65 Hz and the base acceleration of 5 m/s^2 as shown in figure 10. The results showed the output voltages are in harmonic form with the maximum amplitude at around 7.3 V and 7 V for the base excitation on X- and Y-direction, respectively. The output voltages obtained in both cases are totally in similar range. In conclusion, regardless of the base acceleration intensity, the results confirmed that the PCEH may harvest efficiently energy in more than one direction with a proper position of the PVDF layers.

4.3 Parameter optimization

IOP Conf. Series: Earth and Environmental Science **83** (2017) 012010 doi:10.1088/1755-1315/83/1/012010

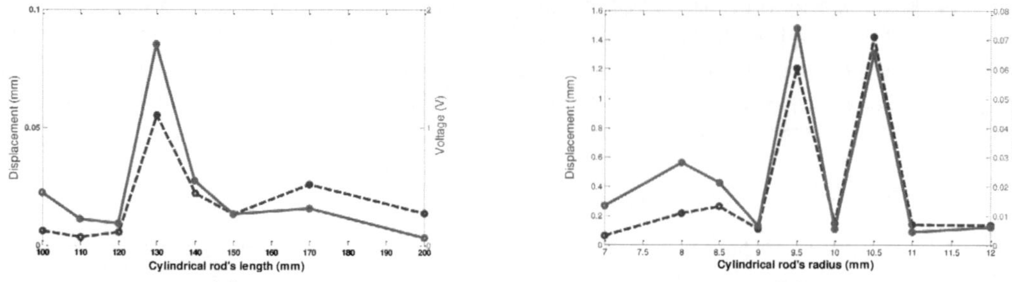

(a) (b)

Figure 11. The variation of the maximum of the tip displacement and the output voltage with the (a) cylindrical length and (b) radius at 0.3 m/s^2.

For the same PVDF layer with the size above, the maximum tip displacement and output voltage are calculated and plotted at different values of the length of the cylindrical rod from 100 mm to 200 mm as shown in figure 11. Notes that in this optimization process, we use the design of 1 PVDF layer. The results demonstrated that there exists a set of values for geometrical parameters with that the optimum energy harvesting performance can be obtained. In particular, the maximum displacement and output voltage have been achieved with the cylindrical rod's length of 130 mm when the radius is fixed at 7.5 mm (as in figure 11(a)) and the optimum radius of 9.5 mm when the length is fixed at 120 mm (as in figure 11(b)). All the results above were obtained with the base acceleration of 0.3 m/s2 and in Y-direction. In addition, it is also worth noting that the numbers of PVDF layer laminated on the surface of the cylindrical rod also affect the energy harvesting performance. The designs of PCEH with 1, 2, 3 and 4 PVDF layers laminated on the spherical surface of the cylindrical rod are presented in figure 12 (a). These designs were tested by conducting the swept sine frequency tests at the same condition (the base acceleration of 0.3 m/s2 and in Y-direction). The frequency response curves for the output voltage of each case are compared in figure 12 (b). The maximum output voltage was obtained with the design using 3 PVDF layers.

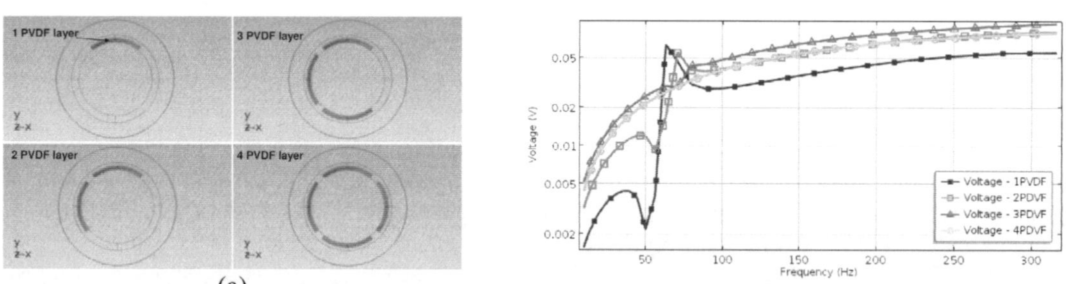

(a) (b)

Figure 12. (a) Schematic designs of PCEH and (b) variation of the output voltage with different numbers of PVDF layer.

To see the effect of resistive load on the performance of the PCEH, a series of parametric study has also been conducted for the design with 1 PVDF layer. The base acceleration was kept at 5 m/s^2 and in Y-direction. Figure 13 showed that while increasing the values of resistive load, the optimum power was determined at $10\,\mathrm{M\Omega}$. Furthermore, from figure 9(a), the output voltage across the load reached the saturated value at $100\,\mathrm{M\Omega}$.

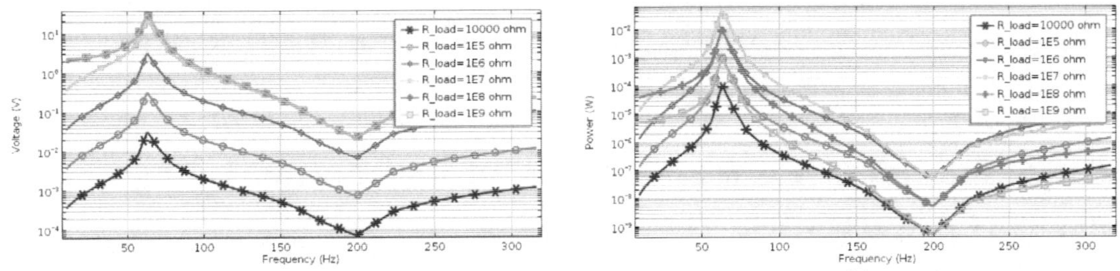

Figure 13. (a) Voltage and (b) power obtained with different values of resistance load.

5. Conclusion

A series of cylindrical-based energy harvesters were proposed in this study. The modeling and simulation study using Finite Element Method (FEM) were performed in COMSOL Multiphysics. The preliminary results pointed out that PVDF layers should be attached at the clamped end of the cylindrical rod for the sake of harvesting efficiency. By now, the PCEH could efficiently harvest energy with multi-directional excitation at low and high excitation intensity with a proper configuration of PVDF layers on the surface of the plastic rod. This finding may help to bypass the challenge of VEH in the multi-directionally vibrating applications. In addition, parametric studies have also been conducted, and the results indicated that there exists an optimal design of the cylindrical structure for PCEH to maximize energy harvesting power (geometry, the number of PVDF layers, and resistive load value). Further study will focus on experimental verification and the design of broadband piezoelectric cylindrical energy harvester.

6. References

[1] Beeby S P, Tudor M J and White N M 2006 *Meas. Sci. Technol.*, vol. **17**, pp. 175-195.
[2] Roundy S, Wright P K and Rabaey J 2003 *Comput. Commun.* **26**, 1131–1144.
[3] Paradiso J A and Starner T 2005 *IEEE Pervasive Comput.*, **4**, pp. 18-27.
[4] Harne R L, Wang K W 2013 *Smart Mater. Struct.* **22**, 023001.
[5] Ottman G K, Hofmann H F, Bhatt A C and Lesieutre G A 2002 *IEEE Trans. Power Electron.*, vol. **17**, no. 5, pp. 669–676.
[6] Christin D, Mogre P S and Hollick M 2010 *Future Internet*, vol. **2**, no. 2, pp. 96–125.
[7] Anton S R and Sodano H A 2007 *Smart Mater. Struct.*, 16R1.
[8] Mei J and Lijie Li 2015 *Sens. Actuators, A*, **233**, 405–413.
[9] Renwen C, Long R, Huakang X, Xingwu Y and Xiangjian L 2015 *Sens. Actuators, A*, vol. **230**, 1–8.
[10] Wei-Jiun S and Jean Z 2013 *Appl. Phys. Lett*, **103**, 203901, doi: 10.1063/1.4830371.

7. Acknowledgments

This research is supported by the Singapore Institute of Manufacturing Technology (SIMTech) under the Agency for Science, Technology and Research (A*STAR).

IOP Conf. Series: Earth and Environmental Science **83** (2017) 012011 doi:10.1088/1755-1315/83/1/012011

Renewable Energy Power Generation Estimation Using Consensus Algorithm

Jehanzeb Ahmad[1], M. Najm-ul-Islam[1] and Salman Ahmed[2]

1 Bahria University Islamabad, Pakistan
2 Sarhad University Peshawar, Pakistan
E-mail: Jehanzeb_ahmad@yahoo.com; sahmedbahria@gmail.com; najam@bui.edu.pk

Abstract. At the small consumer level, Photo Voltaic (PV) panel based grid tied systems are the most common form of Distributed Energy Resources (DER). Unlike wind which is suitable for only selected locations, PV panels can generate electricity almost anywhere. Pakistan is currently one of the most energy deficient countries in the world. In order to mitigate this shortage the Government has recently announced a policy of net-metering for residential consumers. After wide spread adoption of DERs, one of the issues that will be faced by load management centers would be accurate estimate of the amount of electricity being injected in the grid at any given time through these DERs. This becomes a critical issue once the penetration of DER increases beyond a certain limit. Grid stability and management of harmonics becomes an important consideration where electricity is being injected at the distribution level and through solid state controllers instead of rotating machinery. This paper presents a solution using graph theoretic methods for the estimation of total electricity being injected in the grid in a wide spread geographical area. An agent based consensus approach for distributed computation is being used to provide an estimate under varying generation conditions.

1. Introduction

With rapidly decreasing cost of residential PV systems, there has been a rapid increase in the number of households that have shifted to a grid tied PV system. According to Solar Energy Industries Association (SEIA), the residential solar grew by 51% in 2014 compared to 2013 in the United States. [1]. The total installed capacity of residential solar crossed the 1GW barrier in 2014 [1]. We see similar trends in most of the other developing countries as well. The Government owned utility (WAPDA) in Pakistan has not been able to add new generation to keep pace with the increase in demand resulting in massive load shedding throughout the country. The Government approved the policy of net metering in 2015 and it is expected that in the next couple of years a large number of households will shift to grid tied systems. The benefits to consumers and the environment are well known. However since DERs are installed at the distribution level of a power network, it can cause a number of issues related to power quality and reliability (PQR) for the end consumer. For Example, in order to properly maintain voltages at the distribution level according to IEEE C84.1 standard [2], which states that for a 120V line voltage a maximum fluctuation of +- 5% at the service entrance is allowed and up to 8.5% is permissible under special circumstances for short duration of time. In order to maintain such voltages, the location of the Distributed Generators (DGs) and the maximum current injected in the grid become very important. DGs placed very close to the Load Tap Changing (LTC) distribution transformer with line drop compensation will not be able to accurately measure feeder loading, resulting in over voltage at the far end of the feeder [3].

In addition, the various effects on the grid due to DGs increase as the percentage of power injected in the grid increases compared to the maximum current carrying capacity of the feeder. In [4] the authors have shown that the customers close to the distribution transformer were able to inject 100% of their generated power in the grid without causing any PQR issues, whereas the customers at the end of the feeder were able to inject just 35% of their generated power before the limits for PQR were exceeded. Injection of harmonics due to solid state inverters is another issue that needs to be considered. The IEEE 519-1992 [5] standard specifies the maximum amount of harmonics that can be injected before overheating of system equipment occurs. A maximum of 5% Total Harmonic Distortion (THD) and 3% individual harmonic distortion is allowed. In [6] the authors have investigated the maximum percentage of feeder capacity that can be used for DG before the limits of IEEE 519-1992 are exceeded. They have performed calculations for zero and positive sequence harmonics. In their paper it has been shown that for long length feeders of around 10 miles, the allowable limit drops to just 49% of the maximum capacity of the feeder before the 3% individual harmonic limit is exceeded.
In addition to the above mentioned issues, DG can cause voltage flicker and can have an impact on the short circuit currents of the feeder [3]. Keeping in view these problems, it is imperative for the load management centers to have an accurate estimate of the amount of power being injected by these DERs. Since DERs are inherently distributed over a wide geographical area, a central controller is not ideally suited for such conditions. Instead a Multi agent distributed control approach can yield much better and faster results. The distributed technique should also have the ability to cope with rapid changes in the DG topology which can occur due to weather changes or a resource going offline due to some fault conditions. Load management centers typically estimate the power demand and generation on a 15 to 30 minute interval basis. The distributed technique should be fast enough to provide an estimate of the total power every 15 minutes even under changing network conditions. Consensus based graph theoretic methods provide an ideal mathematical foundation to solve such type of distributed networks. Each DG is modeled as an agent with the ability to communicate with other agents in its neighborhood. The advantage of this approach is that each DER just has to communicate in a small geographical area instead of communicating with a central controlled located many miles away. An introduction to graph theory is presented in the section II. A model of the DER has been developed in Matlab and presented in section III. Section IV presents the results of applying the consensus algorithm on the network. Section V concludes the paper showing consensus algorithm to be an effective technique for distributed generation estimation.

2. Consensus algorithm

2.1. Graph theory

A graph is used to model the DER network. Let $\mathcal{G}$ denote a graph such that $\mathcal{G}$ = (V, E, A) where V is the set of elements called vertices or nodes, E is the set of pairs of distinct nodes called edges and A = $[a_{ij}] \in \mathbb{R}^{nxn}$ is the Adjacency matrix. A directed graph has edges with a direction associated with each edge. An undirected graph has no edge directions. The neighborhood $N_i \subseteq$ V of the vertex v_i is the set $\{v_j \in \mathrm{V} \mid v_i v_j \in \mathrm{E}\}$, that is the set of all vertices that are adjacent to v_i. If $v_j \in N_i$ it follows that $v_i \in N_j$ as the edge set in an undirected graph consists of unordered pairs. A graph $\mathcal{G}$ is connected if for every pair of vertices in V ($\mathcal{G}$), there is path that has them as its end vertices.
Different types of matrices can be associated with graphs which represent different properties of the underlying graphs. The three most important matrices associated with graphs are $\Delta(\mathcal{G})$, A($\mathcal{G}$) and L($\mathcal{G}$), where $\Delta(\mathcal{G})$ is the degree matrix i.e a matrix with diagonal entries containing the vertex degrees on the diagonal and remaining entries as 0. The adjacency matrix A ($\mathcal{G}$) is a symmetric nxn matrix encoding of the adjacency relationships in the graph $\mathcal{G}$ such that

$$[A\ (\mathcal{G})]_{ij} = \begin{cases} 1\ if\ v_i v_j \ \ \in E, \\ 0\ otherwise. \end{cases}$$

The L($\mathcal{G}$) represents the Laplacian matrix which plays a very important role in the consensus algorithm. The L($\mathcal{G}$) is defined as

$$L(\mathcal{G}) = \Delta(\mathcal{G}) - A(\mathcal{G})$$

From the definition above it can be inferred that the rows of the Laplacian sum to zero for all types of graphs.

2.2. Consensus algorithm
Let $x_i(k)$ be the kth state of the node i. For a discrete time system the next state x_i (k+1) is given by the equation

$$x_i(k+1) = x_i(k) + u_i(k)$$

Where

$$u_i(k) = \alpha \sum_{j \in N_i} (x_j(k) - x_i(k))$$

When the adjacency is symmetric the equation can be written as

$$u(k) = \alpha[A(\mathcal{G}) - \Delta(\mathcal{G})]x(k)$$

$$u(k) = \alpha\big(-L(\mathcal{G})\big)x(k)$$

Therefore equation (I) becomes

$$x(k+1) = x(k) + \alpha(-L)x(k)$$

$$x(k+1) = x(k)[I - \alpha L]$$

Let

$$A = [I - \alpha L]$$

Then we can write

$$x(k) = A^k x(0)$$

It has been proven that the above equation would converge if A is a Stochastic Indecomposable and Aperiodic matrix (SIA matrix) which has a rank of 1 [7]. The matrix A would be SIA if and only if there exists a spanning tree in the underlying graph of the network.

Consensus is achieved when [8]

$$\lim_{k \to \infty} \{x_i(k) - x_j(k)\} \to 0$$

That is all nodes reach a single value within a specified time. Therefore so long as there is a spanning tree in the graph we can be sure that a consensus value will be reached.

2.3. Estimation of total power using consensus algorithm
A survey of the literature shows that so far graph theory has not been applied in total distributed power generation estimation. In [9] the authors have applied graph theory to solve the economic dispatch problem effectively. In this paper we have used the consensus value reached by the nodes to calculate the total power. This has been achieved by observing that the consensus value reached by the nodes is the average of the initial values of the nodes [10]. Hence if there are n nodes with $p_1, p_2, p_3, \ldots \ldots p_n$ as the initial power being produced by each DER and the consensus power value is $P_{consensus}$ reached after some time (assumption is that the consensus is reached within a 15 minute interval) then total power P_{total} is given by

$$P_{total} = P_{consensus} \times n \text{ [15 minute interval]}$$

The same process can be repeated again with new initial values after 15 minutes to get a new estimate. The interval is small enough that no significant change in power production would occur in that interval from initial values. Taken over a large number of nodes the estimate would be fairly accurate for load dispatch centers.

3. Modeling in matlab

Consensus based algorithm has been implemented in Matlab code. Two different networks have been simulated in Matlab. Fig 1 shows an implementation of 8 nodes representing 8 DERs. The graph is well connected and the edges between the nodes show communication links between the DERs. In figure 2, another topology consisting of 10 nodes has been implemented where node 9 and node 10 have been added to the existing network and node 10 is connected to just node 9 only. This has been done to indicate isolated DERs which communicate with just one other DER in the network.

The initial values of the DERs have been taken as 10KW, 5KW, 5KW, 12KW, 1KW, 15KW, 20KW, 4KW each. Two additional nodes 9 and 10 have been added later with power of 10KW and 7KW each. Simulation time has been set at 20 seconds.

Figure 3 shows the results of consensus algorithm with 8 DERs. It can be seen that within 20 seconds or less the nodes achieves consensus and all nodes acquire the final value of 9 KW. The 9KW is the average of the all initial values of the nodes. In Figure 4 the simulation results of consensus algorithm are shown with the additional nodes (DERs) which are more isolated from the network. As can be seen from the results that consensus is again achieved within 20 seconds and the consensus value if 8.9 KW which is the initial value average of all DERs including node 9 and 10.

In the simulations a time invariant topology has been assumed to simplify the design, however the consensus algorithm is extremely robust and so long as a node is connected to just one other node with the network, consensus is achieved rapidly. Even in case with intermittent communication links and delayed update of values of some of the nodes, consensus can be reached but the mathematical formulation of those conditions is more complex and has not been simulated in this paper.

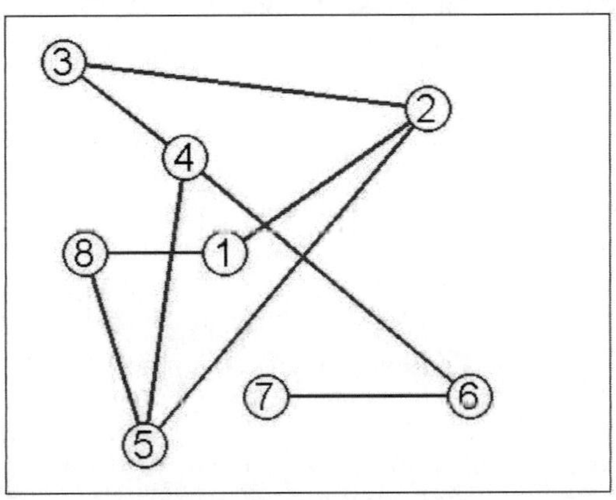

Figure 1. Network of 8 DERs

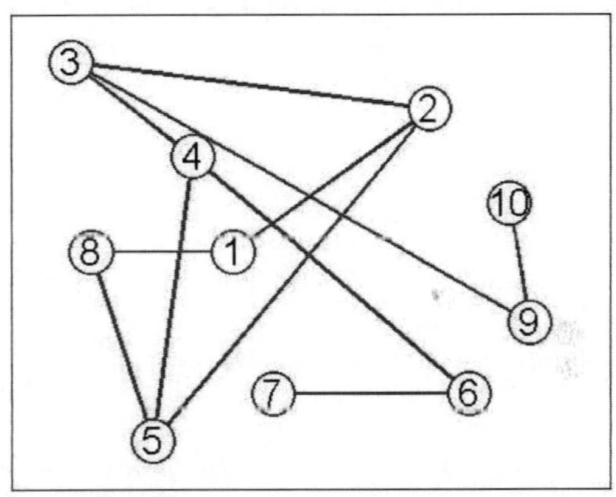

Figure 2. Addition of two isolated DERs

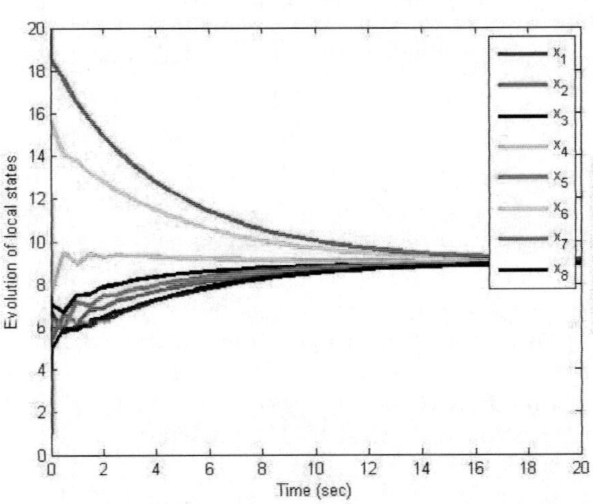

Figure 3. Consensus power with 08 DERs

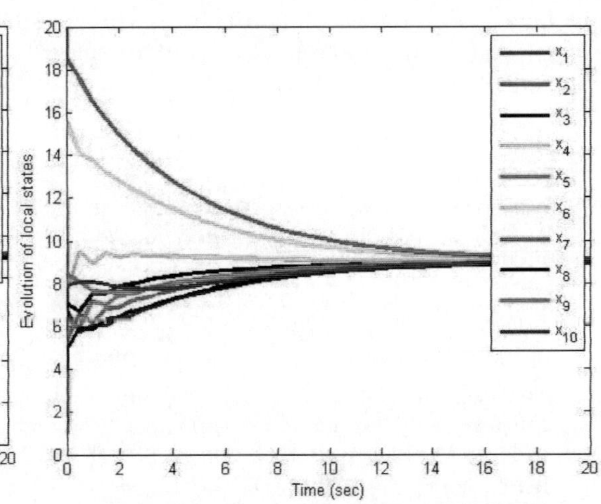

Figure 4. Consensus power with 10 DERs

4. Results

4.1. Summary

As we can see from the simulation results in Matlab, that consensus has been achieved using the standard consensus algorithm. The consensus values are 8.9KW for the 8 nodes network and around 9KW for the 10 nodes network. One of the underlying assumptions in this paper is that the number of nodes representing a DER would be known. Using this knowledge it is a simple process of multiplying the number of nodes with the average consensus value to get an accurate estimate of the total power being injected in the system at any given time interval. In scenario 1 with 8 nodes the total power comes to 72 KWs and in scenario 2 with two additional nodes the total power calculated is 89 KWs. These results can be verified by observing the initial values of the DERs. As, can be seen from the initial values of the power being injected by the DERs, that the algorithm accurately calculates the total power being generated at any given time. The simulation results show that the consensus algorithm is a very power technique to estimate the total power injected in the grid without any central controller.

Assumptions

- The underlying graph of the DERs should have a spanning tree.
- All DERs are able to periodically update their values and share with their neighbors.
- All links are operational within certain time bound intervals.
- An unconnected DER will not be included in the total power estimation.
- No DER is stuck at any given value. In such a case, the algorithm would fail to give the correct value and all nodes would asymptotically acquire the value of the stuck node.
- All nodes are cooperating, and there is no hostile node, remaining stuck at one value or injecting wrong values in the network.

5. Conclusion

This paper has explored the use of consensus algorithm to evaluate the total power injected in the grid through DERs. The method has been simulated with 10 DERs and random communication links between them. It has been shown that the total power is calculated very quickly and accurately. Consensus is achieved in less than 20 seconds. The method is extremely robust and can be scaled to any number of DERs distributed over a large country or a geographical area. The algorithm works without a central controller and the nearest DER to a load management center can convey the total power injected. This reduces the cost of expensive long distance communication links and the links will get progressively shorter as the density of DERs increase in a country or a locality. In our future work we plan to model network with a time varying topology and randomly delayed updates. This will more accurately reflect the real world scenario of nodes going down or not being updated due to some fault in the communication network. In our future work we also plan to increase the number of node to more than a thousand and see if consensus can be reached within a specified time frame needed by the load management centers.

6. References

[1] http://www.seia.org/research-resources/solar-industry-data.

[2] *Voltage Ratings of 60 Hz Electric Power Systems*, ANSVIEEE (284.1-1995, Published by Institute of Electrical and Electronic Engineers, 1995

[3] Barker, P.P.; de Mello, R.W., "*Determining the impact of distributed generation on power systems. I. Radial distribution systems*," Power Engineering Society Summer Meeting, 2000. IEEE , vol.3, no., pp.1645,1656 vol. 3, 2000. doi: 10.1109/PESS.2000.868775

[4] Tonkoski, R.; Lopes, L.A.C.; El-Fouly, T.H.M., "*Coordinated Active Power Curtailment of Grid Connected PV Inverters for Overvoltage Prevention*," Sustainable Energy, IEEE Transactions on , vol.2, no.2, pp.139,147, April 2011. doi: 10.1109/TSTE.2010.2098483

[5] IEEE Draft*: Guide for Applying Harmonic Limits on Power Systems (PAR Withdrawn)," IEEE Std P519.1/D9a* , vol., no., pp.,, 2004 URL: http://ieeexplore.ieee.org/stamp/stamp.jsp?tp=&arnumber=4040444&isnumber=4040443

[6] Bhowmik, A.; Maitra, A.; Halpin, S.M.; Schatz, J.E., "*Determination of allowable penetration levels of distributed generation resources based on harmonic limit considerations*," Power Delivery, IEEE Transactions on , vol.18, no.2, pp.619,624, April 2003. doi: 10.1109/TPWRD.2003.810494

[7] J. Wolfowitz, "*Products of indecomposable, aperiodic, stochastic matrices*," *Proceedings of the American Mathematical Society*, vol. 15, pp. 733–736, 1963.

[8] A. Jadbabaie, J. Lin, and A. S. Morse, "*Coordination of groups of mobile autonomous agents using nearest neighbor rules*," *IEEE Trans. on Automatic Control*, vol. 48, pp. 988–1001, June 2003.

[9] Ziang Zhang; Mo-Yuen Chow, "*Convergence Analysis of the Incremental Cost Consensus Algorithm Under Different Communication Network Topologies in a Smart Grid*," Power Systems, IEEE Transactions on , vol.27, no.4, pp.1761,1768, Nov. 2012 doi: 10.1109/TPWRS.2012.2188912.

[10] Wei Ren; Beard, R.W.; Atkins, E.M., "*A survey of consensus problems in multi-agent coordination*," American Control Conference, 2005. Proceedings of the 2005 , vol., no., pp.1859,1864 vol. 3, 8-10 June 2005 doi: 10.1109/ACC.2005.1470239

IOP Conf. Series: Earth and Environmental Science **83** (2017) 012012 doi:10.1088/1755-1315/83/1/012012

Financing Renewable Energy Projects in Developing Countries: A Critical Review

Donastorg A[1], Renukappa S[1] and Suresh S[1]

1 University of Wolverhampton, U.K.
E-mail: A.donastorg@wlv.ac.uk,

Abstract. Access to clean and stable energy, meeting sustainable development goals, the fossil fuel dependency and depletion are some of the reasons that have impacted developing countries to transform the business as usual economy to a more sustainable economy. However, access and availability of finance is a major challenge for many developing countries. Financing renewable energy projects require access to significant resources, by multiple parties, at varying points in the project life cycles. This research aims to investigate sources and new trends in financing RE projects in developing countries. For this purpose, a detail and in-depth literature review have been conducted to explore the sources and trends of current RE financial investment and projects, to understand the gaps and limitations. This paper concludes that there are various internal and external sources of finance available for RE projects in developing countries.

1. Introduction

It is indisputable that the growing concerns over climate change are provoking a worldwide transformation in the way that governments and industries seek to supply energy while at the same time learning, creating and implementing new measures to aid in minimising greenhouse gas emissions and other environmental impacts. One of the major energy policy strategy applied in many countries worldwide is the employment of renewable energy sources (RES). Because energy is one of the main sectors that fuels the global economic activity. Consequently, the expansion in population growths, which is estimated by [1] to be 2.3 billion more people by 2050 compared to 2016, the living standards and demand, that are estimated by [2] to increase by 21% by 2030, are interlinked and influence the development of a country. The decisions made and implemented for and on the energy sector infrastructure, especially in the investment aspect, would lock the financial strategy at the very least for a few decades. This relates to how efficient the energy sector growth across the economy. The energy sector is directly connected to the sustainability and vibrancy of a countries economy. Any major decision made in the Energy sector will have a ripple effect throughout the economy.

Until recently Renewable energy (RE) investments were treated in the same manner as any other investment, yet RE investments possess certain characteristics that require a high level of understanding, some of the key aspects are:

- The feasibility of the investment (This depends on the policy and regulations and the impact on the economy that such measures can have)
- The durability and real implementation of any subsidies, grants, tradable certificates or tax credits.
- Basic financial analysis.

The governments of many developing countries have embarked on the path to low-carbon development to create and increase energy access, economic opportunities and to reduce carbon

emissions. However, because of scarce public funding and investments, the international community and the private sector engagement will play a major role.
The RE markets is a 19-26 Trillion-dollar investment opportunity, estimated by [3]. The renewable energy industry faces a rapidly increasing opportunity and challenge to grow the business and financial strategies to exploit this investment properly. Current and ongoing cost reductions in renewable power generation technologies can aid developing countries in achieving national and international energy and emissions policy goals, in energy security and reliability and affordable energy; along with promoting access to electricity for all at a lower cost than traditional sources. As Germany, Costa Rica and many more countries have demonstrated. Through the use of Biomass, hydropower, geothermal, solar and onshore wind power for Electricity that shows competitively-levelling cost of RE compared to fossil fuel-fired power generation. Being solar, the most impressive of this Levelized cost of electricity (LCOE) for RE. Solar energy cost generation alone has halved just between 2010 and 2014. Technology enhancements, occurring simultaneity as the reduction in the installed costs of RE sources have augmented the competitiveness of Renewable Energy in the energy market to be equal, and in some cases the LCOE of some RE, like offshore wind even lower than for fossil fuels.

Table 1. New characteristics in renewable energy financing

Cost Reduction	Policy Support	New & Improved Technology
New & improved renewable Systems	Investment	Research and investment
Higher investment	Sustainable goals	Development

The global investment for renewable energy has been exponentially increasing over the years. Due, in part, to the new financial characteristics, that have been implemented in current years (Table 1). The amount of money committed to renewables projects has risen, according to some studies [4]-[6] 5% in 2015, a record high of $285.9 billion, however, according to some research, this drop down 18% in 2016. Along with this record high, 2015 is also the year where investment in renewables in developing countries surpass that in developed countries. Were, China, India, Brazil, South Africa, Mexico, Chile, Morocco, Uruguay, the Philippines, Pakistan and Honduras, together invested over $156 billion in renewable energy investments; developed countries invested only $130 billion. Meanwhile, renewables projects possess attractive advantages for developing countries, such as, the built time, wind farms can take from nine to twelve months, solar parks in three-to-six months, compared to fossil fuel (coal and gas plants) that can take several years to be completed, not to disregard the longer time that nuclear energy takes to installed. Along with the high competitiveness of renewable in current energy markets. However, even with all this improvement policy support for renewables remains fickle.
This study compares main funding mechanisms employed by governments, institutions and businesses to finance renewable energy development programs or projects: feed-in tariffs, tax incentives, and tradable green certificates; in addition to new support mechanisms that have appeared in later years: co-operative funds, hybrid Bonds revolving fund, Tax equity loans, loan loss reserve and crowd funding.

2. Discussion of economic opportunities of financing renewable energy projects

A fundamental concept to understanding the economic competitiveness of any energy project is the real costs of the project versus the benefit achieved throughout its life cycle, regardless if it is fossil or renewable. This concept is the foundation for many investments decisions because it is based on the capital cost, fuel cost and financial costs of the project. These cost at the same time are based on the sum of the cost over the lifetime of the project over the sum of the electrical energy produced over that lifetime. This is subjective to the access to precise, comparable, dependable and current information on the actual costs and life cycle operation of renewable energy technologies and projects. The financial instruments can be created, design and implemented by the private or public sector. As these encompasses the government and non-government own enterprises.

2.1. Private sector

The government of many developing countries has taken measures to commit and ensure a reduction fo carbon emission. For this purpose, renewable energy transformation steps have been taken. However, the financial struggles that many developing countries face is, in some cases, and almost impossible challenge. The private sector engagement is a necessity rather than an option, as this sector can facilitate and aid in the renewable energy investment with strong benefits not just for the country but the non-government businesses as well. As [7] explains an appropriate financial and political framework for the cooperation of the private sector in renewable energy projects. Also, the risk and perceive risk most be address as to attain the engagement of this sector fully. However, as [8], [9] highlight, developing countries use public and concessional resources to attain the aid of the private sector, even though this does not fully address the different challenges or the risk that the private sector experiences. On the other hand, [3], [10], [11] point out that the use of this instruments to engage the private sector has dual a benefit of being more sustainable and minimising the instability that this could bring to the industry; while augmenting the competitiveness of the renewable energy market.

2.2. Public sector

Many developing countries have embarked on a renewable energy transformation, due to the agreement and international commitment to reducing global emission, in addition to creating economic opportunities and increasing energy access of the population; however, most developing countries lack the financial means to achieve this. However, as many researchers have shown[10]-[14], policymakers in developing countries tend to implement a particular financing instrument (Tax incentives, Loans and more) without, in most cases, analysing which instrument or combination of tools would be most effective for the project or country at the time. Affordability is one of the critical challenges that developing countries faced when implementing renewable energy.
Also, the standard process for any energy projects, be it finance by the public or private sector, is first obtaining a power purchase agreement (PPA) or through "yieldco". Although, according to many new researchers [15]-[17], this is also changing along with the funding sources. The new trends that have already been used in 2015 by:

- the UK [17] demonstrated winning bids at 11% below the agreed for onshore wind and offshore a 14% and 18% below the officially set price.
- In South Africa [18] auctions awarded contracts to onshore wind at 41% less in local currency terms than the first auctions
- and Germany [19] the second PV auction in 2015 awarded contracts 7.5% below the previous feed-in tariff level. These auctions allow developed and especially developing countries to offer the generation agreement of energy for keen and in some cases, very reduce prices compared to the traditional PPA. Allowing for the tariff prices to diminish and the competitiveness of the market, particularly the RE, to flourish.

2.3. Investment sources

Traditionally renewable energy (RE) projects had two ways of obtaining funds:

- Borrowing from a bank (Loan). However, Banking institutions will focus on the repayment of the debt and not on the return of the transaction [20]. In such cases, due to the emphasis of the banks, the return for such ventures tends to be smaller than in other funding methods.
- Through equity capital (selling stakes or shares in the business, among others). Equity capital has greater expect returns due to the level of risk that is taken. In some cases, some companies might expect between a 25-35% of return, due to the perceived risk versus the real risk [21]. Also, stakeholders place a greater pressure and expectations on RE projects than traditional financial institutions do. Other ways to finance through equity is by "on balance sheet" funds (from funds drawn internally, treasury department), this is usually done by the utility companies as part of a corporate RE strategy. However, the specific role played by equity may

IOP Conf. Series: Earth and Environmental Science **83** (2017) 012012 doi:10.1088/1755-1315/83/1/012012

change over the lifecycle of the RE project as it is refinanced, depending on existing and new actors that would like to benefit from the project.

Per [22], equity funding is a popular option. As of 2015 financing of RE projects through utility-scale for wind farms and solar parks was of $199 billion in comparison to the 188 billion of the previous year, showing a 6% increase in a years' time. In general, the growth in equity funding in RE in 2015 was between 5.8-12% over previous years, in areas such as roof-top solar projects, reinvestment of equity and acquisition activities.

According to [23]-[25], an important difference between this two methods is also, in which RE projects they are utilised. Loans are usually applied to RE projects with traditional and proven technology and approaches, while equity is used more for new and innovative RE technology projects and methods. Another factor would be the financial options of the country; in some countries, local financing options are plentiful; in others, they are few and far between. The key features of these financial tools have been described in table 2. Other innovations in the funding of RE projects come from Europe; since 2012 new inflation-linked notes have appeared as a way for organisations to access the return flows of RE project, specifically the wind and solar.

Table 2. Key features of traditional funding for RE

	Equity	**Loans**
Source of Funding	Wide range of sources (Insurance companies, pension funds, mutual funds, Stock Market, Real State and more)	Financial Institutions (Banks)
Target	New Technologies, Methods and Markets	Mature Technology and markets
Risk	Low-Medium -High (Depending on the source of funding)	Low risk
Return time	3-10 years (Depending on the RE project and on the Equity that is funded)	2-5 years (Depending on the specific terms of the Loan)
Types	Venture Capital, Private Equity and Funds	Personal, Commercial, Small business and more
Return	Low- Medium-High	Low
Benefits	• Diversity • Liquidity • Public Trading transparency of the market price	• Money guarantee • None involvement in the RE project by the banks • Accessibility and options • Tax Benefits

The Feed-in Tariff (FIT) is another financial tool that is favoured as it is an energy-supply policy or agreement that is the base of supporting the development of new renewable energy power generation. However, more developing countries relay on this method since FIT contracts provides a guarantee of payment (dollars per kilowatt hour, $/kWh) for the full power generated and is a long-term period agreement (between 15-20 usually), this will vary depending on Levelized cost, the renewable technology type (Wind, solar, hydroelectric, etc.), project size, quality, strength and availability of the resource, the PPA and (or) other project/contract/agreement-specific variables. However, not just developing countries implement FIT, according to researchers [13], [22], [23] and much more, over 40 countries implement FIT policies. This policy is attributed the reason, per [15], [23], of the success of the renewable energy markets in German and Spain. Other possible benefits of a properly designed FITs are that it is more cost-effective than standards renewable energy portfolios, also, this makes FIT's more competitive in the general energy market. Taxes and Incentives for Renewable Energy is another go to tool for developing countries. These taxes, are created and implemented to aid in the communication of the government's policies, programs and plans for the development of renewable energy to the energy companies, investors and other entities. As Table 3 shows the main characteristics of Tax incentives and the reason why they have such a successful implementation in developing countries.

Table3. Key features of Tax Incentives (Production tax credit, PTC and investment tax credits, ITC)

Improve funding for renewable energy technologies	Encourage, increase and improve renewable energy market adoption
Job creation in the renewable energy sector	Encouragement of public investment in renewable energy
Increase R&D in local renewable energy technology	Create and encourage renewable energy education and training programs

On the contrary internal and international loans are more than a financial tool for developing countries, they are almost a way of life, especially international loans from global institutions or others countries treaty. This is due to loans being a financial tool (agreement between the actors) implemented in most case by banks as a means of providing finance or capital for companies, investors or individuals to support normal business or project operations. For this, the financial institution will conduct an assessment of the company's or country's financial strength and stability of the project projections and returns, and debt is calculated priced accordingly to the market at the time with a return period and interest rate calculated and agreed upon in the contract. Also, these agreement place few restrictions on how the company or ocuntry can use the funds, provided certain general conditions be met.
Another financial tool that is gaining popularity in developed countries are the Bonds (As can be seen in Table. 4). A bond is a loan or better known as a debt investment that is based on a coupon (fixed or variable interest rate) of an investors loan to an entity (typically corporate or governmental) for a fixed period. In renewable energy, bonds are better known as green bonds, which are any variety of bonds that the incomes will be solely used to finance (re-finance) in full or partly a new and existing renewable energy project. It is often used to raise money to fund an acquisition or a new development. This type of financial tool is rarely used in developing countries.

Table 4. Different types of Green Bonds
Green Use of Proceeds Bond
Green Use of Proceeds Revenue Bond
Green Use of Proceeds Project Bond
Green Use of Proceeds Securitized Bond
Pure Play
Hybrid Bond
Other types will appear as the renewable energy market matures

A cooperative fund or trust is a self-governing alliance of people that have united voluntarily to fulfil a common economic, social and (or) cultural need(s) through a jointly and democratically owned and controlled business. Crowdfunding is an alternative finance form that has become a popular financing tool in the last couple of years [25]. This tool allows a project, an organisation or a company to gain its funds from the general and global public using open calls on the internet, on a dedicated and detailed web page and platform. This is a result of the progressiveness of the information and communication technology (ICT), the increased use and popularity social networks and the rapid advance and use of interactive technology. Crowdfunding can indeed be defined as an 'economic superstructure' of social networks and crowdsourcing.

2.4. Risk and return

An important area of understanding any finance and investment subject is a risk and return. The purpose of any business is at its core the generation of income. In this same way, financial institutions are based in on the return of the investment versus the risk that has or will be undertaken. This is directly proportionate; a higher risk of investment equals a greater return. The RE sector utilises finance from across the entire risk-reward spectrum (Table 5).

Table 5. Different risks of investing in RE

Country	Governed by the stability, status, seriousness and transparency of the government, it is the legal system, business practices and links to the risk encore by currency.
Economic	Depends on the inflation and future projection of the currency, local financial regulation, market growth and GDP
Financial	Based on the coupons or interest rates, refinement of agreements or projects, insurance of business, projects and companies along with asset liquidation and shares in the company
Currency	Subject to the exchange rate fluctuation, currency controls, devaluation, currency flow
Political	Influenced by the changes in the legal framework, directly link to the countries risk. However, more dependent of the policies and implementation of them.
Security	Refers to the asset insurance and robust legal framework in correspond to the ownership and operations of a project
Perceived	Based on the popular perception and social knowledge of the actors involved in the project and financing. Influenced by a subjective understanding of the individuals based on past cases or projects.

3. Conclusion

Financing Renewable energy in developing countries has become a revolution and necessity, as the global initiatives to mitigate climate change grow more critical. Renewable energy targets have been implemented in at least 164 countries. Developing countries, especially, have a challenging future when it comes to the transformation to renewable. This is due to the core challenges that developing countries face: (1)Financing Renewable Energy, (2) Knowledge Management of renewable energy, (3) Legal framework and implementation of this context for renewable energy and (4) Political leadership and transparency. In this paper, the focus is on the economic understanding, as developing countries lack the means to implement renewable energy projects on their own successfully. This usually leads developing countries to maintain old and unfitting financial methods and tools for the renewable energy projects. This is in part due to a perceived but unfounded risk of the new financial methods. As many developed countries around the world step up efforts in the financial direction, a better understanding of the economic impact will support informed policy and decision makers. Also, this will maximise the benefits of the implementation of renewable energy. Due to the specific situation of developing countries, it is hard, if not impossible, for the governments alone to accurately evaluate and determine which renewable energy technologies and financing tool are the most appropriate.

Developed countries have anchored the investor's confidence by creating credible and time sensitive renewable energy targets, this also helps in mapping the trajectory of the renewable energy development in the current energy sector and provides with a future projection for financial tools. Consequently, these renewable energy targets must be based and backed by comprehensive and dedicated policies, regulations and legal frameworks; along with in-depth education programs and knowledge management. As the deployment of policies is not enough, due to the situation of developing countries, a mix of policies is the best approach for developing countries. These measures

ensure an expectable income stream for renewable energy projects. While aiding in creating a stable investment environment, that will help in overcoming noneconomic barriers (social, environmental and more.). Effective deployment of policies has supported the growth of renewables globally (point in case Germany and Spain). As of 2015, more than 145 countries have introduced regulatory (e.g. feed-in tariff, net metering, auctions), and fiscal incentives and public cooperatives as financial tools, that have effectively transformed the economy to a low carbon-based economy.
Financing (e.g. capital subsidy, investment or production tax credit). This number has increased nearly ten times over the past decade, due to the new trends in financing renewable energy. Renewable energy now forms part of a high share of the energy mix for power generation of several countries, with substantial growth projected for this sector in the coming decades, due to the lowering of cost and continues innovation of the renewable technology and the outreach of the knowledge management is this area. This has led to a dramatic shift in the challenges that once plague renewable, and has shifted the perception of policy makers worldwide, and presenting a growth opportunity for developing countries. Accordingly, governments are adapting existing policies to ensure that incentives are appropriate while increasing transparency and stability. Renewable energy policies have focused mainly on the electricity sector. There is already a trend towards greater adoption of policies for the heating/cooling and transportation sectors, but further attention will be required for the end use sectors. One of the biggest challenges that new financial trends face in developing countries is the perceived risk due to the inaccurate and sometimes misleading knowledge that some developing countries have regarding the cost and power generation of renewable energy and that policy makers based decisions. This is due to the lack of transparency in the government and the methodology along with the lack of reliable data, resources and assumptions used to make cost calculations. However, many ways exist to measure the cost of power generation technologies, and each way brings its insights, benefits and challenges. To strengthen and empower institutions, it is crucial to identify, assess and address existing barriers to their modes of operation and development. Developing countries find themselves at a crossroad when it comes to investments in the clean technology sector, as this is always combined with a capital intensity and new technologies, that the countries lack. Therefore, securing project financing from local or international sources can prove to be a critical step in the creation and growth of the renewable energy sector. PPA have proven to work best in developing countries. However, a new trend of Dutch auctions are making progress through developed and hopefully developing countries would follow; as this allows for the cost of electricity to diminish while creating long-term agreements that benefit all impacted actors.
An investment-friendly environment is essential to overcome financing barriers and attract investors. To double the share of renewables by 2030, global annual investments in the renewable power sector need to be in the range of USD 500 billion to USD 750 billion between now and 2030 (compared to over USD 270 billion in 2014). The greatest shares of new investments will need to come from private sources. As deployment grows and new markets emerge, developers are improving in forecasting cash flows, while financiers can more accurately assess risk and design financing products suited for renewable energy projects. Nevertheless, actual and perceived risks continue to slow down investment growth in renewable energy, especially in new markets. Policymakers and international financial institutions must deploy the right policy and fiscal tools to address these risks and mobilise private sector investment. Public funding will continue to remain a major catalyst and will need to increase. Implementing tax incentives or Green Bond trading mechanism could be considered ideal policies to aid in the growth of the renewable energy market and the countries economy. However, the discussion regarding the new and standard financial renewable portfolio policies showed that the firsts are successful when used for new and innovative technologies and projects, but are linked to a high risk that many developing countries are not willing or possess the knowledge on how to handle. Also, the latter are appropriate financial policies when it is applied by the government, and have a low risk for the investors. Finally, a review of funding mechanism for each country is needed as each case differs to others. Also, financial tools alone cannot transform a country to the use of renewable energy, it most is a joint effort, of not only the actors (Government, industry and population) but of the different sectors, business plan, technology, education and resources.

IOP Conf. Series: Earth and Environmental Science **83** (2017) 012012 doi:10.1088/1755-1315/83/1/012012

4. References

[1]. UNEP, 2012. Financing renewable energy in developing countries. UNEP finance initiative – innovative financing for sustainability. Feb-12.
[2]. Taylor M, Daniel K, Ilas A and Young So E, 2014. Renewable Power Generation Costs In 2014. IRENA
[3]. UNEP, 2009. Private financing of renewable energy -A guide for policymakers-.
[4]. Chirambo, D, 2016. Addressing the renewable energy financing gap in Africa to promote universal energy access: Integrated renewable energy financing in Malawi. Renewable and Sustainable Energy Reviews, 62, pp.793–803.
[5]. Lee, C W & Zhong, J, 2015. Financing and risk management of renewable energy projects with a hybrid bond. Renewable Energy, 75, pp.779–787.
[6]. UNEP, 2016. Green Finance For Developing Countries. Inquiry: Design of a sustainable financial system. May-16
[7]. Abolhosseini, S & Heshmati, A, 2014. The main support mechanisms to finance renewable energy development. Renewable and Sustainable Energy Reviews, 40, pp.876–885.
[8]. UNEP and SEFI, 2007. Analysis of Trends and Issues in the Financing of Renewable Energy and Energy Efficiency in OECD and Developing Countries. Global Trends In Sustainable Energy Investment 2007
[9]. UNEP and SEFI. 2004. Financial Risk Management Instruments for Renewable Energy Projects
[10]. Cicea, C et al., 2014. Environmental efficiency of investments in renewable energy: Comparative analysis at macroeconomic level. Renewable and Sustainable Energy Reviews, 30, pp.555–564.
[11]. Ferroukhi R, Lopez-Peña A, Kieffer G, Nagpal D, Hawila D, Khalid A, El-Katiri L, Vinci S and Fernandez A, 2014. Renewable Energy Benefits: Measuring the Economics. IRENA
[12]. Malik, I A et al., 2014. Turn on the lights: Macroeconomic factors affecting renewable energy in Pakistan. Renewable and Sustainable Energy Reviews, 38, pp.277–284.
[13]. [13]. Ming, Z et al., 2014. Review of renewable energy investment and financing in China: Status, mode, issues and countermeasures. Renewable and Sustainable Energy Reviews, 31, pp.23–37.
[14]. Griffith-Jones S, Ocampo J A, Spratt S, 2012. Financing Renewable Energy In Developing Countries: Mechanisms And Responsibilities. European Report on development
[15]. Ng, T H & Tao, J Y, 2016. Bond financing for renewable energy in Asia. Energy Policy, 95, pp.509 517.
[16]. Wang, T, Gong, Y & Jiang, C, 2014. A review on promoting share of renewable energy by green-trading mechanisms in power system. Renewable and Sustainable Energy Reviews, 40, pp.923–929.
[17]. Repowering London, 2013. Case study of renewable energy co-operatives: Brixton Energy Solar 1, Solar 2 & Solar 3
[18]. Cory K, Couture T, and Kreycik C, 2009. Feed-in Tariff Policy: Design, Implementation, and RPS Policy Interactions. Technical Report NREL/TP-6A2-45549 March.
[19]. Blazejczak, J et al., 2014. Economic effects of renewable energy expansion: A model-based analysis for Germany. Renewable and Sustainable Energy Reviews, 40, pp.1070–1080.
[20]. Bobinaite, V & Tarvydas, D, 2014. Financing instruments and channels for the increasing production and consumption of renewable energy: Lithuanian case. Renewable and Sustainable Energy Reviews, 38, pp.259–276.
[21]. Omri, A, 2014. An international literature survey on energy-economic growth nexus: Evidence from country-specific studies. Renewable and Sustainable Energy Reviews, 38, pp.951–959.
[22]. Couture T D, Cory K, Kreycik C and Williams E, 2010. A Policymaker's Guide to Feed-in Tariff Policy Design. Technical Report NREL/TP-6A2-44849 July.
[23]. World Bank and CIF, 2012. Financing renewable energy Options for Developing Financing Instruments Using Public Funds
[24]. Justice S, 2009. Private Financing of Renewable Energy- A Guide for Policymakers. UNEP
[25]. CEDRO, 2015. Crowdfunding and the Energy Sector, Empowering Lebanon With Renewable Energy. August-15

Comparison of Iron and Tungsten Based Oxygen Carriers for Hydrogen Production Using Chemical Looping Reforming

M N Khan[1] and T Shamim[1]
1 Institute Center for Energy (iEnergy), Department of Mechanical and Materials Engineering, Masdar Institute of Science and Technology, P.O. Box 54224, Abu Dhabi, United Arab Emirates
E-mail: tshamim@masdar.ac.ae

Abstract. Hydrogen production by using a three reactor chemical looping reforming (TRCLR) technology is an innovative and attractive process. Fossil fuels such as methane are the feedstocks used. This process is similar to a conventional steam-methane reforming but occurs in three steps utilizing an oxygen carrier. As the oxygen carrier plays an important role, its selection should be done carefully. In this study, two oxygen carrier materials of base metal iron (Fe) and tungsten (W) are analysed using a thermodynamic model of a three reactor chemical looping reforming plant in Aspen plus. The results indicate that iron oxide has moderate oxygen carrying capacity and is cheaper since it is abundantly available. In terms of hydrogen production efficiency, tungsten oxide gives 4% better efficiency than iron oxide. While in terms of electrical power efficiency, iron oxide gives 4.6% better results than tungsten oxide. Overall, a TRCLR system with iron oxide is 2.6% more efficient and is cost effective than the TRCLR system with tungsten oxide.

1. Introduction

The interest in hydrogen (H_2) as an energy carrier and decarbonized fuel is growing due to concerns about rising greenhouse gas emissions caused by the use of fossil fuels. A common large scale H_2 production method is steam-methane reforming (SMR). However, CO_2 is also produced during the process. To reduce the environmental impact of this process requires the capturing of CO_2 by using the costly and energy intensive technologies such as pressure-swing adsorption and amine absorption [1]. The use of CO_2 capture method increases the cost of H_2 production. In this regard, chemical looping reforming (CLR) offers a cost effective method of H_2 production with inherent CO_2 capture and no additional air separation unit.

A three reactor chemical looping reforming (TRCLR) process is considered as the most promising CL technology for H_2 production [2]. In this process, an oxygen carrier (OC) is used to transfer oxygen from air to the fuel. This process is economical and the cost of H_2 produced is in parity with other H_2 production technologies [3]. It employs three reactors: fuel reactor, steam reactor and air reactor. In the

Fuel Reactor: $$MeO_x + CH_4 \rightarrow MeO_{x-a} + CO_2 + 2H_2O \quad (1)$$

Steam Reactor: $$MeO_{x-a} + H_2O \rightarrow MeO_{x-a+1} + H_2 \quad (2)$$

Air Reactor: $$MeO_{x-a+1} + (a-1)O_2 \rightarrow MeO_x \quad (3)$$

Overall Reaction: $$CH_4 + \frac{2}{3}H_2O + \frac{2}{3}O_2 \rightarrow CO_2 + \frac{8}{3}H_2 \quad (4)$$

fuel reactor (FR), the fuel (e.g., methane (CH_4)) is oxidized to CO_2 and water (H_2O) by taking up the oxygen (O_2) from the OC (eq. (1)). The reduced OC goes into the steam reactor (SR) where it is oxidized by the steam to produce H_2 (eq. (2)). Finally, in the air reactor (AR), the OC is fully oxidized by the air (eq. (3)). The O_2 needed for the reactions is circulated among the three reactors by means of a transition metal OC. The overall reaction is shown in eq. (4).

The selection of OC is very critical task of the TRCLR process. The information about suitable OCs for the CLR process is limited since the range of options for OCs is quite narrow. Zafar et al. [4] investigated different OCs such as CuO, Mn_2O_3, NiO and Fe_2O_3 supported on SiO_2 for H_2 production. They concluded that NiO is the most feasible OC to be used in the process. However, their finding was based on the use of a two reactor CLR system. Kang et al. [5] performed a comprehensive thermal analysis to identify the most suitable OC for the reduction reaction with CH_4 and the oxidation reaction with steam. Oxides of iron (Fe), tungsten (W) and cerium (Ce) were selected for the H_2 production application. Cormos [6] presented the plant concepts and methodology to evaluate the plant performance. Iron based CL systems with different power generation schemes using natural gas and syngas were evaluated. Kathe et al. [7] investigated the thermodynamic limits for full conversion of natural gas through an iron-based CLR system. They reported higher thermal efficiency than that from a conventional SMR process.

In the current study, the best reaction pathways of the OCs of base metal Fe and W are used in a thermodynamic model of a CH_4-fueled TRCLR plant. The model is similar to that used by Khan and Shamim [3] in their techno-economic assessment. All the sub-models and assumptions used are similar. However, in the current study, the plant input fuel capacity is 500 MW as opposed to 350 MW in the aforementioned study. The plant performance based on the electrical, hydrogen and exergetic efficiencies is compared for both OC materials. In addition, the reactor temperatures and the emissions such as CH_4, CO and NO are also compared. It is worth mentioning here that the main objective of the present study is to demonstrate the competitiveness and viability of the CLR technology. This study on TRCLR plant will assist in complying with the urgent need of reduction in CO_2 emissions and transferring the dependence on hydrocarbon fuels to de-carbonized fuels like H_2 to mitigate the threat of climate change.

2. Thermodynamic modeling

The thermodynamic model was developed by employing conservation of mass and energy for all the cases in Aspen plus. The model specifications and the assumptions used can be found in Khan and Shamim [2]. The process flow diagram of the plant with the components used are shown in Figure 1. RGIBBS reactor was used for FR, SR and AR, which assumes chemical and phase equilibrium based on the Gibbs energy minimization concept. Separation of solids and the product gases were assumed to be perfect and was done by using cyclones. A turbocharger was used for compression of incoming air into the AR to a desired pressure and power production through a gas turbine by the outgoing vitiated air from the AR. The exhaust of the AR was used to generate intermediate pressure (IP) and low pressure (LP) steam in a heat recovery steam generator (HRSG). The exhausts from the FR and the SR were used to produce a high pressure (HP) steam in a parallel HRSG. The HP and IP steam were supplied to the steam turbines (ST) to produce power. Some of the LP steam was compressed to IPs to fulfill the steam requirement of the plant by using a steam compressor (SC). HRSG and ST unit constitute a steam cycle. The AR exhaust was released into the atmosphere while the FR and SR exhausts were compressed to the desired pressures in a two stage intercooled compression system (COMP).

Thermodynamic equilibrium has been assumed for the calculations. The properties were evaluated using the property method Redlich-Kwong-Soave (RKS) equation of state with the Boston-Mathias modifications. For the steam cycle the property method STEAM-TA has been used. The base case model was validated with a good agreement in our previous work [2]. The circulating OCs used in the present study were oxides of iron (Fe_2O_3, FeO, Fe_3O_4) and tungsten (WO_3, WO_2, $WO_{2.72}$) with 70% (by weight) inert material $MgAl_2O_4$ according to the reaction pathways shown in Table 1. The operating conditions and assumptions are given in Table 2. All the cases were performed at the stoichiometric OC mass flow rate based on the fuel flow rate. However, in actual systems, excess OC is supplied to ensure full conversion of fuel in the FR. The excess amount is usually 2% of the total flow rate [8]. Hence, the effect of excess OC is insignificant on the plant performance.

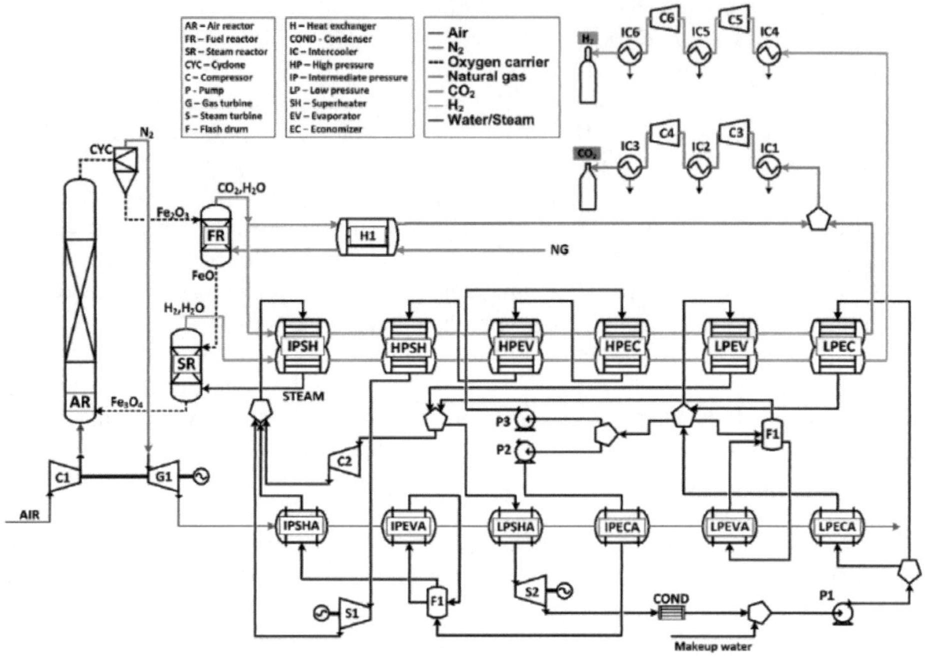

Figure 1. Process flow diagram of TRCLR plant

Table 1. Selected pathway reactions in three reactors

OC	Reaction
	$4\ Fe_2O_3 + CH_4 \rightarrow CO_2 + 2\ H_2O + 8\ FeO$
Iron	$3\ FeO + H_2O \rightarrow H_2 + Fe_3O_4$
	$4\ Fe_3O_4 + O_2 \rightarrow 6\ Fe_2O_3$
	$4\ WO_3 + CH_4 \rightarrow CO_2 + 2\ H_2O + 4\ WO_2$
Tungsten	$1.39\ WO_2 + H_2O \rightarrow H_2 + 1.39\ WO_{2.72}$
	$7.14\ WO_{2.72} + O_2 \rightarrow 7.14\ WO_3$

Table 2. Assumptions and operating conditions

Item	Value
Ambient conditions	15 °C, 1.013 bar
Natural gas heating value (MJ/kg)	45.467
CL reactors operating pressures (bar)	AR-16, FR-20, SR-18
CL reactors thermal losses	0.2% of thermal input
CL reactors thermal losses	0.2% of thermal input
Heat exchangers - Pressure loss	2%
GT/TC compressor polytropic efficiency	0.924
GT/TC turbine polytropic efficiency	0.926
Steam cycle pressures (HP, IP, LP, Condensation)	90/22/3/0.04 bar

HRSG pinch, approach temperature	10/25℃
Max. Turbine inlet temperature	500℃
Mechanical efficiency (pumps, compressors, turbines)	98%
Isentropic efficiency (pumps, compressors, turbines)	85%
Liquid CO_2 condition	25 °C, 120 bar
H_2 condition	25 °C, 60 bar

For thermodynamic assessment, the technical indicators used are the electrical and H_2 production efficiencies and which are defined as follows

$$\eta_E = \frac{\dot{W}}{Q_{NG}};\ \eta_H = \frac{Q_H}{Q_{NG}} \tag{5}$$

where Q_H and Q_{NG} are thermal power (product of mass flow rate and LHV of the gas) of the H_2 output and natural gas input, respectively, and $\dot{W}$ is the net power output. As the main product of this plant is H_2 and the byproduct is electricity, it is difficult to define a single performance indicator for a co-production plant. However, a single index called global efficiency can be defined which includes both the efficiencies. It can be called as the equivalent H_2 production efficiency as it is calculated based on the fuel used for only H_2 production i.e. the fuel used for electricity production is deducted from the total fuel input.

$$\eta_{GL} = \frac{Q_H}{Q_{NG} - \dot{W}/\eta_{E,REF}} = \frac{\eta_H}{1 - \eta_E/\eta_{E,REF}} \tag{6}$$

where $\eta_{E,\ REF}$ is the reference electrical efficiency of an alternative power generation process. Such a reference efficiency is 50% and is taken from the study on a natural gas fired combined cycle plant. The reference efficiency is used to calculate the equivalent natural gas consumption to produce the same amount of electricity as actually produced by the plant.

The overall exergetic efficiency is defined as the ratio of the exergy of the product obtained from the operation of the components to the exergy of the fuel expended in that operation as shown in equation (7).

$$\varepsilon_{tot} = \frac{\dot{E}_{H_2} + \dot{W}_{net}}{\dot{E}_{Air} + \dot{E}_{Water} + \dot{E}_{Fuel}} \tag{7}$$

3. Results and discussions

As mentioned earlier, the thermodynamic model was validated from the data available in the literature [8]. The results show that the endothermic reaction in the FR forms the products namely CO_2, H_2O, H_2, CO, N_2 and traces of CH_4 while the highly exothermic reaction in the AR gives out argon, CO_2, H_2O, N_2, O_2 and traces of NO. The thermodynamic performance indicators for Fe- and W-based plants are shown in Figure 2. The results show that the H_2 production efficiency, in the case of W-based oxides, is 4 percentage points higher than that of Fe-based oxides while the electrical power efficiency is 4.6 percentage points lower than that of Fe-based oxides. The global efficiency, however, is 2.6 percentage points higher in the case of Fe-based oxide. In terms of exergetic efficiency, W-based plant showed a better performance with 5.3 percentage points higher efficiency than a Fe-based plant. The gross power output obtained in both the scenarios are 59.4 and 31.3 MW, respectively. Of which 65/61 % comes from GT and 35 and 39 % comes from ST, respectively. About 59/68 % and 10/44 % of the gross power output is consumed in air compression and steam compression, respectively. The CO_2 and

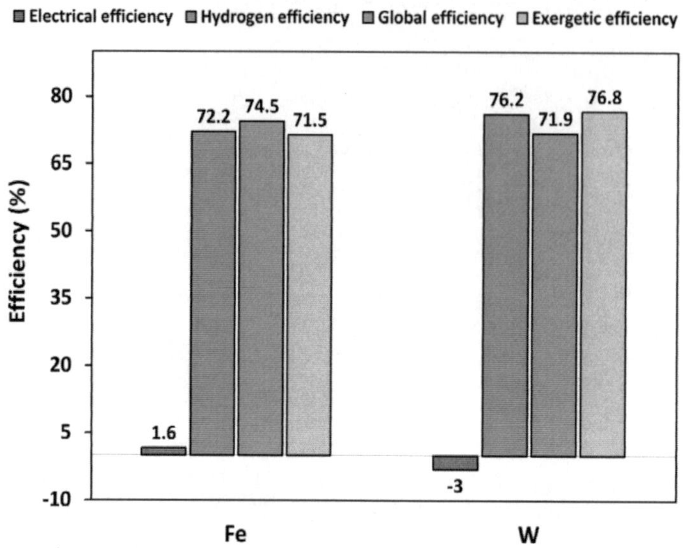

Figure 2. Plant performance with iron and tungsten oxides as oxygen carriers

H_2 compression units consume about 7.7/16 % and 9.3/19 %, respectively. The power output in W-based plant is lower because of the lower reactor temperatures. In the FR, the temperature difference between the Fe-based and W-based plants is about 125 K, whereas, in the SR and the AR, the temperature differences are about 131 K and 156 K, respectively, as shown in Figure 3. Table 3 compares performance data and the emissions of unburnt CH_4, CO and NO. Fe-based plant produces net power whereas there is no net power in W-based plant since extra power is required by the auxiliaries of the plant. The CH_4 emissions from Fe-based plant is very low, about 17 ppm as opposed to 2778 ppm from W-based plant. The CO emissions are similar for both the plants. In terms of NO emissions, W-based plant performs better. This is because of the low reactor temperatures.

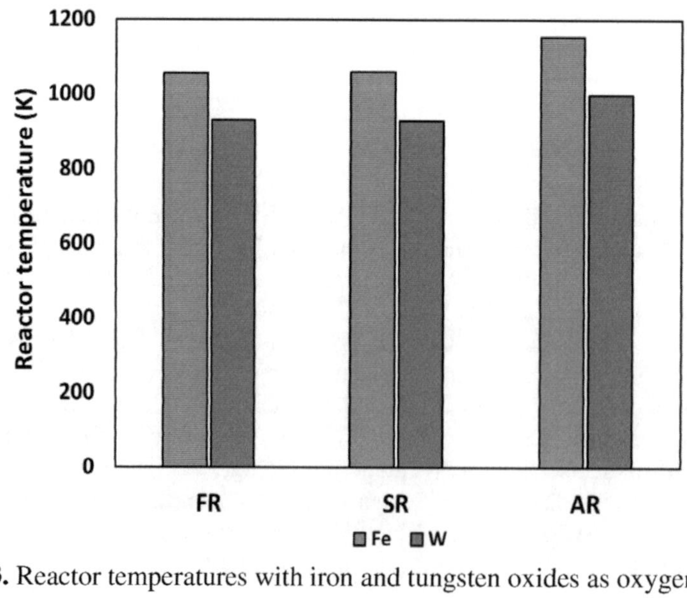

Figure 3. Reactor temperatures with iron and tungsten oxides as oxygen carriers

Table 3. Performance data and emissions from the TRCLR plants

	Fe-based	W-based
Fuel input (kg/s)	11	11
Gross/net output (MW)	59.36/7.9	31.27/-
Auxiliary loads (MW)	16.6	25
OC mass flow rate (kg/s)	1346	1954
Steam mass flow rate (kg/s)	45	48
Air mass flow rate (kg/s)	77	46
Emissions (ppm)		
CH_4	17	2778
CO	25620	26687
NO	87	8E-07

4. Conclusions

The current study investigated the performance of Fe- and W-based OCs in a TRCLR plant. A thermodynamic study has been performed by using these OC materials in an Aspen based model. The results in the form of H_2 and power production efficiencies were compared. The results show that the H_2 production efficiency, is 4 percentage points higher for W-based oxides were than that of Fe-based oxides. The electrical power efficiency is 4.6 percentage points lower for W-based oxides than that of Fe-based oxides. Furthermore, Fe is much cheaper than W. Therefore, an Fe-based plant is the most suitable option for H_2 generation and it has the potential to be used commercially.

5. Acknowledgment

This research was supported by the Government of Abu Dhabi to help fulfil the vision of the late President Sheikh Zayed Bin Sultan Al Nahyan for sustainable development and empowerment of the UAE and humankind.

6. References

[1] Barelli L, Bidini G, Gallorini F and Servili S 2008 *Energy.* **33** 554-70.
[2] Khan M N and Shamim T 2016 *Appl. Energy.* **162** 1186-94.
[3] Khan M N and Shamim T 2016 *Int. J. Hydrogen Energy.* **41** 22677-88.
[4] Zafar Q, Mattisson T and Gevert B 2005 *Ind. Eng. Chem. Res.* **44** 3485-96.
[5] Kang K S, Kim C H, Bae K K, Cho W C, Kim S H and Park C S 2010 *Int. J. Hydrogen Energy.* **35** 12246-54.
[6] Cormos C C 2011 *Int. J. Hydrogen Energy.* **36** 5960-71.
[7] Kathe M V, Empfield A, Na J, Blair E and Fan L S 2016 *Appl. Energy.* **165** 183-201.
[8] Chiesa P, Lozza G, Malandrino A, Romano M and Piccolo V 2008 *Int. J. Hydrogen Energy.* **33** 2233–45.

IOP Conf. Series: Earth and Environmental Science **83** (2017) 012014 doi:10.1088/1755-1315/83/1/012014

Green Technology for Smart Cities

M Casini[1]

1 Eng. PhD, Professor of Architecture Technology and Environmental Certification of Buildings - Department of Planning, Design and Architecture Technology, Sapienza University of Rome.
E-mail: marco.casini@uniroma1.it

Abstract. In view of the enormous social and environmental changes at the global level, more and more cities worldwide have directed their development strategies towards smart policies aimed at sustainable mobility, energy upgrading of the building stock, increase of energy production from renewable sources, improvement of waste management and implementation of ICT infrastructures. The goal is to turn into Smart Cities, able to improve the quality of life of their inhabitants by offering a lasting opportunity for cultural, economic and social growth within a healthy, safe, stimulating and dynamic environment. After an overview of the role of cities in climate changes and environmental pollution worldwide, the article provides an up to date definition of Smart City and of its main expected features, focussing on technology innovation, smart governance and main financing and support programs. An analysis of the most interesting initiatives at the international level pursued by cities investigating the three main areas of Green Buildings, Smart grid-Smart lighting, and Smart mobility is given, with the objective to offer a broad reference for the identification of development sustainable plans and programs at the urban level within the current legislative framework.

1. Introduction

Global population increase, progressive decrease of energy sources and their consequent higher cost, climate change and air pollution are some of the main problems that the cities of the future will have to cope with to survive, transforming into Smart Cities and focusing on Green Building and Smart Mobility.
Because of the low energy efficiency of buildings and transportation systems, the cities of today are responsible, on average, for 70% of greenhouse gases emissions and over 60% of energy consumed worldwide [1].
The global increase of carbon dioxide emissions, whose values, equal to over 32 Gigatonnes per year in 2016, exceeded over 50% of those in 1990, caused the increase of CO_2 concentration levels in the atmosphere, by now stably higher than 400 parts per million (which has not been occurring for 300 millions of years) [2].
2016 was also the hottest year ever recorded in NOAA's 137-year series, since measurements began in 1880. Remarkably, this is the third consecutive year a new global annual temperature record has been set [3]. The average global temperature across land and ocean surface areas for 2016 was 0.94°C (1.69°F) above the 20th century average of 13.9°C (57.0°F), surpassing the previous record warmth of 2015 by 0.04°C (0.07°F). This marks the 40th consecutive year (since 1977) that the annual temperature has been above the 20th century average. To date, all 16 years of the 21st century rank among the seventeen warmest on record (1998 is currently the eighth warmest). 2017 lends itself to being a record year too. In fact, the global land and ocean surface temperature during January-April 2017 was 0.95°C (1.71°F) above the 20th century average of 12.6°C (54.8°F). This was the second highest such period since records began in 1880, behind 2016 by 0.19°C (0.34°F) and ahead of 2015 by 0.10°C (0.18°F).
The issues of acoustic pollution and air quality typical of urban centres go in addition to the climate changes. In the EU, buildings alone are responsible for 40% of the final energy use, 36% of CO_2 emissions and above 40% of Particulate Matter emissions (PM_{10} and $PM_{2.5}$). Current mobility systems based on fossil fuel, besides being responsible for above 25% of polluting emissions, are unsuitable to the needs of urban areas, making movements

difficult especially during rush hours, with journey speeds around 7-8 km/h (the same speeds recorded in 1700) [4]

Every year worldwide 12.6 million people die because of environmental pollution, equal to one fourth of the total deaths. Air, water and soil pollution, chemical exposure, climate changes and ultraviolet radiation contribute to the increasing of over 100 illnesses and health damages [5].

Atmospheric pollution is the fourth risk factor for deaths on a global level, and undoubtedly the main environmental risk factor for lungs and heart diseases: over 5.5 million people die every year all over the world because of air pollution, more than Finland, Slovacchia and Sicily inhabitants. Italy hits the record of dead from smog with 59,500 premature deceases from $PM_{2.5}$, 3,300 from Ozone and 21,600 from NO_x only in 2012 [6].

These issues are going to increase with the progressive decrease of resources, the consequent increase of energy cost and the population development that is estimated to reach 9 billions of individuals in 2050 (from current 7.4 billions) of which over two thirds will live in the urban centres. These will produce the 80% of global GDP and will consume the 75% of global resources, contributing to create a model of urban-centric development.

The economic resources that the worldwide cities have addressed to adaptation measures to climate changes like protective barriers against inondations, more resilient infrastructures and better draining systems (around the 0.22% of GDP for the developed countries compared to the 0,15% for the cities of developing countries) are already relevant.

Looking at this scenario, cities have to be ready and capable of handling enormous social and environmental mutations, becoming the fulcrum of the fight against global warming and catalyzing investments and policies oriented to sustainability and efficiency in a Smart vision.

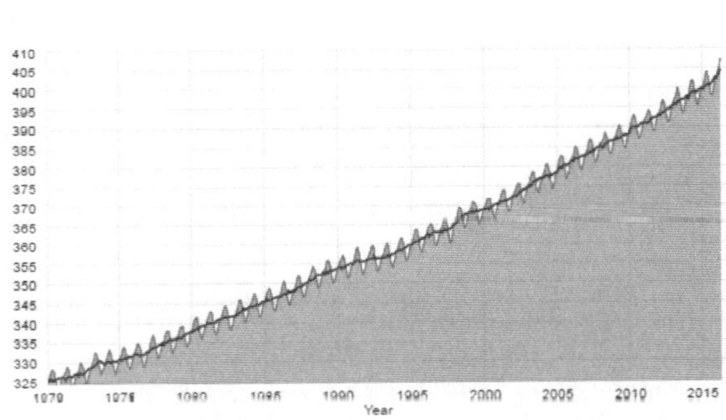

N	Year	Anomaly	N	Year	Anomaly
1	**2016**	**0.94 °C**	6	2005	0.66 °C
2	2015	0.90 °C	7	2009	0.64 °C
3	2014	0.74 °C	8	1998	0.63 °C
4	2010	0.70 °C	9	2012	0.62 °C
5	2013	0.67 °C	10	2003, 06, 07	0.61 °C

Figure 1. World carbon dioxide concentration (ppm) and temperature ranking (1880-2016) [3]

2. Becoming a smart city

A Smart City, or intelligent city, is a city capable of improving its citizens' life quality, offering a lasting opportunity for cultural, economic and social growth in a healthy, safe, stimulating and dynamic environment [7].

A Smart City is a city that guarantees:

- economic competitivity (smart economy), innovation, enterprise, economic image and brands spirit, productivity, job market flexibility, international integration, transformation capacity;
- training and social interaction of citizens (smart people), qualification level, long-term training, social and ethnic plurality, flexibility, creativity, cosmopolitanism and mental opening, participation to public life;
- administration functioning and services (smart governance), participation to decisional processes, public and social services, transparent government activity, politics strategies and perspectives;
- availability of information and communication technologies and modern and sustainable transportation systems (smart mobility), local accessibility, international accessibility, availability of IT infrastructures, sustainable, innovative and safe transportation systems;
- high environmental quality (smart environment), attractiveness of natural conditions, pollution, environment protection, sustainable management of resources;
- life, culture, health and safety quality (smart living), social structures, health conditions, individual security, dwellings quality, educational structures, touristic attractiveness, social cohesion.

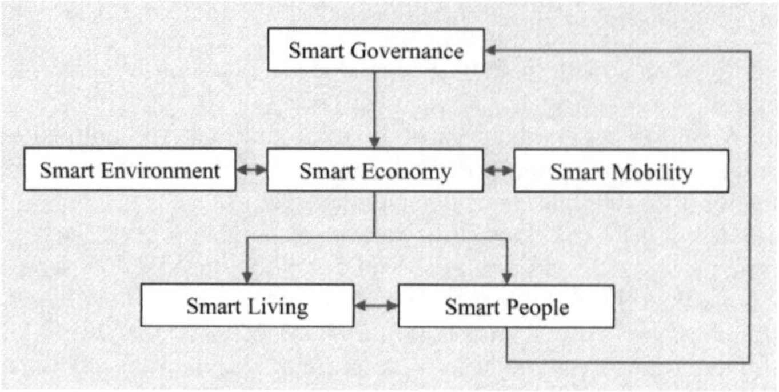

Figure 2. Smart City model

2.1 Technologic innovation

Today, reaching these strategic goals is possible also thanks to the availability of cutting-edge Technologies that are changing the aspect of the city, functioning of services and users' behaviour: renewable energies, advanced materials, innovative transportation systems, ICT, broadband, geolocation systems, Internet of Things, smartphones and tablets, social networks, city apps and urban data.

The long-term predictions by Navigant Research track, for the next eight years, a consistent growth of the market related to Smart cities that in 2023 will be valued for about 25.3 billion euros. This really consistent amount means that the 85% of urban projects will focus on the use of technology to produce digitalized services and improve the efficiency in the city. The segments of the Smart city world that will lead the growth include:

- smart grids, the intelligent grids capable of exchanging information with the administrations and regulating the energy flows;
- latest generation public illumination, LED lamps monitoring traffic and urban pollution levels, improving safety;
- smart mobility, hybrid and electric vehicles, smart parking, bike and car sharing;
- technologies related to the resilience in the city to cope with the meteorological phenomena related to climate change;
- smart systems to avoid hydric resources waste;
- all the devices linked to the urban data, constituted by all the data that the city produces daily, measured and translated into facts, figures and views.

In particular, regarding the urban data, according to Eric Schmidt, Executive Chairman and ex CEO at Google, the entire amount of data collected from the dawn of humanity to 2003 is equivalent to that today is produced in two days. This phenomenon of enormous proliferation of information, to which we often refer as "Big Data", involves the need of filtering and making this new asset of information accessible.

By 2020, the users connected to mobile devices will be 5 billions and a half, around the 70% of global population. There will be more people with a phone (5.4 billions) than those with electricity (5.3 billions), water (3 billions and a half) and cars (2.8 billions). Growth will be also lead by fourth generation web and, mostly, by videos: these will represent alone the three fourths of mobile traffic. The increasing of users in the five years from 2015 to 2020 will be twice more rapid than that of global population.

Internet traffic from mobile devices will reach a volume of 367 exabyte per year. A quantity of data that is not easy to visualize: it is equal to seven thousand billions of videoclips on YouTube, or to sending or receiving 28 images per inhabitant of the world, per day and for a year.

Regarding the Urban Data in Chicago, currently there is a highly innovative project called "Array of Things", promoted by the Computation Institute of Chicago University with the purpose of extending the Internet of Things to the urban scale.

The project involves the construction of a network of 500 sensors placed at strategic points in the city to measure all "vital parameters" and make it more efficient, livable and safe. In particular, the sensors will have the task to measure and make available, in real time, data such as: temperature and air pressure, wind, precipitation, lighting level, vibrations, pollutants (CO, NO_2, SO_2, Ozone), noise levels, pedestrian and vehicular traffic, surface temperature.

2.2. Smart governance

Particular importance in the development of a Smart City is given to the Smart Governance tools such as: Digital Democracy, Open governance, Citizen Empowerment, Participated Urbanism and Urban Data.

Digital Democracy is based on the implementation of information technologies, communication and social media in the service of political and governmental processes. If applied well, it produces a wider participation of citizens and a more transparent administration. Digital democracy also serves to increase the common responsibility between government agencies and the participant public.
Open governance refers to accessibility of citizens to information, data and governmental processes to favor the widespread participation and collaboration in the city's decision-making processes. This commitment often exploits technology to make it easier to facilitate a more active and open communication between the citizens and government, leading to a more efficient use of funds and a better quality of life for the inhabitants. Open governance aims to make the urban administrations more responsible to their citizens, increasing the legitimacy of the rulers.
Citizen Empowerment measures the consciousness of the citizens that their actions can contribute actively to decision-making processes and to changes in the city. By creating communication opportunities and participation, citizens feel called to contribute to the city with their own time, their own energy and ideas.
Finally, participating urbanism is the condition, often facilitated by technology, in which citizens have the ability to collect and share data, ideas and proposals with the city's governing bodies. The idea is based on the fact that members of the community are experts in the related urban situations and they already have knowledge and solutions for multiple issues. Participating urbanism allows professionals to identify actual needs and utilize knowledge and local human resources, rather than imposing the change from above.

3. Financing and support programs

Several cities in Europe (London, Amsterdam, Vienna, Barcelona, Stockholm) and the world (New York, Los Angeles, Seattle, Seoul, Melbourne, Vancouver, Shenzhen) have started in the last ten years to address their own efforts towards the development of sustainable mobility, the energy regeneration of building stock, the increase in energy production from renewable sources, the improved management of waste and the implementation of ICT infrastructures. There is not an existing city, big or small, that has not launched at least one project to be able to enjoy the title of "smart", focusing essentially on digital technology, environmental sustainability, civic initiatives, mobility and businesses.
In Europe, the Covenant of Mayors program was launched in 2008 (and updated in 2015), being the main European movement brought forward by local and regional authorities with the goal of increasing energy efficiency and the use of renewable energy sources in their territories. Through their commitment, the signatories of the Pact aim to reach and go beyond the European objectives of reduction in CO_2 emissions by 2030. In order to translate their political commitment into measures and tangible projects, the signatories of the Covenant undertake to present a Sustainable Energy and Climate Action Plan (SECAP) which outlines the main actions they intend to launch until 2030 for climate change mitigation and adaptation activities.
There are also numerous initiatives and programs promoted on a European and international level to stimulate the transition to an "intelligent, sustainable and inclusive urban model". Concerning Europe, worth noting are the two initiatives started by the European Commission "Smart Cities and Communities European Innovation Partnership (SCC EIP)" and “Stakeholder Platform Smart Cities”, with the goal to launch innovative projects in urban areas integrated in the fields of energy, transport and information and communication technologies (ICT), as well as identifying and disseminating needs and information among all interested parties, public and private.
In addition to these, financing programs Horizon 2020 (€ 6 billion), Connecting Europe Facility (€ 6 billion) and the Cohesion Funds 2014-2020 (€ 23 billion for renewable energies, energy efficiency, and smart grids and mobility) are provided.
In particular, the Horizon program through the Call "Smart Cities and Communities" finances innovative projects on an urban scale regarding energy efficiency of buildings, the development of intelligent networks (electricity, district heating, telecommunications, water, etc.), energy storage, electrical mobility and intelligent charging infrastructures, the use of the latest generation of ICT platforms (130 million euros in 2016 alone).
As for Italy, in March 2016 through the Smart City Address Act, the Ministry of Economic Development (MiSE) has launched its first intervention program for the Smart cities by allocating 65 million euros. In particular, the program includes both promotion in urban areas of more efficient energy infrastructures and services and the activation of large pre-commercial contracts, aimed at responding to the most innovative needs expressed by the local administrations.

4. The main areas of intervention

Among the areas of intervention that characterize the transformation process of cities into Smart Cities, of particular interest are those addressed to energy and environmental efficiency of existing buildings, the introduction of renewable energy sources on an urban scale and the launch of smart mobility plans. These areas of action are in fact the most effective way to reconcile environmental objectives (reduction of energy consumption and polluting emissions), economic goals (reduction of management costs for citizens and for public administration, development of businesses and rising employment levels) and social goals (improvement of the welfare and quality of the services).

According to International Renewable Energy Agency (IRENA), more than 9.8 million people were employed in the renewable energy sector in 2016 [8]. Renewables are directly supporting broader socio-economic objectives, with employment creation increasingly recognised as a central component of the global energy transition. Solar photovoltaic (PV) was the largest employer in 2016, with 3.1 million jobs - up 12 per cent from 2015 - mainly in China, the United States and India. In the United States, jobs in the solar industry increased 17 times faster than the overall economy, growing 24.5 per cent from the previous year to over 260,000. New wind installations contributed to a 7 per cent increase in global wind employment, raising it up to 1.2 million jobs. Brazil, China, the United States and India also proved to be key bioenergy job markets, with biofuels accounting for 1.7 million jobs, biomass 0.7 million, and biogas 0.3 million. The number of people working in the renewables sector is expected to reach 24 million by 2030, more than offsetting fossil-fuel job losses and becoming a major economic driver around the world.

In the European Union, the renewable energy sector records in 2015 over one million employees, with a turnover of about 143.6 billion euros. In Italy, renewable energies give work to 82,500 people. Italy is fourth in the European green job ranking behind Germany (347,400), France (169,630) and Great Britain (92,850). Also, in Italy there are 20,000 employees in the wind, 19,000 in biomass, 10,000 in photovoltaics, 8.500 in the field of heat pumps, 5.500 in geothermal, 4,500 in hydroelectric, 5,500 in biofuels and 5 thousand in biogas.

4.1 Green buildings

The construction industry is one of the cornerstones of priority intervention for the achievement of the objective of a "smart, sustainable and inclusive" growth and for a transition to an economy based on efficient use of resources and low carbon emissions. Currently, in fact, over 40% of the final energy consumption in the EU-27 is absorbed by houses, public and private offices, shops and other buildings (43% is used by households, 44% by industry and remaining 13% from the services). In residential homes 67% of consumption is used for heating of environments, 15% for lighting and the use of electric equipment, 14% for heating of water and the remaining 4% for kitchen appliances [9].

From numerous studies emerge that the building industry presents a potential for improvement of energy efficiency that can be estimated, in 2020, in about 30% of current consumption and which can be exploited with effective interventions also in terms of costs [10]. There is also an enormous existing building potential of unused space usable for the integration of renewable energy sources. 40% of the total electricity demand of the European Union in 2020 could be satisfied if all roofs and the appropriate facades of European buildings were covered with photovoltaic panels [11].

In addition, in Europe, the population spends almost 90% of their time inside buildings: a bad design of buildings or inadequate construction methods may have a negative effect on the health of the occupants and can make maintenance and management of the buildings themselves extremely more expensive, starting from heating and cooling, with strong repercussions especially on older people and the most disadvantaged groups of the population.

As for buildings, the goal is therefore to promote the energy upgrading of existing buildings, and public ones in particular (communal buildings and primary schools), through: measures to reduce dispersion through the building envelope (energy sufficiency measures); the promotion of the use of cleaner energy sources (replacement of diesel fuel with natural gas) and renewable energy sources (solar thermal, solar photovoltaic, heat pumps) for air conditioning and hot domestic water production (clean energy measures); the efficiency of heating systems and a constant monitoring of their emissions (energy efficiency measures). Various cities have provided rewarding systems for condominiums that adopt more efficient solutions. The interventions aiming to reduce energy consumption and atmospheric emissions also concern the public lighting systems.

More and more numerous in Europe are examples of solar retrofit of existing historic buildings, from photovoltaic coverage of the historic bridge of Blackfriars (completed in January 2014) until the eco-renovation in 2014 on the first floor of the Eiffel Tower in Paris, which includes the integration of wind and photovoltaic technologies. The City of Paris also launched in 2014 the maxi urban redevelopment project wanted by the Mayor Anne Hidalgo representing the deepest transformation of Ville Lumiere for 150 years. The project "Reinventing Paris" is all about sustainable technologies and green buildings: green roofs, urban farms, urban forests, biofaçades, aquaponic farming, photocatalytic concrete are just some of the features of the 22 winning projects that will transform the French capital.

The city of Seoul has instead started the construction of an urban park of 10,000 square meters in place of a highway now dismantled. Seoul Skygarden designed by MVRDV will be a botanical encyclopedia with over 250 different species of trees sorted according to the Korean alphabet. The conversion, which will allow citizens to take a shortcut to get to the rail station, will allow you to travel in 11 minutes instead of 25, walking through trees and shrubs and looking at the city by 17 feet height.

4.2. Smart grid and smart lighting

Interventions for reducing energy consumption on an urban scale driven by Smart Cities are focused also on the improvement of the municipal electricity grid to make it a 'smart grid', and on increasing efficiency of public lighting with the replacement of outdated lamps with last generation LED lamps, the introduction of photovoltaic or wind power plants, to the introduction of sensors for real-time data detection.

The city's power grid needs a smart system able to manage, in a dynamic way and in real-time, both the inversion of flows of energy, from peripheral nodes distributed in the territory towards the center of the system (distributed generation), and any local energy surpluses due to renewable sources, balancing demand and supply.

In this vision, with Resolution 87/2016/R/eel, the Italian Energy Authority (AEEGSI) presented its minimum features for new smart electricity counters (version 2.0) that, according to distributor choices (the subjects holding the measurement activity) will progressively take the place of the first generation installed since 2001 and whose useful life (15 years) ended at the end of 2016. The new generation of smart counters promises to be more smart and user-friendly compared to the previous one. In particular, electric gauges will be able to provide a detailed report to the users on their consumption to facilitate efficiency and savings choices.

Particularly interesting in the field of smart grids are the projects for the realization of microgrids of communities aiming to sharing clean energy such as that carried out by New York start-up TransActive Grid in two districts of Brooklyn in New York (Park Slope and Gowanus). The project envisages the realization of a network of transactive grid, meaning that it is based on a transaction and active in both directions instead that one way. Within this network each individual, with his/her own home, constitutes a hub. To get involved, in addition to physical connection, it is necessary to enter into special contracts with the house neighbors, giving or buying an amount of energy. The difference with the traditional energy supplies, where the private buys from the government or from a huge company, is the relationship with the other network hubs. Transaction security, and warranty that the computer system (Ethereum) that regulates the whole microgrid can not be tampered, is given by the use of a blockchain system, the same that regulates the transactions of the virtual coin bitcoin.

For what concerns public lighting, particularly interesting is the smart cities project by Sap and Philips that kicked off in 2016 in Buenos Aires, where intelligent urban lighting systems will be installed on 91,000 street lamps, controlled by the Philips Citytouch system interfaced with the platform Hana by Sap and with which it is expected a reduction of over 50% of consumption.

Real time data capture by the connected lamps will show the public administration information useful on the traffic, the outflow of the waters after rains, crowding at the public transport stops and the availability of parking lots. The control of road lighting will also help increase security levels.

Other innovative systems such as the one patented by the company EnGoPLANET and installed in 2016 in the city of Las Vegas expect integration of lighting with floor piezoelectric tiles able to transform the mechanical energy of the pedestrian passage into electric energy. The system is completed by sensors of motion to adjust the levels of illumination, USB charging devices or wireless pads and additional sensors which measure air quality, temperature, precipitations, in addition to video surveillance cameras for vehicle traffic analysis.

4.3. Smart mobility

Intelligent mobility is one of the key aspects of a Smart City, towards which cities worldwide are starting the most interesting transformations, either through large structural investments, and low cost initiatives that act on social innovation and on raising public awareness.

In particular, the interventions for a Smart mobility brought forward by the Cities concern:

- enhancement and efficiency of the system of public transport and a modernization of the related means of transport with vehicles that use low-emission combustion engines, electric or hydrogen motors, up to the introduction of driverless vehicles;
- promotion of the use by citizens of electric and hybrid vehicles, even with the installation of new charging columns (as required by the EU) and the activation of electric cars rental services, and the introduction of smart charging systems (vehicle to grid and vehicle to building);
- enhancing bike sharing, car sharing and car pooling policies;
- implementation of early warning systems for conveying traffic and of parking addressing systems and the management via smartphones of the access to restricted traffic areas and pay parking;
- digitalization of the public transport system with the introduction of smart palettes and panels with a variable message at public transport stops and applications dedicated to info-mobility that can provide useful information about urban lines, waiting times, possible criticalities and atmospheric disturbances directly to the users' smartphones;

- introduction of interchange parking spaces where to leave own's car to continue with other lower environmental impact means;
- promotion and development of pedestrian traffic activating policies to encourage walking even through the retraining of the paths, the improvement of lighting and the introduction of dedicated signage;
- introduction of intelligent traffic lights taking real-time count of car flows as now happens in several cities in the United States;
- introduction of intelligent streetlights capable of automatedly modulating lighting according to the intensity of the transit and that, through a survey in real time of the detected data, are able to provide public administration useful information;
- realization, as in Netherlands and France, of solar photovoltaic cycle paths.

Concerning the electric car market, in 2015 there was a global growth of 65% compared with 2014 (of which 36% of the total in the United States, 27% in Europe and 24% in China) [12]. Today the United States hold 46% of the market, Japan, second country, is included in the RoW (Rest of the World) which has 27%. Still behind European Union and especially China, which have respectively 18% and 8% of total sold electric cars. For the year 2021 the situation will be completely different: the European Union should represent 37% of the market and China, in the strongest growth, would be around 30%. In 2025, 8% of cars in the world will be electric, against the current 0.6% of 500 thousand in 2015.

In 2015, Australia inaugurated an electric bus with 1000 km of autonomy able to travel from Sidney to Melbourne without ever charging the battery. The vehicles, built by Brighsun in Melbourne, mount a high performance lithium ion battery, combined with an electric motor and a regenerative braking system. On the other hand, the world's first hydrogen tram is Chinese. Produced by a public company, it is the first to mount fuel cells and can travel 100 kilometers with nearly 400 people on board. In 2015 it began to run on the tracks of Chinese city Tsingtao. It reaches a maximum speed of 70 km/h and thanks to the fuel cells it incorporates, the public vehicle only releases water. Refilling takes only three minutes, and once the tank is loaded the tram can travel for up to 100 kilometers. In the city of Tsingtao, that means it can cover the urban path from terminal to terminal three times in a row. Inside it offers 60 seats and 320 standing, for a total of about 380 passengers.

In the Netherlands, instead, the world's first solar bicycle path was inaugurated in 2014 and has produced in 6 months 3000 kWh electricity. Just 70 meters long and with integrated photovoltaic panels, the cycle track connects Krommenie and Wormerveer, two suburbs of Amsterdam.

France, on the other hand, announced the realization of 1000 km of photovoltaic road in 5 years that will provide energy to 5 million people, 8% of the French population. Works began in spring 2016. The infrastructure will be paid with an increase in taxes on fossil fuels contributing for around 300 million euros to the project.

According to the French Agency for the environment and management of energy, 4 meters of "solarized" road are sufficient to meet one family's electrical needs, without counting the heating. One kilometer of these panels is able to give electricity to a community of 5 thousand inhabitants.

5. The virtuous examples of London and New York

Among the cities that are most engaged in programming in smart mobility and energy retrofit of existing buildings, the first places definitely belong to London and New York, the latter winner of the City Climate Leadership Awards 2014 for the category "Energy Efficient Built Environment". Both cities are characterized by the fact that over 75% of their greenhouse gas emissions are due to buildings, 80% of which will be still standing in 2030 (2050 in London). Their renovation is therefore an indispensable goal to face the challenge of climate change and ensure a high quality of life for citizens.

With the two RE:FIT programs, reserved to commercial buildings and public bodies, and RE:NEW, addressed to housing, the city of London has set the goal to reduce carbon emissions by 60% by 2025. The two programs are alongside other important initiatives oriented to a timely monitoring of emissions (London's GHG inventory) and to the promotion of sustainable mobility for the improvement of air quality through the replacement with electric vehicles (New Taxi for London) of all taxis in the city, responsible alone for about 35% of PM_{10} emissions and more than 15% of city NO_x emissions. London is also committed to investing more than £ 300 millions so that by 2020, all 300 one-storey buses working in central London will be electric and the over 3,000 two-storey buses will be hybrid (to date they are already over 1,300). In March 2016 the city also inaugurated the first five entirely electric two-storey buses. Manufactured and equipped by the Chinese company BYD, the new EV Buses, with 10.2 m length and a capacity of 81 passengers, have an autonomy of 305 km and will be in service on Route 98.

After an inquiry involving millions of citizens, London capital also launched the London Infrastructure Plan 2050 (LIP 2050), a series of interventions for a London Smart City that will surpass 1800 billion euros [13].

The plan, approved by the Citizens Council, will affect the adaptation of infrastructures, building heritage, schools, energy efficiency of buildings, the promotion of green technologies and the expansion of green areas:

42% of funding (£ 547 billion) will be allocated to the building industry, to improve the structures of the capital; 35% (466 billion pounds) will serve to make the urban mobility system smarter and more efficient - among the various interventions, a wider and more efficient cycle path system; 11% will be intended to improve the energetic system and to pursue the goal of reducing by 80% carbon emissions in the atmosphere by 2050.
In addition, 68 billions will be used to improve school buildings, 49 billion are destined to the improvement of water and energy efficiency, 22 billion will go to expand the green areas, 14 billion are destined to make the waste management smarter, 8 billion will make London an interconnected and digital Smart City.
Thanks to the London Infrastructure Plan, it is expected that in 2050 the UK capital will be greener, cleaner and more efficient and according to estimates it will go from 8.6 millions of inhabitants to 11, who will enjoy advanced services in a sustainable and digitized environment.
Finally, in 2016 London started a remarkable project of Smart city. The British capital has studied an interesting green finance plan to support innovation and sustainability projects in the urban sphere. The financial market can be crucial to accessing new economic resources useful for the development of smart city projects, environmental sustainability, energy efficiency and resilience. This initiative could yield 100 billion a year of green bonds to be put on the market to give birth to new projects in the field of sustainable mobility, intelligent services, the use of big data, monitoring traffic flows and urban efficiency.
To put in practice the interesting green design three strategies have been identified. The first envisages increasing trust in green bonds through the improvement of transparency and the accreditation of operators. The second one will see an increase of information addressed to operators of the industry, businesses and citizens. The latest strategy aims at promoting the green finance with every means available and the development of a new financing model studied ad hoc for Smart City projects.
London, the capital of world finance, dues its success to fast economic development, quality labor, cost of living, and infrastructures optimization, international popularity and the technology sector.
The city of New York has adopted in 2007 a global sustainability plan, the PlaNYC, to reduce urban greenhouse gas emissions by 30% by 2030. To achieve this, the city has issued under the "Greener, Greater Buildings Plan" (GGBP), a set of energy efficiency standards in particular addressed to the properties of larger dimensions (over 50,000 square meters) which, though being only 2% of all the properties of the Big Apple, are responsible for the 45% of carbon emissions. In 2007 the city has also launched the volunteer program NYC Carbon Challenge, initially reserved to Universities and Hospitals and later extended from 2013 to multi-family residential and commercial buildings, aimed at reducing emissions by 30% in 10 years.
These and other virtuous examples carried out in other urban realities show how cities can play a key role in combating climate change and how the challenge of efficiency improvement and reducing emissions can be a strong change engine that can improve economic competitiveness, environmental and social quality.

6. Conclusions

Resilient, attractive and competitive cities make up an indispensable prerequisite to face the challenges that the society of the third millennium will have to face in the next years.
From the analysis of the numerous initiatives carried forward by worldwide cities, it emerges that many smart city projects are today still mostly focused on individual areas such as energy efficiency in buildings, flexible public transport services, digital infrastructures, smart grids, etc.
To turn cities into smart cities it is indispensable that such initiatives are carried forward within a systemic and integrated approach, capable of making the best use of infrastructures and encouraging the interoperability and scalability of solutions.
The challenge for cities is to be able to integrate the new enabling infrastructures and sensors technologies with existing structures on the territory, exploiting synergies and interoperability between systems in order to deliver added value services for citizens, thus contributing to improve their quality of life.

7. References

[1] International Energy Agency 2017 Tracking Clean Energy Progress 2017 (Paris: OECD/IEA).
[2] International Energy Agency 2017 *World Energy-related CO_2 Emissions* (Paris: IEA)
[3] National Oceanic and Atmospheric Administration 2017 State of the climate Global Climate Report 2017 (Silver Spring: NOOA)
[4] International Energy Agency 2013 Transition to Sustainable Buildings: Strategies and Opportunities to 2050 (Paris: OECD/IEA).
[5] World Health Organization 2016 Preventing disease through healthy environments A global assessment of the burden of disease from environmental risks (Geneva: WHO).
[6] European Environment Agency 2016 Air quality in Europe - 2016 report (Copenhagen: EEA)

[7] Smart Cities Council 2015 *Smart Cities Readiness Guide* (Reston, VA: Smart Cities Council)
[8] International Renewable Energy Agency 2017 *Renewable Energy and Jobs Annual review 2017* (Abu Dhabi: IRENA)
[9] International Energy Agency 2013 Technology Roadmap: Energy Efficient Building Envelopes (Paris: OECD/IEA).
[10] International Energy Agency 2017 Tracking Clean Energy Progress 2017 (Paris: OECD/IEA).
[11] European Photovoltaic Industry Association. 2011 Solar Generation 6: Solar Photovoltaic Electricity Empowering the World (Brussels: EPIA).
[12] International Energy Agency 2016 Global EV Outlook 2016 (Paris: OECD/IEA).
[13] Mayor of London 2015 *London Infrastructure Plan 2050 - Update* (London: Mayor of London)

Key Barriers to the Implementation of Solar Energy in Nigeria: A Critical Analysis

D Abdullahi[1], S Suresh, S Renukappa and D Oloke

1 School of Architecture and Built Environment, Faculty of Science and Engineering, University of Wolverhampton, WV1 1LY
E-mail: D.Abdullahi@wlv.ac.uk; S.Subashini@wlv.ac.uk; Suresh.Renukappa@wlv.ac.uk; D.A.Oloke@wlv.ac.uk

Abstract: Nigeria, potentially, has abundant sunshine throughout the year, making it full thirst for solar energy generation. Even though, the country's solar energy projects have not realised a fair result over the years, due to many barriers associated with initiatives implementation. Therefore, the entire power sector remains incapacitated to generate, transmit and distribute a clean, affordable and sustainable energy to assist economic growth. The research integrated five African counterpart's solar energy initiatives, barriers, policies and strategies adopted as a lesson learned to Nigeria. Inadequate solar initiative's research, lack of technological know-how, short-term policies, lack of awareness and political instability are the major barriers that made the implementation of solar initiatives almost impossible in Nigeria. The shock of the barriers therefore, constitutes a major negative contribution to the crippling of the power sector in the state. Future research will concentrate on initiatives for mitigating solar and other renewable energy barriers.

1. Introduction

Solar energy contributes a tiny fraction of power generation in Nigeria, although, the country receives maximum sunlight exposure, potential for solar PV. Nigeria is in the solar belt, which heightens its solar potential but sadly, the opportunities for harnessing the renewable energies remained narrow and unpractical compare to the conventional electricity [1]

Traditionally, solar energy in the country has predominantly been used for various activities by employing the open to the sun method, mostly in the rural communities. Solar electricity generation, on the other hand, emerged approximately two decades ago and it has seen a steady growth [2]. The primary drivers for the solar power uptake in the country are attributed to research centres and energy institutions, initiatives apart from the huge electricity production deficits. Besides the learning and research institutions, government and international agencies' activities have led to the development of the largest solar-PV plants across the country [3]. The major aims for the various solar electricity generation initiatives include the enhancement of water supply and the maintenance of clinics and healthcare facilities in the rural areas [4]

2. Solar photovoltaic (PV)

Nigeria is blessed with the abundance of solar radiation throughout the year. However, according to [5] only the exact utilisable solar resource base, formerly available in the state is unknown despite the huge financial commitment towards ensuring sustainable electricity in the country. They further added that Nigeria lacks the proper equipment and infrastructure to conduct solar radiation measurement. At the moment, Nigeria operates about thirty (30) measuring stations which are managed by the Nigerian

Meteorological Agency (NMA). These are airport based and the data from these stations are used to calculate an estimate probability for solar radiation in the country [6]. They are little local research and development activities related to solar energy in Nigeria. According to surveys conducted by the Energy Commission of Nigeria (ECN) in 1999 and, it was gathered that there are about 44 companies and research centres responsible for the importation and installation of photovoltaic systems in Nigeria. Out of these, only one company (Exide Batteries Nigeria Limited) produces batteries that are used for photovoltaic solar systems [6].

3. Solar energy potential

Nigeria is located in the solar belt with an average sunshine of up to 9 hours per day, equivalent to 5.5 kWhm-2 days-1 degrees of solar radiation are seen almost throughout the year [1]; [7]. The availability of massive radiations combined with the developments in the photovoltaic technologies makes it clear that huge amounts of electricity can be generated and utilised to combat the country's electricity crises that has become an obstacle to economic development [8]. [8], argued that the solar PV prices have been in steady decline in the recent years, dropping to approximately 50% compared to when it was first introduced years ago. This phenomenon was attributed to ramping up of solar PV productions in China and the technological breakthroughs experienced in the field [2].

Given the large amounts of incident radiation, installing solar panels on at least one percent of the country's geography can easily generate enough electricity to outdo the amount currently produced in the country [9]. Annually, the state receives up to 115, 000 times its total energy production through solar energy. This fact highlights the immense amount of energy which the country can benefit from the solar energy, bearing in mind that currently, solar energy contributes less than one percent of the total power generation [7]

Solar energy is the most stable and reliable source of energy that can be harnessed for the benefit of domestic and commercial purposes [9]. They further argued that, in line with the objectives of cutting down global warming through limiting the use of carbon fossils, solar energy has emerged to be a strategic component for achieving the sustainable development goals. Not only it is environmentally friendly, but it is also free and available in Nigeria. Electricity is not only produced cheaply, but also environmentally friendly [10].

4. Policies adopted

The national energy policy of 2003 approved by the federal government of Nigeria was aimed at leveraging the latent potential in power generation [11]. The plan highlights the need to develop energy using sustainable and environmentally friendly sources. The policy emphasised on electricity generation with the use of solar power among other renewable energy sources [12]. Regarding stakeholder composition in the energy production, the policy is much oriented towards the integration of public private partnership (PPP). Essentially, the policy based on hope that the private sector will take up the production, acquisition and distribution of renewable energy sources in the country [3]. The broad objectives of the policy include the generation of electricity in the rural areas, the creation of employment opportunities, environmental protection and mitigation of climate change and the diversification of energy production in the country. Besides other objectives, is to develop the country's solar utilisation capabilities [1]; [13].

The electric power sector reform Act of 2005 is the second policy formulated with aims to overcome the historic energy crisis in Nigeria [14]. The plan's principal objective was the liberalisation of the electricity generation industry to allow private investors to take part in energy generation, transmission and distribution [15]. The country's planners envision that by 2020 the country will be the 20th largest economy in the world. Bearing in mind the fact that power generation comes second to political goodwill in a country's development, the planners had to align energy production to their vision. Solar energy was therefore slated to be the major boost to the country's developments [16].

5. Impact of the policies

The policies highlighted above have a huge impact as far as solar energy generation in Nigeria is concerned. For instance, the Power Reform Act of 2005 has allowed private investors to get involved in the power generation exercises [17]. This measure has enabled the country to achieve some of its short-term energy objectives with rural areas being the primary beneficiaries [16]. The aligning of solar power generation with the country's Vision 2020 has also ensured that the program features in most government plans. For example, in parliamentary procedure to bring down the cost of the new textiles, the Nigerian government has removed duty on solar modules being imported into the country [13]

6. Lessons from five African countries

Apart from Nigeria, the other countries in Africa which have successfully adopted solar PV Panels include Morocco, South Africa, Ghana, Chad, and Kenya. Morocco so far has one of the most ambitious solar projects in the world [18] The Moroccan government sets some policies to direct their energy industry: The National plan of priority actions (PNAP) and the National Energy Strategy (NES). The NES major objective is to ensure that 20% of the country's energy needs is met through renewable energies [19]. Similarly, other related goal was to ensure that the country attains an energy efficiency of 20% by 2020 and a further 15 % by the dawn of 2030 [20] The PNAP policy on the other hand aimed at diversifying the fuel varieties and their sources. Subsequently, other policy, is the provision of energy in all sectors and at relatively competitive prices. The renewable promotion was another major objective aimed at heightening energy efficiency [20]. Finally, a policy was also aimed at integrating markets from the euro-Mediterranean region.

Morocco is on track towards the achievement of its energy objectives with the generation of 500MW of solar electricity having been initiated. The beginning stage of the project amounting to 160MW was completed in February 2016 with the other phases being scheduled to be finalised in 2018 [21].

South Africa, on the other hand, has also developed some policies to direct its renewable energy generating sector. South Africa's most important energy policy is the integrated resource plan [22] The plan outlines the country's aims to develop up to 9600 MW of solar power by the dawn of 2030. Resultantly, the solar electricity generation in the country has moved from nil to a point where it has provided a significant contribution to the national grid [23]. Furthermore, the rapid developments in the industry can be imputed to the friendly policies which contributed to the liberalisation of the sector. In combination with the relatively affordable Solar PVs, the industry has witnessed tremendous growth [23].

Ghana has also made strides towards the development of its solar energy generation capacity. The government's resolution is captured in its policies which include the national energy policy of 2010, aimed at introducing public-private partnership in the development of infrastructure [24]. The other policy was the Ghana sustainable energy for all action plan 2012 and was aimed at the development of renewable energy technologies. The government also developed a strategic national energy plan for 2006 to 2020 [24]. The plan as the title insinuates, was to outline the role of renewables in the national grid. The various policies set by the government have had a positive impact on the development of renewable in the country and specifically solar power generation. For instance, the country could add 20MW of solar energy to its national grid [25].

To achieve the SDGs, Chad initiated a myriad of measures towards the generation of power. One important renewable energy policy formulated by the country's government is the system of renewable [26]. The policy is aimed at providing the necessary framework for developing renewable energy schemes. There has been a great success as far as the attainment of renewable energy goals is concerned. For instance, recently, the country benefited from a sustainable energy fund for Africa loan aimed at the development of a 40MW solar plant in the country [27].

Kenya is optimally located along the equator to access maximum sunlight throughout the year. According to [28], the country experiences up to 300 days of sunlight within a year this highlighting the potential that it has regard to the generation of solar energy. The government has put in place some policies to ensure the seamless development of the renewable energy industry [28]. The policies include scaling up renewable energy program put into effects the elimination of import duties and value added tax (VAT) on all renewable energy-related imports. The government has also introduced a feed-in tariff to help the uptake of the energy produced from renewable sources such as solar power. The measures taken by the government have generated positive results as attributed to the proliferation of solar firms in the country [29]. Despite the potential of renewable energy, especially solar PV, in Africa, particularly Nigeria, the implementation suffers from various challenges which is further discussed.

7. Key barriers to solar initiatives in Nigeria

Despite the many benefits that can be accrued from the use of solar energy potential and uses is still very low. The major impediment to the technology adoption are series of barriers which makes it hard to implement.

Table 1: Barriers to Solar energy implementation in Nigeria

Barriers category	Barriers	Remarks	Sources
Technical Barriers	Lack of skilled personnel, lack of code of standard, lack of maintenance and operation, lack of training facilities and entrepreneur's development mechanism, lack of Reliability.	The barriers lead to poor plans, poor standard, constraints of the competitive market, inadequate knowledge to know-how about the technology and risk acceptance. All these barriers resulted in technology locked -up.	[1]; [30]; [31]
Social; Cultural Behaviour	Lack of consumer awareness about the product, lack of understanding of benefit of solar PV and public resistance to chance for new technology.	The barrier, affect the market projection negatively, cultural and religious faith controversies towards economic development and sustainability.	[32]; [33]; [34]
Economic/ Financial Barriers	Lack of access to capital, credit to consumers and financial instrument. Lack of support to R & D, high interest rate, import duties subsidies to support local manufacturing.	At the early stage, solar projects need incentives to encourage entrepreneurs. The barriers make it difficult to adopt and sustain due to financial constraints.	[13]; [17]; [35]
Institution al/ Legal barriers	Institutional barriers, legal framework, regulatory issues, non-integration of energy mix, non-participation of private sector, poor R & D culture and stakeholder's non-interference.	The barriers cause risk of uncertainty in support of solar energy, lobbies against RET, poor communication mechanism to reach the institutional policy makers for improvement and negative perception about the technology.	[7]; [36]
Political/ Policies Issues	Lack of long term policies, lack of political will to diversify into clean energy, constantly changing of government and re-shuffling of institutions.	These barriers serve as a deterrent to future planning for solar and other renewable energy adoption and sustainability. There is the fear of uncertainty in government.	[1]; [3]; [30]
Market Distortions Issues	Trade barrier for new product, energy sector controlled, lack of access to diversified technology, lack of facilities and backup technology, non-market oriented research for solar energy technology and application.	The barriers cause hindrance to market penetration and hence new technology failed at some point.	[3] [37]

Apart from awareness, the workforce to support and technical know-how to install the solar panels is still lacking. The necessary expertise to assemble and manage the solar facilities in existence has further composed the issue. Currently, Nigeria relies on experts from other developed countries to run its solar generation plants. This lack of capacity increases the maintenance cost of the projects. This fact impedes the technology's uptake even in cases where the end users are aware of its adverse effect. Despite the major reduction in the price of solar PVs, the lack of financial and fiscal incentives present a challenge to the stakeholders. Essentially, the development of a solar firm requires massive amounts of money which may not be easily available in Nigerian situation [16].
Solar generation also suffers from the insufficient institutional framework. Apparently, the country's energy regulation Commission (NERC), is also mandated to oversee the licensing of electricity plants that generate at least 1MW. Incidentally, renewable energy plants which produce less than 1MW are not adequately regulated. The intermittent nature of resources leaves the stakeholders with operational challenges. Renewable energy sources are cyclic in character, with seasons of high supply and others of low supply. Therefore, the lack of storage makes the operation of solar systems, complex [17].

8. Findings

It is evident that Africa in general and Nigeria to be precise have huge potentials as far as the generation of clean energy is concerned. It is, however, disappointing to note that solar power which is the most efficient and readily available resource is least utilised with most countries covered having below one percent solar power generation. The major impediments to the uptake include the lack of sufficient finance, technical know-how and general awareness [38].

9. Recommendations

As far as the improvement of solar energy production in Nigeria is concerned, creating awareness concerning its advantages will go a long way towards its adoption. The Nigerian government should, therefore, put in place a policy that encourages the sensitisation and awareness initiatives. Besides creating awareness, the government should also introduce in the country's learning institutions, courses on the handling of solar panels. Such a policy will aid in the development of the necessary workforce capacity. A special fund should also be put in place to subsidise the raw materials needed for the installation and production of solar panels [39]. Future work of this research will proffer a framework to mitigate challenges for solar energy initiatives in Nigeria.

10. Conclusion

Nigeria has not taken advantage of its potential in the renewable energy generation. Given the large amounts of incident radiation, the country can greatly benefit if the necessary resources are committed towards subsidised opportunity for socioeconomic development. It is, therefore, imperative upon the stakeholders to put in place policies that will create an enabling environment for solar energy production. The stakeholders for energies, especially solar, can learn from other African countries such as South Africa, Kenya and Morocco who to some extent have yielded positive result. The Nigerian government needs to address the challenges also by integrating research recommendations from various research more especially the institutional and technical barriers.

11. References

[1] Sambo, A.S. and Bala, E.J., 2012. Penetration of Solar Photovoltaic into Nigeria's Energy Supply Mix. In World Renewable Energy Forum (WREF).
[2] Osinowo, A.A., Okogbue, E.C., Ogungbenro, S.B. & Fashanu, O. 2015, "Analysis of Global Solar Irradiance over Climatic Zones in Nigeria for Solar Energy Applications", Journal of Solar Energy, 2015, pp. 1-9.
[3] Ohunakin, O.S., Adaramola, M.S., Oyewola, O.M. & Fagbenle, R.O. 2014, "Solar energy applications and development in Nigeria: Drivers and barriers", Renewable and Sustainable Energy Reviews, 32, pp. 294-301.

[4] Oseni, M.O., 2012. Improving households' access to electricity and energy consumption pattern in Nigeria: Renewable energy alternative. Renewable and Sustainable Energy Reviews, 16 (6), pp.3967-3974.

[5] Emodi, N.V. 2015, "Policy Scenarios for Low Carbon Energy Development in Nigeria", afore, pp. 237-237.

[6] Udoakah N. Y.-O. 2014. Sustainably meeting the energy needs of Nigeria: The renewable options. IEEE International Conference, IEEE, pp. 326-332.

[7] Aliyu, A.S., Dada, J.O. & Adam, I.K. 2015, "Current status and future prospects of renewable energy in Nigeria", Renewable and Sustainable Energy Reviews, (48), pp. 336-346.

[8] Amankwah‐Amoah, J., 2015. Solar Energy in Sub‐Saharan Africa: The Challenges and Opportunities of Technological Leapfrogging. Thunderbird International Business Review, 57 (1), pp.15-31.

[9] Ozoegwu, C.G., Mgbemene, C.A. & Ozor, P.A. 2017, "The status of solar energy integration and policy in Nigeria", Renewable and Sustainable Energy Reviews, 70, pp. 457-471.

[10] Masini, A. and Menichetti, E., 2012. The impact of behavioural factors in the renewable energy investment decision making process: Conceptual framework and empirical findings. Energy Policy, 40, pp.28-38.

[11] Adeyanju, A.A. 2011, "Solar Thermal Energy Technologies in Nigeria", Research Journal of Applied Sciences, 6 (7), pp. 451-456.

[12] Ajayi, O.O., Ohijeagbon, O.D., Nwadialo, C.E. & Olasope, O. 2014, "New model to estimate daily global solar radiation over Nigeria", Sustainable Energy Technologies and Assessments, 5, pp. 28-36.

[13] Shaaban M, Petinrin, J. O. 2014. Renewable energy potentials in Nigeria: Meeting rural energy needs. Renewable and Sustainable Energy Reviews, 29:72-84.

[14] Kieran, P. 2014. Review of the reform and the Privatization of Power Sector in Nigeria: Solution for Growing Economies. Available at: http://www.energynet.co.uk/webfm_send/427. [Accessed 14 February 2017].

[15] Emetere, M.E. & Akinyemi, M.L. 2016, "Prospects of solar energy in the coastal areas of Nigeria", AIP Conference Proceedings, 1705 (1).

[16] Oyedepo, S.O., 2012. Energy and sustainable development in Nigeria: the way forward. Energy, Sustainability and Society, 2 (1), p.15.

[17] Emodi, N.V. and Boo, K.J., 2015. Sustainable energy development in Nigeria: Current status and policy options. Renewable and Sustainable Energy Reviews, 51, pp.356-381.

[18] Humphreys, G. 2014, "Harnessing Africa's untapped solar energy potential for health", Bulletin of the World Health Organization, 92 (2), pp. 82.

[19] Writer, G. 2016, Solar energy in Africa comes of age as costs reach grid parity, The Financial Times Limited, London.

[20] Kousksou, T., Allouhi, A., Belattar, M., Jamil, A., El Rhafiki, T., Arid, A. and Zeraouli, Y., 2015. Renewable energy potential and national policy directions for sustainable development in Morocco. Renewable and Sustainable Energy Reviews, 47, pp.46-57.

[21] Nfaoui, H. and Sayigh, A., 2017. Renewable Energy in South of Morocco and Prospects. In Mediterranean Green Buildings & Renewable Energy (pp. 667-679). Springer International Publishing.

[22] Donev, G., van Sark, W.G.J.H.M., Blok, K. & Dintchev, O. 2012, "Solar water heating potential in South Africa in dynamic energy market conditions", Renewable and Sustainable Energy Reviews, 16 (5), pp. 3002.

[23] Dekker, J., Nthontho, M., Chowdhury, S. and Chowdhury, S.P., 2012. Economic analysis of PV/diesel hybrid power systems in different climatic zones of South Africa. International Journal of Electrical Power & Energy Systems, 40 (1), pp.104-112.

[24] Atsu, D., Agyemang, E.O. & Tsike, S.A.K. 2016, "Solar electricity development and policy support in Ghana", Renewable and Sustainable Energy Reviews, 53, pp. 792-800.

[25] Asumadu-Sarkodie, S. and Owusu, P.A., 2016. The potential and economic viability of solar photovoltaic power in Ghana. Energy sources, Part A: Recovery, utilization, and environmental effects, 38 (5), pp.709-716.

[26] AfDB supports development of Chad's Starsol PV 2015, Euromoney Institutional Investor PLC, London.

[27] Komendantova, N., Patt, A., Barras, L. and Battaglini, A., 2012. Perception of risks in renewable energy projects: The case of concentrated solar power in North Africa. Energy Policy, 40, pp.103-109.

[28] Ondraczek, J., 2014. Are we there yet? Improving solar PV economics and power planning in developing countries: The case of Kenya. Renewable and Sustainable Energy Reviews, 30, pp.604-615.

[29] Newell, P., Phillips, J., Pueyo, A., Kirumba, E., Ozor, N. and Urama, K., 2014. The political economy of low carbon energy in Kenya. IDS Working Papers, 2014 (445), pp.1-38.

[30] Painuly, J.P. 2001, "Barriers to renewable energy penetration; a framework for analysis", Renewable Energy, vol. 24, no. 1, pp. 73-89.

[30] Luthra, S., Kumar, S., Garg, D. & Haleem, A. 2015, "Barriers to renewable/sustainable energy technologies adoption: Indian perspective", Renewable and Sustainable Energy Reviews, vol. 41, pp. 762-776.

[31] Pasqualetti, M.J. 2011, "Social Barriers to Renewable Energy Landscapes", Geographical Review, vol. 101, no. 2, pp. 201-223.

[32] Pollmann, O., Podruzsik, S. and Fehér, O., 2014. Social acceptance of renewable energy: Some examples from Europe and Developing Africa. Society and Economy, 36 (2), pp. 217-231.

[33] Akinwale, Y.O., Ogundari, I.O., Ilevbare, O.E. and Adepoju, A.O., 2014. A Descriptive Analysis of Public Understanding and Attitudes of Renewable Energy Resources towards Energy Access and Development in Nigeria. International Journal of Energy Economics and Policy, 4 (4), pp. 636-646.

[34] Kar, S.K. & Sharma, A. 2015, "Wind power developments in India", Renewable and Sustainable Energy Reviews, 48 pp. 264-275.

[35] Charles, A., 2014. How is 100% renewable energy possible in Nigeria? Global Energy Network Institute (GENI), 1 (619), pp. 595-0139.

[36] Fagbenle, R.O., Katende, J., Ajayi, O.O. & Okeniyi, J.O. 2011, "Assessment of wind energy potential of two sites in North-East, Nigeria", Renewable Energy, 36 (4) pp. 1277-1283.

[37] Bazilian, M., Nussbaumer, P., Rogner, H.H., Brew-Hammond, A., Foster, V., Pachauri, S., Williams, E., Howells, M., Niyongabo, P., Musaba, L. and Gallachóir, B.Ó., 2012. Energy access scenarios to 2030 for the power sector in sub-Saharan Africa. Utilities Policy, 20 (1), pp.1-16.

[38] Timilsina, G.R., Kurdgelashvili, L. and Narbel, P.A., 2012. Solar energy: Markets, economics and policies. Renewable and Sustainable Energy Reviews, 16 (1), pp.449-465.

IOP Conf. Series: Earth and Environmental Science **83** (2017) 012016 doi:10.1088/1755-1315/83/1/012016

MOF-reduced Graphene Oxide Composites with Enhanced Electrocatalytic Activity for Oxygen Reduction Reaction

Yuan Zhao[1], Rong Fan[1], Chuanxiang Zhang[2], Haijun Tao[1], Jianjun Xue[1]

1 College of Material Science and Science Technology, Nanjing University of Aeronautics and Astronautics, Nanjing 211106, PR China

2 College of Materials Engineering, Nanjing Institute of Technology, Nanjing 211167, PR China

E-mail: zy871817440@163.com

Abstract. Development of inexpensive and scalable cathode catalysts that can efficiently catalyze the oxygen reduction reaction (ORR) is of significance in practical application of fuel cells. The oxygen reduction activity of the MOF-based catalyst is much lower than that of Pt, which is mainly due to the high overpotential. In this work, we designed a superior composite named Co@Co_3O_4-reduced graphene oxide (Co@Co_3O_4-rGO) derived from MOF-rGO by an in-situ synthetic method which gathered both the advantages of MOF and rGO. The Co^{2+} which belongs to the MOF provides the metal source, while the N sources are supplied by the organic ligands benzimidazole. With the combination of rGO, Co@Co_3O_4-rGO has got a higher specific surface area and much better transport pathways for oxygen and the electrolyte than Co@Co_3O_4-C derived from the pure MOF. The half-wave potential, onset potential of Co@Co_3O_4-rGO are close to the superior commercial Pt/C catalyst. The number of electron transfer in the process of catalytic oxygen reduction is close to 4, the excellent properties benefited from the synergistic effect of rGO and MOF.

1. Inrtoduction

Global resource shortages and environmental pollution have forced us to seek alternative energy conversion and storage systems with the advantages of high efficiency, low cost and environmental benignity [1]. With these advantages, Direct Methanol Fuel Cells (DMFC) has been concerned widely, the oxygen reduction reaction (ORR) is one of the most basically and the core technology of electrochemical reactions for DMFC [2]-[4]. Generally, Pt-based nanomaterials have been applied on promoting the oxygen reduction reaction as catalysts for their dominant electro-catalysts. Nevertheless, resource starvation, high price, poor stability and environmentally hazardous hinders Pt-based catalysts being widely applied to practical application in fuel cell devices. To overcome these bottlenecks, non-precious metal-based nanomaterials electro-catalysts have been intensively studied for substituting the high cost electro –catalysts Pt for the oxygen reduction reaction [5]-[8].

Metal-organic frameworks (MOFs) are built from the reaction of metal ions and organic ligands, which have the advantage of particular porous structures and high specific surface area [9]-[11]. It has caught attention of plenty of researchers in gas adsorption[12]-[14], storage [15-17], sensors [10] and catalysis [18], [19]. Moreover, zeolitic imidazolate framework (ZIFs) could be designed with abundant metal-N_x coordinate moieties that can improve electro-catalytic performance [20]. As a result, most

MOF-derived electrocatalysts for ORR were based on porous carbon materials prepared by pyrolysis of ZIFs or ZIF-supported precursors [18]. Unfortunately, the oxygen reduction activity of the MOF-based catalyst is much lower than that of Pt, which is mainly due to the high overpotential effect of the MOF-based catalyst, which can be divided into two vital aspects. On the one hand, nitrogen-doped carbon or M-N-C structure which supplies active sites plays a vital role on ORR catalysts but not effective. On the other hand, the transmission of O_2 and the interaction between electrolytes with catalysts are limited .In this regard, in order to obtain more efficient catalysts for ORR, the most vital factor is how to get rapid mass transport properties of ordered carbon nanomaterials [21]. Graphene, a two-dimensional sheet of sp^2 conjugated carbon atoms, can be considered as a giant polyaromatic platform for performing chemistry due to its open ended structure [21], [22]. The combination of high specific surface area and high electrical conductivity makes graphene sheets high promising as an electrocatalyst platform. Loh et al.[23] studied the electro-catalytic action of the composites of MOFs and graphene oxide (GO) on the hydrogen evolution reaction (HER), oxygen evolution reaction (OER), and oxygen reduction reaction (ORR). Their studies showed that the composites exhibited the better electrochemical performance and stability than that of pure MOFs due to the synergistic effect of GO and MOF. Therefore, in-depth study of MOF-graphene composites is necessary.
In this work, we report a ZIF9-based nanomaterial which was inserted into reduced graphene oxide by an in situ method. In the process of Benzimidazole and $Co(NO_3)_2{\cdot}6H_2O$ were used as precursors to synthesis ZIF9, at the same time reduced graphene oxide was added into the mixture, MOF-rGO composite was obtained in one step. Then after thermal treatment under the atmosphere of N_2, Co@Co_3O_4-rGo composite was synthesized finally. During the prepared composite is pyrolyzed into Co@Co_3O_4-rGO, the Co^{2+} which was surrounded by benzimidazole ligands converted into Co@Co_3O_4 and it was wrap by N-doped porous carbon derived from benzimidazole. And the rGO can enhancing the electron transfer between metal oxide nanoparticles and carbon support. Due to the improvement mentioned above, the Co@Co_3O_4-rGO which prepared showed higher ORR catalyst performance than Co@Co_3O_4,which is approach to the commercial Pt/C catalyst. And the Co@Co_3O_4 -rGO electrocatalyst also revealed excellent stability and methanol tolerance in comparison to the Pt/C catalyst at the same time.

2. Experiment

2.1. Synthesis of MOF-rGO
Graphite oxide(GO) was prepared by oxidation of graphite powder via modified Hummers' method [24]. The reduced graphene is obtained by hydrazine hydrate reduction [25]. In order to obtain the MOF-rGO, $Co(NO_3)_2{\cdot}6H_2O$ (0.42g,1.44 mmol) was dissolved into DMF(20 mL), 0.1g r-GO was added to the solution above and sonicated for 30 min. In other hand, benzimidazole (0.17 g) was dissolved into 30 mL DMF, and this solution was stirred for 10 min. Then reflux unit was equipped, and the vessel was heated at 130 ℃ for 12 h. After cooling to room temperature the as-formed MOF-rGO composite was obtain by filtration and washed with water and ethanol several times. The filter residue was dried by freeze-drying.

2.2 Synthesis of Co@Co_3O_4 and Co@Co_3O_4-rGO
The above prepared MOF and MOF-rGO composite was added into porcelain crucible and then pyrolyzed at 800 ℃ for 2 h under the atmosphere of Ar_2 before cooling down to room temperature. The composite was heated to 500℃ at a heating rate of 10℃ min^{-1}, then obtained material was heated to 800 ℃ at a heating rate of 5 ℃ min^{-1}. At last, via thermal treatment the purple MOF and atropurpureus MOF-rGO composite was converted to black powder.

2.3 Characterization
The crystal structure of the electro-catalyst was identified by X-ray diffraction (XRD) 2θ ranges from 10° to 90°. X-ray spectroscopy (XPS) was performed by. The morphological structure of the electro-catalyst was investigated by scanning electron microscopy measurements and transmission electron

IOP Conf. Series: Earth and Environmental Science **83** (2017) 012016 doi:10.1088/1755-1315/83/1/012016

microscopy. Raman spectra were characterized by Micro-Raman Spectroscopy System (Renishaw inVia).

2.4 Electrochemical measurements
All the electrochemical measurements were carried through CHI660D electrochemical workstation which was supplied by CHENHUA Shanghai via a standard three-electrode system. The working electrode was a glassy carbon rotating disk electrode (GC) which was coated by the catalyst with a diameter 5 mm. First, the working electrode was polished by α-Al_2O_3 powders in order to get a smooth, neat and clean surface, and then washed with deionized water and ethanol before using. To prepare the working electrode,5mg of the as-prepared electro-catalyst which was prepared with different doping content and different calcination temperature and 50 μ L Nafion solution were mixed and dispersed in 1mL 1:4 (v/v) water/ethanol mixed solvent and the mixture were treated with sonication for 1h at least in order to obtain a homogeneous suspension solution. The 20% Pt/C catalyst was also prepared by the method mentioned above. Then 25 μ L of the mixed suspension were dropwise added onto the surface of the GC electrode and dried in air with natural drying. An Hg/Hg_2Cl_2 electrode with KCl solution was used for reference electrode and Pt sheet was applied as counter electrode. All potentials here are converted to a reversible hydrogen electrode (RHE) scale.
Cyclic voltammetry (CV) and linear sweep voltammetry (LSV) were performed in a N_2-saturated and an O_2-saturated 0.1 M KOH in the potential range of 0-1.2 V (*vs.* RHE). All the electrochemical measurements were performed in a 0.1 mol/L KOH aqueous solution which was used as electrolyte at room temperature with pumping oxygen or nitrogen into the electrolyte.
Koutechy-Levich equation.

$$\frac{1}{J}=\frac{1}{J_L}+\frac{1}{J_K}=\frac{1}{B\omega^{1/2}}+\frac{1}{J_K} \qquad (1)$$

$$B=0.2nFC_0(D_0)^{2/3}v^{-1/6} \qquad (2)$$

Here J represents the measured electric current density, J_L and J_K are the diffusion and kinetic-limiting electric current densities, respectively. ω means the angular velocity , n is the electrons number which was transferred during the ORR, F is faraday constant (96485 C mol^{-1}) , C_0 is the bulk concentration of O_2 in 0.1 M KOH (1.26×10^{-3}mol L^{-1}), D_0 is the oxygen diffusion coefficient in 0.1 M KOH (D_0 = 1.9×10^{-5} cm^2 s^{-1}) and V is the kinetic viscosity of the electrolyte (n=0.01cm^2 s^{-1}). Chronoamperometric measurements were also performed in an O_2-saturated KOH electrolyte adding the methanol with the purpose of testing the methanol tolerance performance of the electro-catalyst [26].

3. Results and discussion
Morphological and microstructure were investigated by scanning electron microscopy (SEM) and transmission electron microscopy (TEM). As shown in fig 1a, the pure MOF have got a purple appearance, respectively. However, the MOF-rGO composite in fig.1c displays a comprehensive appearance between purple and black which can defined as modena. The morphologies and structural features of the as-prepared electro-catalyst, including MOF, rGO and Co@Co_3O_4-rGO can be observed clearly and directly from SEM and TEM images. SEM images of the pure MOF and MOF-rGO are shown in fig.2b and fig.2d. As we can see from fig.2d, a layer of rGO with parts of drapes tiled in the image.Fig.2b displays the nubbly morphology of the pure MOF, fig.2d reveals the nubbly MOF was inserted into the carbon layer of rGO forming MOF-rGO. MOF-rGO was pyrolyzed under N_2 atmosphere in order to prepare Co@Co_3O_4-rGO, as it is shown in fig.2e, the rGO is still remained via thermal treatment and Co@Co_3O_4 nanopaticles are uniformly distributed and anchored in the layer of rGO.

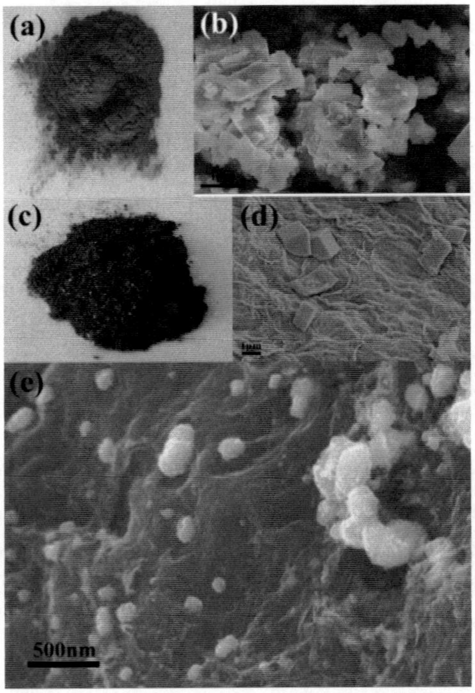

Figure 1. Photographs of (a) ZIF9 (c) ZIF9-rGO; SEM images of (b) ZIF9, (d) ZIF9-rGO and (e) Co@Co_3O_4-rGO

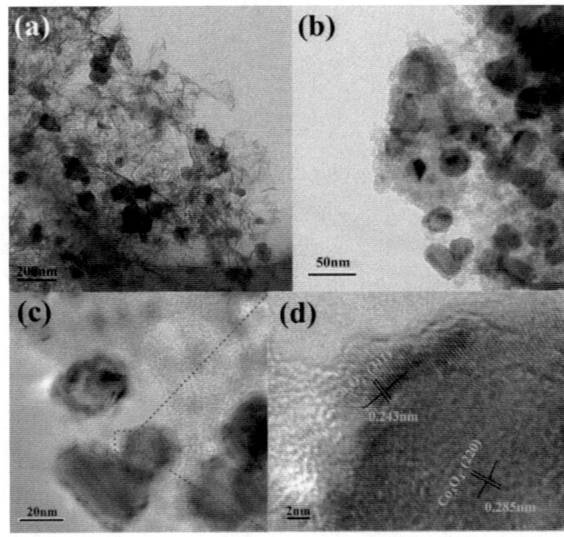

Figure 2. TEM image of (a) ZIF9-rGO, and (b) Co@Co_3O_4-rGO. HRTEM images of (c) (d) Co@Co_3O_4-rGO

The tulle-like graphene can be seen in the TEM spectrum clearly (fig 2(a)), after high temperature pyrolysis, the graphene transition to C, and Co^{2+} that exists in the layer is also converted into Co_3O_4. As shown in fig 2(d) the high-resolution TEM (HRTEM) image reveals that the lattice spacings of Co_3O_4 nanocrystala on rGO are 0.243nm and 0.285nm, and which mathes well with the lattice spcings of Co_3O_4 (311) and (220). This result is consistent with the XRD spectrum.

In order to further analyze the structure and chemical composition of the Co@Co_3O_4-rGO composites, XRD and XPS analysis were performed.. From fig 3a, the characteristic peaks of Co, Co_3O_4, C can be observed obviously. The prominently sharp carbon peak at 26.2° came from the carbon which was pyrolyzed from benzimidazole after combined with rGO. The peak at 44.3° is assigned to the (111) lattice plane of Co, also the peaks at 51.3° and 76.2° are assigned to the (200) and (220) lattice plane Co_3O_4 respectly. The lattice planes mentioned above further testify the existence of Co. The rest peaks corresponding to the (111),(220),(311),(511) and (440) lattice planes [27].

The full XPS spectrum of the Co@Co_3O_4-rGo fig.3b demonstrates the existence of the element of carbon, cobalt, nitrogen and oxygen. Among the elements mentioned above, the nitrogen element is derived from the benzimidazolate ligands. We can see from the high-resolution XPS spectrum of N peaks, pyridinic N (398 eV), pyrrolic N (400 eV) and graphitic N (401 eV) can be observed in fig.3c. Pyridine nitrogen can provide effective oxygen reduction activity site [28]. The regional Co 2p spectrum displayed in fig.3d indicates that there are two energy bands, the higher energy band is at 795 eV(Co $2p_{1/2}$), the lower one is at 780 eV(Co $2p_{3/2}$).

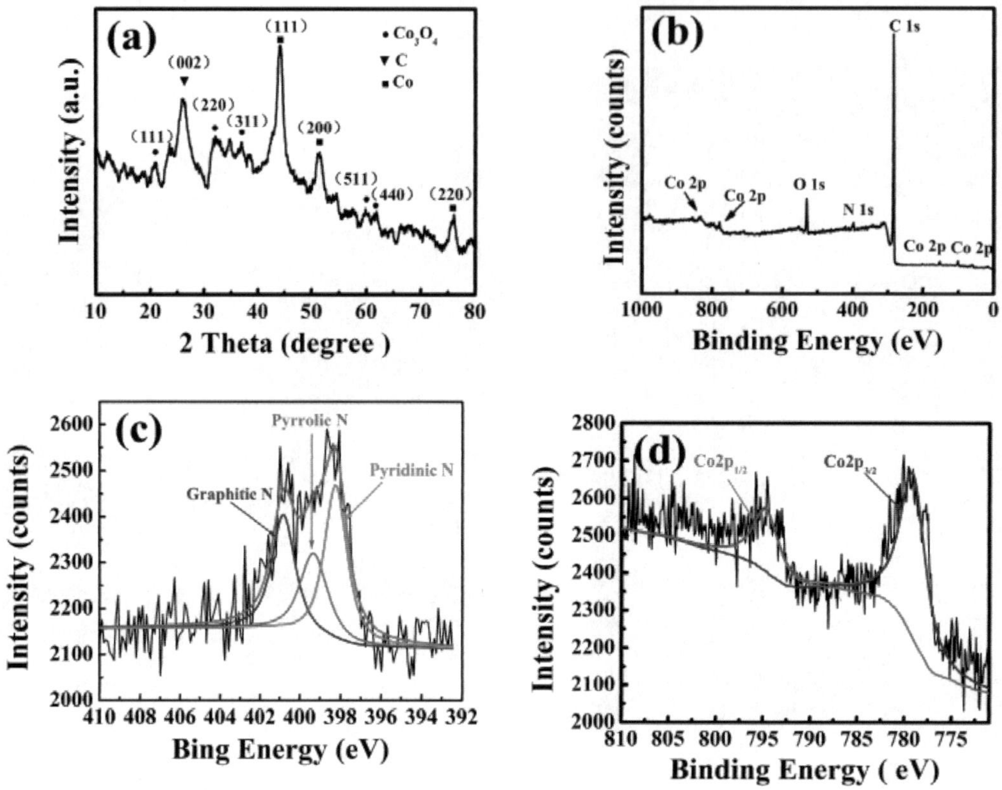

Figure 3. (a) XRD pattern, (b) XPS full spectrum, (c) N1s spectrum and (d) Co 2p XPS spectrum.

Co@Co_3O_4-rGO shows type IV isotherms (mesoprous capillart condensation) with a H3-type hysteresis loop (P/P_0 >0.4), indicating the nanocatalyst is a mesoprous. As we can see from the Fig 4b, MOF-rGO also shows type IV isotherms while it shows H4-type hysteresis loop. Also from the BJH method we obtained the pore size is about 4.36 nm and the Brunauer–Emmett–Teller (BET) surface area is 305.4 m^2g^{-1},as well as the cumulative pore volumes is 0.31cm^3g^{-1}. Contrasted to the pure MOF Fig 4a, the composite has much higher specific surface area. The results displayed upon demonstrated that the high surface areas and pores of the pyrolyzed pure MOF remained after thermal treatment. The pyrolyzed composite shows better ORR catalytic activity than the pyrolyzed pure MOF as a result of the improvement on specific surface area and pore size.

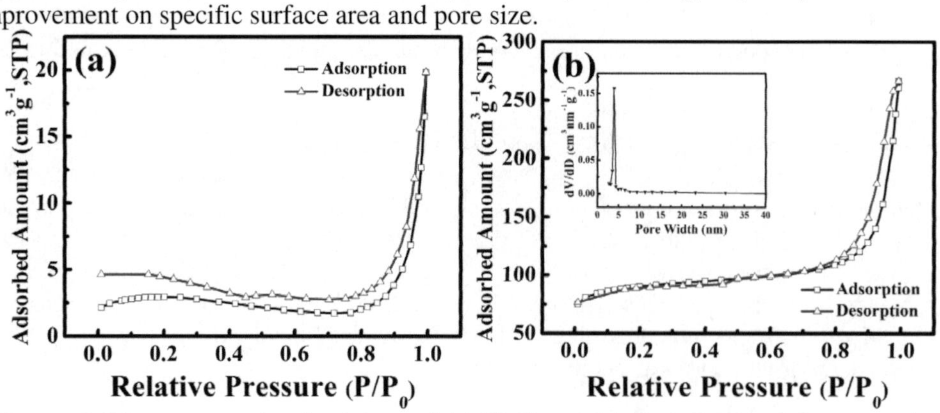

Figure 4. Nitrogen sorption isotherms of (a) ZIF9 and (b) Co@Co_3O_4-rGO composite.

The inset shows corresponding pore size distribution based on the Quenched Solid Density Functional Theory (QSDFT) model using the adsorption branch.

In electrochemical measurements section, the CV of Co@Co_3O_4-rGO catalyst take place in 0.1M KOH solution. As we can see from fig.5a, the Co@Co_3O_4-rGO catalyst shows a common curve with no characteristic in N_2-saturated solution. Nevertheless, the Co@Co_3O_4-rGO catalyst reveals obvious cathodic peaks for the ORR in O_2-saturated 0.1M KOH solution. The fig.5b displays the ORR polarization curves of Co@Co_3O_4-rGO,Co@Co_3O_4,rGO and commercial Pt/C via linear sweep voltammetry. It is not difficult to see that the catalysts obtained from MOF-rGO has got a higher onset potential of 0.90 V and a higher half-wave potential of 0.85 V than the catalyst derived from pure MOF with the onset potential of 0.84 V and the half-wave potential of 0.79 V. It is worth mentioning that the half-wave potential, onset potential of Co@Co_3O_4-rGO are close to the superior commercial Pt/C catalyst. As we can see from the comparison curves, the catalyst derived from composite material with the addition of rGO reveals obvious advantage on onset potential and limited current than the catalyst derived from the pure MOF. The rGO improves the electrical conductivity. In a word, the outstanding performance on ORR activity makes the Co@Co_3O_4-rGO be the most high-efficiency catalyst among the non-precious metal under the same loading capacity. In addition, the Co@Co_3O_4-rGO catalyst possesses a higher limit current density than that of Co@Co_3O_4 and the rGO, which is approaching to the limit current density of superior commercial Pt/C.

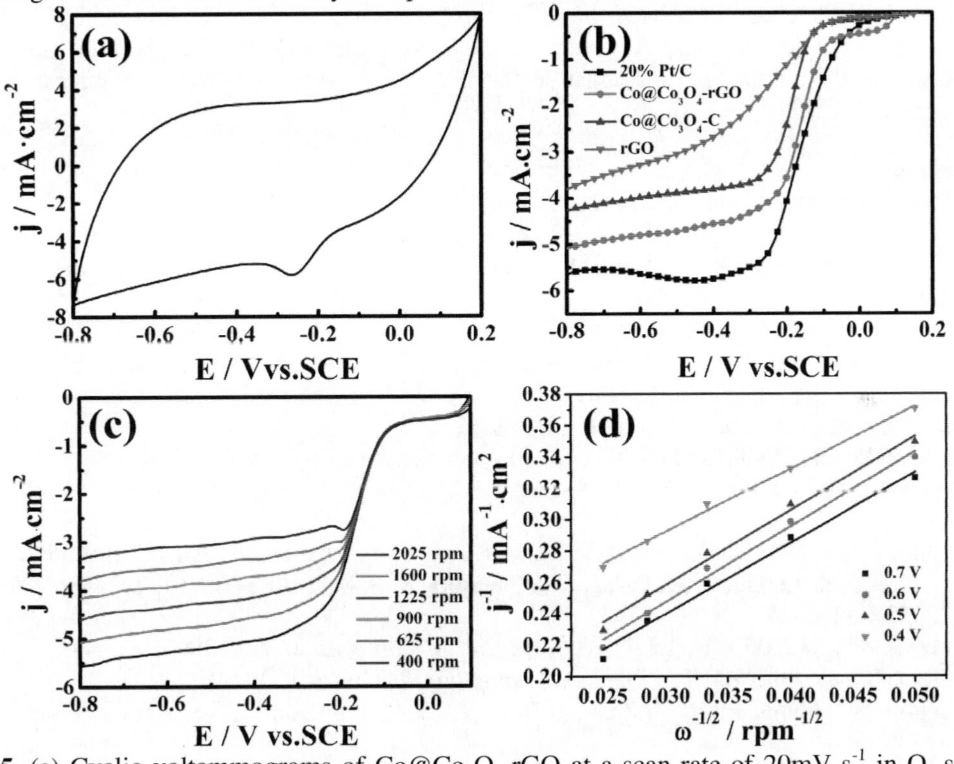

Figure 5. (a) Cyclic voltammograms of Co@Co_3O_4-rGO at a scan rate of 20mV s^{-1} in O_2-saturated aqueous solution of 0.1M KOH; (b)rotating disk electrode (RDE) voltammograms of 20% Pt/C, Co@Co_3O_4-rGO, Co@Co_3O_4-C, and rGO in an O_2-saturated 0.1M aqueous KOH solution with a scan rate of 5 mV s^{-1} at a constant rotation rate of 1600 rpm; (c) RDE voltammograms of Co@Co_3O_4-rGO at different rotation rates from 400 to 2025 rpm and (d) Koutecky-Levich plot for the electrode materials at -0.3V.

The linear sweep voltammetry curves at various rotating rates and corresponding Koutechy-Levich plots of Co@Co_3O_4-rGO are obtained and calculated which are shown in fig.5c and fig.5d. A good linearity of the plots and near parallelism of the fitting lines at various potentials were observed, suggesting that the reactions on Co@Co_3O_4-rGO are first-order over the potential range. To assess the ORR property of the catalysts fairly, the preferred electron transfer number should four[29]. As we obtained from the Koutechy-Levich calculation, the electron transfer number on Co@Co_3O_4-rGO ranges from 3.8 to 3.9, which is very close to 4, indicating that the catalyst we synthesized suffers 4-electron during the ORR process.

4. Conclusions

In summary, this work offered an in situ method to prepare MOF-rGO by refluxing treatment of benzimidazole, cobaltous nitrate hexahydrate and reduced graphene oxide. And then after a step pyrolysis get Co@Co_3O_4-rGO directly. With the combination of rGO, Co@Co_3O_4-rGO has got a higher specific surface area and much better transport pathways for oxygen and the electrolyte than Co@Co_3O_4-C derived from the pure MOF. By analyzing the properties of as-obtained product, the Co@Co_3O_4-rGO catalyst demonstrated superior catalytic activity compared to Co@Co_3O_4-C and rGO. The excellent properties benefited from the synergistic effect of rGO and MOF. It is believed that the simple but efficient way could be further developed for the production of metal-free electrocatalyst to replace Pt/C in the area of fuel cell and metal-air battery in the future.

5. Acknowledgments

The authors gratefully acknowledge the Jiangsu Innovation Program for Graduate Education(Grant no. KYLX15-0305), Fundamental Research Fund for the Central Universities (Grant no.NS2015061), the Foundation of Graduate Innovation Center in NUAA (Grant no. KFJJ20160611), and Priority Academic Program Development of Jiangsu Higher Education Institutions for the financial support of this work.

6. References

[1] S. Chu and A. Majumdar, *Nature*, 2012, **488**, 294-303.
[2] S. K. Kamarudin, F. Achmad and W. R. W. Daud, *Int. J. Hydrogen Energy*, 2009, **34**, 6902-16.
[3] S. K. Kamarudin, W. R. W. Daud, S. L. Ho and U. A. Hasran, *J. Power Sources*, 2007, **163**, 743-54.
[4] R. Dillon, S. Srinivasan, A. S. Aricò and V. Antonucci, *J. Power Sources*, 2004, **127**, 112-26.
[5] M. Shao, Q. Chang, J. P. Dodelet and R. Chenitz, *Chem. Rev.*, 2016, 116, 3594-657.
[6] D. Li, H. Lv, Y. Kang, N. M. Markovic and V. R. Stamenkovic, *Annual Review of Chemical and Biomolecular Engineering*, 2016, 7, 509-32.
[7] J. Liu, P. Song, Z. Ning and W. Xu, *Electrocatalysis*, 2015, 6, 132-47.
[8] G. L. Tian, M. Q. Zhao, D. Yu, X. Y. Kong, J. Q. Huang, Q. Zhang and F. Wei, *Small*, 2014, **10**, 2251-59.
[9] W. Xia, J. Zhu, W. Guo, L. An, D. Xia and R. Zou, *J Mater Chem A*, 2014, **2**, 11606.
[10] L. E. Kreno, K. Leong, O. K. Farha, M. Allendorf, R. P. Van Duyne and J. T. Hupp, *Chem. Rev.*, 2012, **112**, 1105-25.
[11] J. L. C. Rowsell and O. M. Yaghi, *Microporous Mesoporous Mater.*, 2004, **73**, 3-14.
[12] C. Petit, B. Mendoza and T. J. Bandosz, *Langmuir*, 2010, **26**, 15302-9.
[13] X. Zhou, W. Huang, J. Shi, Z. Zhao, Q. Xia, Y. Li, H. Wang and Z. Li, *J Mater Chem A*, 2014, **2**, 4722.
[14] K. Sumida, D. L. Rogow, J. A. Mason, T. M. McDonald, E. D. Bloch, Z. R. Herm, T. H. Bae and J. R. Long, *Chem. Rev.*, 2012, **112**, 724-781.
[15] S. Liu, L. Sun, F. Xu, J. Zhang, C. Jiao, F. Li, Z. Li, S. Wang, Z. Wang, X. Jiang, H. Zhou, L. Yang and C. Schick, *Energ Environ Sci*, 2013, **6**, 818.
[16] Z. Li and L. Yin, *J. Mater. Chem. A*, 2015, **3**, 21569-77.
[17] D. Ji, H. Zhou, Y. Tong, J. Wang, M. Zhu, T. Chen and A. Yuan, *Chem. Eng. J.*, 2017, **313**, 1623-32.
[18] H. Wang, J. Wei, Y. Hu, Y. Liang, B. Kong, Z. Zheng, J. I. N. Zhang, S. P. Jiang and Y. Zhao, *J. Mater. Chem. A*, 2017, DOI: 10.1039/c7ta00276a.
[19] S. Ma, G. A. Goenaga, A. V. Call and D. J. Liu, *Chemistry*, 2011, **17**, 2063-7.
[20] W. Xia, R. Zou, L. An, D. Xia and S. Guo, *Energy Environ. Sci.*, 2015, **8**, 568-76.
[21] M. Jahan, Q. Bao and K. P. Loh, *J. Am. Chem. Soc.*, 2012, **134**, 6707-13.
[22] C. Huang, C. Li and G. Shi, *Energ Environ Sci*, 2012, **5**, 8848.

[23] M. Jahan, Z. Liu and K. P. Loh, *Adv. Funct. Mater.*, 2013, **23**, 5363-72.
[24] D. C. Marcano, D. V. Kosynkin, J. M. Berlin, A. Sinitskii, Z. Sun, A. Slesarev, L. B. Alemany, W. Lu and J. M. Tour, *ACS Nano*, 2010, **4**, 4806-14.
[25] D. Li, M. B. Muller, S. Gilje, R. B. Kaner and G. G. Wallace, *Nat Nanotechnol*, 2008, **3**, 101-5.
[26] W. N. Yan, X. C. Cao, J. H. Tian, C. Jin, K. Ke and R. Z. Yang, *Carbon*, 2016, **99**, 195-202.
[27] Y. Liang, Y. Li, H. Wang, J. Zhou, J. Wang, T. Regier and H. Dai, *Nat Mater*, 2011, **10**, 780-6.
[28] D. H. Guo, R. Shibuya, C. Akiba, S. Saji, T. Kondo and J. Nakamura, *Science*, 2016, **351**, 361-5.
[29] A. M. Gomez-Marin, R. Rizo and J. M. Feliu, *Catalysis Science & Technology*, 2014, **4**, 1685-98.

IOP Conf. Series: Earth and Environmental Science **83** (2017) 012017 doi:10.1088/1755-1315/83/1/012017

Laccase Spontaneous Adsorption Immobilization: Experimental Studies and Mathematical Modeling at Enzymatic Fuel Cell Cathode Construction

I N Arkadeva[1], V A Bogdanovskaya[2], V A Vasilenko[1], E A Fokina[1] and E M Koltsova[1]

[1] Department of informational computing technologies, D. Mendeleev University of Chemical Technology of Russia (MUCTR), Miusskaya sq., 9, 125047, Moscow, Russia
[2] Laboratory of Electrocatalysis and Fuel Cells, Russian academy of sciences
A.N. Frumkin Institute of Physical chemistry and Electrochemistry (IPCE RAS), Leninsky prospect, 31, Moscow, 199071, Russia
E-mail: iarkadyeva@gmail.com; kolts@muctr.ru

Abstract. The activity of a bioeletrode is largely determined by the amount of enzyme adsorbed on its active layer, including the distribution of enzyme along thickness in the carrier layer. The distribution of enzyme is also required for calculations of the characteristics of bioelectrocatalysis process using a mathematical model. In the present article, on the basis of conducted experimental research a mathematical model of laccase immobilization by spontaneous adsorption on carbon-based sorbentsof different nature was developed. The model can be used to predict adsorption value and enzyme distribution in the layer of an adsorbent. The model includes the equations of the enzyme concentration changing due to its adsorption on the surface of the carbon material (CM) and the enzyme diffusion over the thickness of CM. The diffusion equation is based on the second Fick's law and contains fractional derivatives instead of the first oder derivative.

1. Introduction

More recently, enzymatic electrodes have increasingly been considered by researches as a possible alternative to electrodes based on noble metals. Enzyme-based systems operate under mild conditions, such as ambient temperature and neutral pH, and with renewable biocatalysts and wide variety of fuels. But, on the other hand, electrochemical characteristics of such systems are not high enough and due to this fact the question of development of highly active biocatalyst systems still remains of current interest. The capability of predicting bioelectrocatalysis characteristics and output for the devices based on bioelectrocatalysis as well as reducing the number of laboratory studies drives research interest to a mathematical modelling of such systems.

By analyzing the researches available at the moment it can be noticed that the activity of immobilized enzymes largely depends on the method of immobilization and on the nature of the adsorbent. While constructing a biofuel cell electrode highly disperse carbon based materials (CM) as carriers [1]-[3] and physical adsorption,covalent binding and entrapment as immobolization methods [4]-[11] are widely used.

Lager, comparing to nanoparticles of metals, molecules of enzymes provide lower values of current density, thus making it nessesary to find a method which would increase the degree of filling of carrier's surface by an enzyme. In addition to that, the electrocatalytic activity of enzymes largely depends on its orientation, so studying possible factors that would increase a percentage of active enzymes on a carrier is important [12]. Immobilization by spontaneous adsorption is of particular interest, since it is assumed that in this case immobilized enzyme is being attached to the electrode in orientation which is favourable for the direct electron transfer from the electrode to the active centers (i.e. enzyme is being attached in active state).

Basic blocks which are used in extended mathematical models of biofuel cells [13]-[21] can be subdivided into the following parts: describing kinetics of the reactions that are taking place in the system, mass and charge balances, initial and boundary conditions. Most of mathematical models available in relevant literature operate with the value of surface concentration of enzyme, but the value is assumed to the equal to the total amount of enzyme deposited on the electrode. With this assumption, it is impossible to define the factors that affect final characteristics of enzymatic electrode.

Considering everything stated above, the main task of this study was to develop a mathematical model which describes the process of spontaneous adsorption of immobilized enzyme on carbon-based materials and to define the factors which affect adsorption process.

2. Materials and metods

This work is a continuation of earlier studies on the influence of the nature of CM on the laccase activity in the reaction of electroreduction of oxygen to water [12].

In this study the Trametesversicolor laccase obtained according to the procedure of Shteinberg et al [22] was used. The initial concentration of enzyme was 19 mg/ml with activity 0.189×10^{-3}mmol ABTS$\cdot$min$^{-1}\cdot$mg^{-1}. 0.2 M phosphate-acetate buffer solution (pH 4.0-4.5) prepared from $Na_2HPO_4\cdot2H_2O$, acetic acid (CP type) and deionized water were used.

Figure 1. Working electrode with CM deposited on its surface.

The study was carried out on a floating electrode (figure 1) in the form of a tablet made from hydrophobized carbon black, which contained 35% mas. fluoroplastic and was obtained by heat treatment (300 °C) of mixture of acetylene black and fluoroplastic emulsion (containing surfactant),which was placed in 1 cm diameter mold and pressed at a pressure of 5 MPa/cm^2. At the same time, platinum mesh was pressed into the tablet to be used as current collector. Electrode thickness was equal to 2 mm, the surface area of contact with electrolyte was equal to 1 cm^2. Electrode of this design by simulating gas diffusion electrode provides efficient supply of oxygen when studying highly dispersed catalytic materials [22] and allows to eliminate the use of binders in electrode-catalyst system which are required in case rotating disc electrode is used.

Commercial carbon black (XC-72) and carbon nanotubes (CNT) provided by D.I. Mendeleyev University of Chemical Technology of Russia were used as CMs. CM was deposited on the surface of the floating electrode, the electrode then was pressed at 5 MPafor 30 s. After that, for the immobilization by spontaneous adsorption the prepared electrode was held for 2 hours on the surface of electrolyte containing laccase in concentration of 0.076 mg/ml. The measurements were conducted after rinsing the electrode in the buffer solution in order to remove the enzyme molecules which are not attached to the electrode surface. Polarization curves were obtained in oxygen atmosphere.

Electrochemical measurements were conducted in a three-electrode cell with separated electrode spaces. A saturated Ag/AgCl electrode was used as a reference electrode and platinum wire was used as a counter electrode. Polarization curves were registered by potentiostat IPC-Pro 3A.
The amount of laccase in solution was determined spectrophotometrically by Specord M40 spectrophotometer with the help of calibration plot of the dependence of the optical density of the ABTS solution on the concentration of laccase solution added. To create a calibration line, 500 μl of ABTS solution was placed into 3mm cuvette, 1μl of enzyme solution of known concentration was added and the change in optical density was registered at the wavelenght of λ = 420 nm (ε_{ABTS} – 36 $mM^{-1}cm^{-1}$) for 6 minutes (time required to reach maximum optical density). The amount of laccase adsorbed on the electrode was calculated as the difference between the initial amount of laccase in the solution and in the solution after enzyme adsorption on CM.

3. Mathematical simulation

3.1. Model of laccaseadsorbtion from solution
The developed mathematical model assumes two modeling domains. The first modeling domain is the volume of the laccase solution and the second one is the active layer of the carbon material (figure 2).

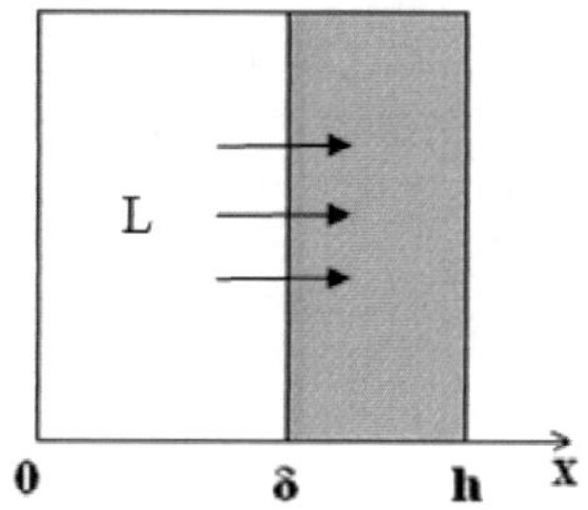

Figure 2. Schematic representation of modeling domains:$0<x<\delta$ – volume of the laccase solution; $\delta<x<h$– active layer of CM.

The area of the active layer is CM, which, as it's expected, due to its roughness, porosity and other irregularities can be described using a mathematical apparatus developed for fractal structures.
Laccase outflow from the electrolyte phase occurs only due to the departure of adsorbed molecules and can be described by the model of ideal mixing, i.e. there are no concentration gradients in this region, and the loss of the enzyme occurs only due to the adsorption:

$$V_l\,\mathrm{d}c_l/\mathrm{d}t = -aW_{ads} \qquad (1)$$

where V_l, m^3 – volume of solution from adsorption is carried out; c_l, $mol \cdot m^{-3}$– enzyme concentration in the solution; a, m^2–surface area of CM, available for the enzyme adsorption; W_{ads}, $mol \cdot m^{-2} \cdot s^{-1}$, – rate of the adsorption process, expressed as in [23]:

$$W_{ads} = pJ_{ads} = pc_l\left(\frac{k_BT}{2\pi m}\right)^{1/2} \qquad (2)$$

where p– adsorption probability; J_{ads}, $mol \cdot m^{-2} \cdot s^{-1}$– adsorbate flow to the surface; k_B, J/K– Boltzmann constant; T, K – temperature; m, kg – mass of adsorbate molecule.

Equation (1) was solved analytically. The expression to calculate adsorbate concentration in the solution vs adsorption time has the following form:

$$c_l(t) = c_l^0 \exp\left[-ap\left(\frac{k_BT}{2\pi m}\right)^{1/2} V_s^{-1}t\right] \qquad (3)$$

where c_l^0 – the concentration of laccase in the solution at the initial time. The determination of the surface area available for adsorption is based on known data of the material specific surface area measured by Brunauer–Emmett–Teller (BET) method S_{BET}, $m^2 \cdot g^{-1}$ and CM loading m_{CM}, g:

$$a = S_{BET} \cdot m_{CM}. \qquad (4)$$

The adsorption probability p is fitted parameter, its value was found from the conditions of calculated data correspondence to the experimental one.

3.2. Model of laccase diffusion in carbon material

In order to carry out further calculations of oxygen bioelectrocatalytic reducing process it is nessesary to obtain the information of enzyme distribution over the CM layer thickness. At modeling the process of impregnation of CM by enzyme, it was considered that the loss of substance from the local volume of the material occurs only due to its diffusion.The diffusion equation is based on the second Fick's law and contains fractional derivatives instead of the usual ones due to the fractal structure of the porous CM, the fractional derivative order γ characterizes the porosity of the material and corresponds to the fraction of the channels available to the flow, $0 < \gamma < 1$.

As it has been shown [24] in such structuresultra-slow transfer processes can be realized:

$$\frac{\partial^\gamma c}{\partial t^\gamma} = D_\gamma \frac{\partial^2 c}{\partial x^2} \tag{5}$$

where c, mol·m^{-3}– enzyme concentration in CM; x, m – the coordinate of active layer thickness; D_γ, m^2·s$^{-\gamma}$— the coefficient of enzyme diffusion in a porous medium, is determined by the relation:

$$D_\gamma = D^\gamma * S^{(1-\gamma)} \tag{6}$$

where$D$, m^2·s^{-1}– the coefficient of enzyme diffusion; S, m^2 – the cross-sectional area of the channel available for the flow, calculated from the data of the pore diameter d_p, m.

The value of the fraction of the channels available for the flow was determined from the ratio of the surface area free for the adsorption ($S_{external}$) to the total surface area of the material determined by the BET method (S_{BET}):

$$\gamma = S_{external} S_{BET}^{-1} \tag{7}$$

The equation of enzyme distribution over the active layer thickness (5) was solved by the finite-difference method using the implicit scheme (8) [24]:

$$\frac{c_j^{n+1} - \gamma c_j^n}{\Gamma(1-\gamma)(1-\gamma)\Delta t^\gamma} = D_\gamma \frac{c_{j+1}^{n+1} - 2c_j^{n+1} + c_{j-1}^{n+1}}{\Delta x^2} \tag{8}$$

whereΓ– gamma function, n– time step, j – coordinate step.

The scheme (8) is absolutely stable and approximates equation (5) with (2-γ) time order and 2-th coordinate order [24]. The approximation time order (2-γ) shows that the fractional derivative equation describes the mass transfer process more accurately than the partial differential equation, where the given order is 1.

The implicit scheme (8) was solved by the numerical methodusing following boundary conditions:

the concentration of the adsorbed enzyme at the left border: solution/CM is equal to the laccase concentration in the solution:

$$c(x=\delta, t) = c_l(t) \tag{9}$$

The absence of a concentration gradient along the thickness at the right boundary:

$$D_\gamma \, \partial c(x=h,t)/\partial x = 0 \tag{10}$$

The solution of equation (5) was carried out using the "floating" coordinate x in order to determine the time-varying adsorbtion thickness H, m. The thickness H was determined from the condition of the material balance: the mass of the enzyme left the solution is equal to the mass adsorbed in the CM.

As a result the mathematical model allows to obainthe dependence of the enzyme concentration in the solution on the time of adsorption, the values of the impregnation thickness of LM with laccase, the distribution of the enzyme concentration vs CM layer thickness and time.

4. Results and discussion

This study was carried out to establish the dependence of the laccase-based electrode activity on the catalytic layer thickness and the type of carbon material. Figure 3 shows that the adsorption values for the two materials (XC-72 and CNT) are approximately the same, but its efficiency decreases with increasing thickness of the catalytic layer for the two types of CM. Apparently, this is due to the limitations of access of the enzyme to the deeper layers of the CM. In addition, figure 3 shows the data calculated with the model equation. One can see a good agreementbetween calculated (lines) and experimental (dots) data.

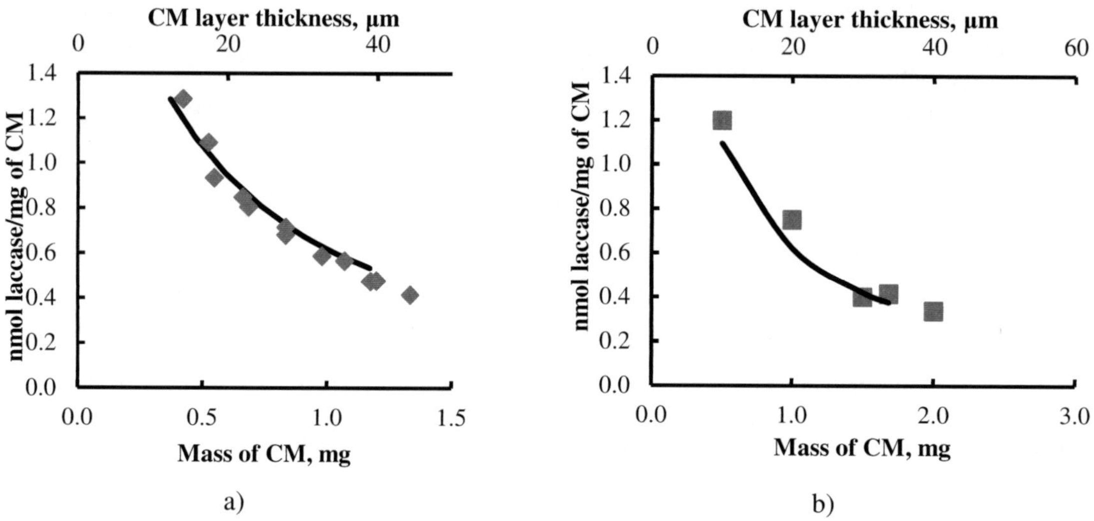

Figure 3. Dependence of the efficiency of laccase adsorption on the mass (thickness) of the active layer for: a) CNT; b) XC-72.

However, the specific activity of bioelectrodes based on laccase, adsorbed on CM of two types (figure 4), varies considerably. For carbon nanotubes, it sharply increases with increasing mass (thickness) of the CM layer to 1 mg, and then begins to decrease gradually. Obviously, only part of the adsorbed on the CM enzyme takes part in current-forming reactions, and the fraction of laccase molecules adsorbed in the electrochemically active state on nanotubes is much larger in comparison with carbon black. Probably, the enzyme, due to its size, does not penetrate deep into the CM layer but adsorbs only on the accessible surface.

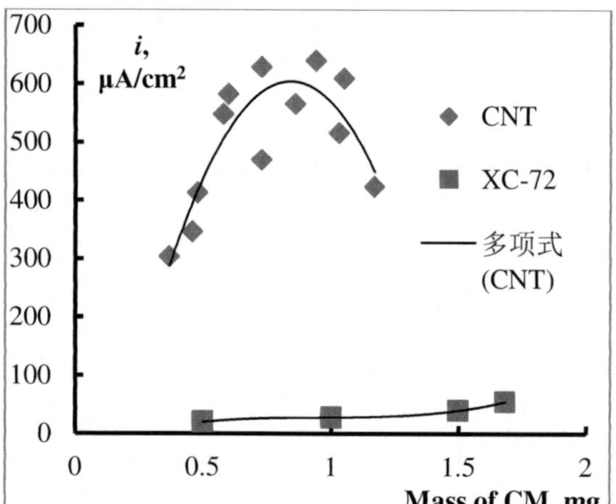

Figure 4. Dependence of the specific activity of the electrode on the mass (thickness) of the active layer (at 0.4 V, sweep rate 1 mV/s;oxygen).

Table 1. The structural characteristics of the materials

Parameter	XC-72	CNT
S_{BET}, m^2/g	230	216
V_{pores}, cm^3/g	2	3.8
$S_{external}$, m^2/g	150	198

Figure 5 shows photomicrographs made in the centre of collective use of the D. Mendeleev University of Chemical Technology of Russia, which depicts electrodes with CM on its surface. The photos clearly show mesopores of materials (pores between agglomerates of nanotubes or particles of carbon black). Obviously, the mesoporosity of CNTs is higher than carbon blacks, which corresponds to the data on the structural characteristics of the materials presented in table 1.

a) b)

Figure 5. Photomicrographs of electrodes surface with CM on it: a) CNT, b) carbon blackXC-72

Possibly, the mesoporosity of the material allows the enzyme to penetrate deep into the layer, surrounding it and increasing the possibility of its attachment by the binding centre of the second substrate responsible for the transfer of electrons from the electrode surface to the active centers of laccase. Figure5 shows the carbon blacks maximum mesopore size is much smaller than the size of mesopores of nanotubes, so the enzyme is forced to adsorb in the surface layer of the material, where the probability of attachment in a favourable position for direct electron transfer is much smaller.

In the result of simulation following dependences were obtained: the distribution of the enzyme concentration on CM layer thickness at different process times (figure 6). As shown in figure 6, the penetration depth of the enzyme increases with time, and its concentration in the CM impregnated layers decreases insignificantly. Obtained data will be used at the next stage of mathematical modeling to estimate the dependencies of oxygen bioelectrocatalytic reduction by laccase.

Input, measuredandcalculated parameters are listed in table 2.

Table 2. List of parameters values used in the simulation

Parameter	Symbol	Value	Reference
Laccase diffusion coefficient	D, $m^2 \cdot s^{-1}$	1.49×10^{-10}	[25]
Temperature	T, K	298	Experimental
Volume of solution	V_l, m^3	5×10^{-7}	Experimental
CM mass on electrode	m_{CM}, kg	62×10^{-6}	Experimental
Enzyme initial concentration in solution	c_l^0,$kg \cdot m^{-3}$	0.154	Experimental
Average CNT pore diameter	d_{pore}, m	11×10^{-7}	[12]
Fraction of channels available to electrolyte flow	γ	0.94	Calculated (Eq. 7)
Laccase diffusion coefficient in active layer	D_γ, $m^2 \cdot s^{-\gamma}$	6.59×10^{-11}	Calculated (Eq. 6)
Adsorption probability	p	3.4×10^{-10}	Fitted

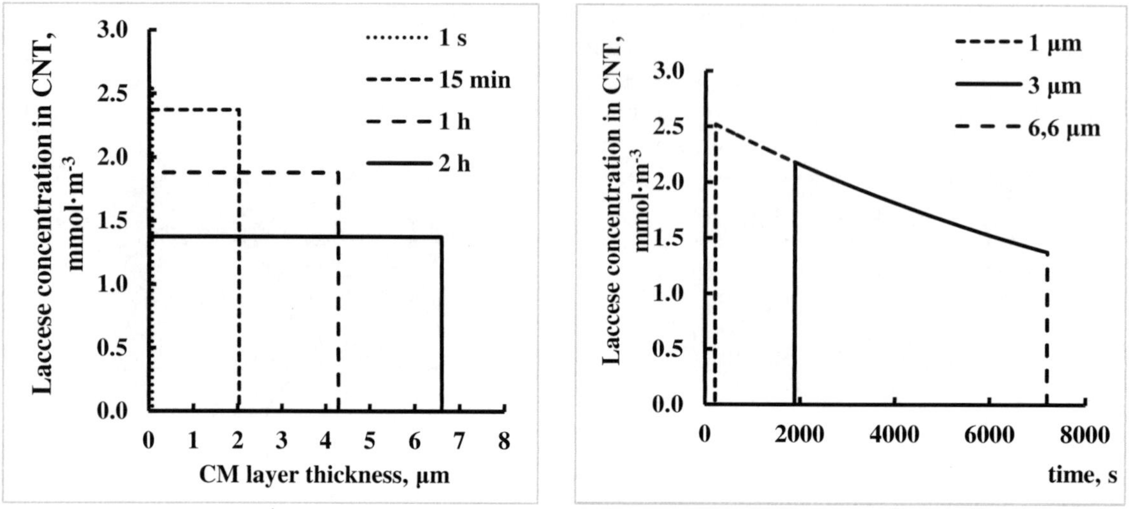

a) b)

Figure 6. Distribution of the concentration of adsorbed laccases: a) vs CM layer thickness; b) vs adsorption time.

5. Conclusions

We have carried out a series of experimental studies of establishing the main dependencies of the values of laccase adsorption on carbon materials from their structural characteristics. We have developed a mathematical model of adsorption process, which predicts the amount of adsorbed enzyme on CM of different dispersity, as well as enzyme distribution over the thickness of the CM layer. The model includes the equations for enzyme concentration changing in the solution as a result of its adsorption on the surface of the CM and due to the enzyme diffusion within the thickness of the CM. We have developed an algorithm for calculating equations of the mathematical model. As a result, the dependencies of the distribution of the enzyme amount over the thickness of the CM layer, the accumulation of the substance by different layers of CM as a function of the time of adsorption, were obtained. The parameters of the laccase adsorption process at CMs were determined. Calculations for the amount of adsorbed at CM laccase for different amounts of CM and different concentrations of the initial enzyme solution are in good agreement with the experimental data. The next stage of this scientific research is the development of a mathematical model of direct electron transfer bioelectrocatalysis, which will use the obtained in this paper data and which can predict the electrochemical characteristics of a laccase-based electrode.

Acknowledgements

Authors gratefully acknowledge Doctor of chemical sciences, Prof. Gorshina N.N. and Doctor of chemical sciences, Prof. Yaropolov A.I. for providing the laccase and research assistance.
This study is partially supported by state contract with Russian Foundation for Basic Research (RFBR) № 16-08-01140.

References:

[1] Holzinger M, Le Goff A and Cosnier S 2012 *Electrochim. Acta* 82 179
[2] Jensen U B, Lurcher S, Vagin M, Chevallier J, Shipovskov S, Koroleva O, Besenbacher F and Ferapontova E 2012 *Electrochim. Acta* 62 218
[3] Stolarczyk K, Kizling M, Majdecka D, Zelechowska K, Biernat J F, Rogalski J and Bilewicz R 2014 *J. Power Sources* 249 263
[4] Kim R E, Hong S G, Hab S and Kim J 2014 *Enzyme and Microbial Technology* 66 35
[5] Dagys M, Lamberg P, Shleev S, Niaura G, Bachmatova I, Marcinkeviciene L, Meskys R, Kulys J, Arnebrant T and Ruzgas T 2014 *Electrochim. Acta* 130 141

[6] Gupta G, Lau C, Branch B, Rajendran V, Ivnitski D and Atanassov P 2011 *Electrochim. Acta* 56 10767
[7] Deng L, Shang L, Wang Y, Wang T, Chen H and Dong S 2008 *Electrochem.Commun.*10 1012
[8] Stolarczyk K, Nazaruk E, Rogalski J and Bilewicz R 2008 *Electrochim. Acta* 53 3983
[9] Pöller S, Beyl Y, Vivekananthan J, Guschin D A and Schuhmann W 2012 *Bioelectrochemistry* 87 178
[10] Karaskiewicz M, Nazaruk E, Zelechowska K, Biernat J F, Rogalski J and Bilewicz R 2012 *Electrochem.Commun.* 20 124
[11] Gutiurrez-Sanchez C, Jia W, Bey l Y, Pita M, Schuhmann W, De Lacey A L and Stoica L 2012 *Electrochim. Acta* 82 218
[12] Arkadeva I N, Bogdanovskaya V A 2015 Trudy 10 Konferentsii molodyh uchyonyh, aspirantov i studentov IFHE im. A.N. Frumkina RAN "Fizikohimiya-2015"[*Proc. 10 th Conf. of young scientists, post-graduate students and students at A.N. Frumkin IPCE RAS "Physicochemistry-2015"*].(Moscow) p 119–120 (in Russian)
[13] Osman M H , Shah A A, Wills R G A and Walsh F C 2013 *Electrochim. Acta* 112 386
[14] Do T Q N, Varnicicr M, Hanke-Rauschenbach R, Vidakovicr-Koch T and Sundmacher K 2014 *Electrochim. Acta* 137 616
[15] Der-Sheng Chan, Der-Jong Dai and Ho-Shing Wu 2012 *Energies* 5 2524
[16] Osman M H, Shah A A and Wills R G A 2013 *J.Electrochem. Soc.*160 (8) F806-14
[17] Piyush Kar, Hao Wen, Hanzi Li, Shelley D. Minteerand Scott Calabrese Barton 2011 *J. Electrochem. Soc.*158 (5) B580
[18] Tamaki T, Ito T and Yamaguchi T 2009 *Fuel Cells* 1 37
[19] Zeev Rubin V and Mor L 2013 *Adv. Chem. Eng. Sci.*3 218
[20] Kjeang E, Sinton D and Harrington D A 2006 *J. Power Sources* 158 1
[21] Song Y, Penmatsa V, Wang C 2014 *Energies*74 694
[22] Shteinberg G V, Kukushkina I A, Bagotskii V S and Tarasevich M R1979 *Russian Journal of Electrochemistry* 15 6517
[23] JungLS and Campbell CT2000 *Physicalreviewletters*84(22) 5164-7
[24] Kol'tsova E M, Vasilenko V AandTarasov V V 2000*Russian Journal of Physical Chemistry A* 74 (5) 848
[25] Gelo-Pujic M, Hyug-Han Kim, Butlin N G, Tayhas G and Palmore R 1999 *Applied and environmental microbiology* 65 (12) 5515

IOP Conf. Series: Earth and Environmental Science **83** (2017) 012018 doi:10.1088/1755-1315/83/1/012018

Optimal Renewable Energy Integration into Refinery with CO2 Emissions Consideration: *An Economic Feasibility Study*

M Alnifro[1], S T Taqvi[1], M S Ahmad[1], K Bensaida[1,2], and A Elkamel[1,3]

1 Department of Chemical Engineering, University of Waterloo, 200 University Avenue W, Waterloo, ON

2 Department of Mechanical Engineering, National Engineering School of Sfax, Tunisa

3 Department of Chemical Engineering, The Petroleum Institute, Khalifa University, UAE

E-mail: jacky.mucklow@iop.org

Abstract. With increasing global energy demand and declining energy return on energy invested (EROEI) of crude oil, global energy consumption by the O&G industry has increased drastically over the past few years. In addition, this energy increase has led to an increase GHG emissions, resulting in adverse environmental effects. On the other hand, electricity generation through renewable resources have become relatively cost competitive to fossil based energy sources in a much 'cleaner' way. In this study, renewable energy is integrated optimally into a refinery considering costs and CO_2 emissions. Using Aspen HYSYS, a refinery in the Middle East was simulated to estimate the energy demand by different processing units. An LP problem was formulated based on existing solar energy systems and wind potential in the region. The multi- objective function, minimizing cost as well as CO_2 emissions, was solved using GAMS to determine optimal energy distribution from each energy source to units within the refinery. Additionally, an economic feasibility study was carried out to determine the viability of renewable energy technology project implementation to overcome energy requirement of the refinery. Electricity generation through all renewable energy sources considered (i.e. solar PV, solar CSP and wind) were found feasible based on their low levelized cost of electricity (LCOE). The payback period for a Solar CSP project, with an annual capacity of about 411 GWh and a lifetime of 30 years, was found to be 10 years. In contrast, the payback period for Solar PV and Wind were calculated to be 7 and 6 years, respectively. This opens up possibilities for integrating renewables into the refining sector as well as optimizing multiple energy carrier systems within the crude oil industry

1. Introduction

Renewable energy is defined as "the energy generated from natural resources that can be renewed naturally in the environment" by sustainable energy resources. These resources include hydropower, wind, biomass, geothermal, and solar [1]. The depletion of fossil fuel reserves has caused an increase in demand and price of petroleum compounds. Fossil fuel accounts for 88% of total primary energy consumption share with oil (35%), coal (29%) and natural gas (24%) as the major fuels [2]. In addition, 28% of the world's primary energy is being consumed in transportation sector. Moreover, transportation fuel demand is predicted to increase up to 40 % by 2040 [3], [4]. However, the fact remains that fossil fuels are non-renewable scarce resources of energy [5].

IOP Conf. Series: Earth and Environmental Science **83** (2017) 012018 doi:10.1088/1755-1315/83/1/012018

Generally speaking, production of petroleum yields involve enormous amounts of energy. Petroleum refining is one of the most complex processes in the oil and gas industry. It includes many unit processes and subsidiary facilities. Most refineries are different from each other and have a unique combination and arrangement of units. They are highly energy intensive due to their large production capacity. The capacity of modern petroleum refineries usually range from 800,000 to 900,000 barrels of crude oil feed per day [6], [7]. Since the mid of 20th century, petroleum products have become a dominant energy source, surpassing coal demand.

Present scenario focuses on meeting future challenges to cope with the energy demand throughout the world in developed countries. Renewable energy resources have been utilized in pursuit of overcoming this problem, since the past several decades. They play an important role in producing 'clean' energy that reduces greenhouse gas (GHG) emissions, specifically CO2, as compared to fossil fuels. The aim of this study is to determine the feasibility of optimal renewable energy integration into a refinery within the Middle East. Specifically, simulating a refinery to define energy demand by various units within the environment. Additionally, developing a model to find optimal distribution of energy. Lastly, investigating the economic feasibility of having such an integration.

2. Methodology

2.1 Superstructure

A superstructure was designed, outlining the available energy resources as well as the energy required by each unit, within the refinery, as seem in figure 1. These energy sources are limited due to their availability within the Middle East region. Moreover, electricity provided via the grid is assumed to be produced by natural gas.

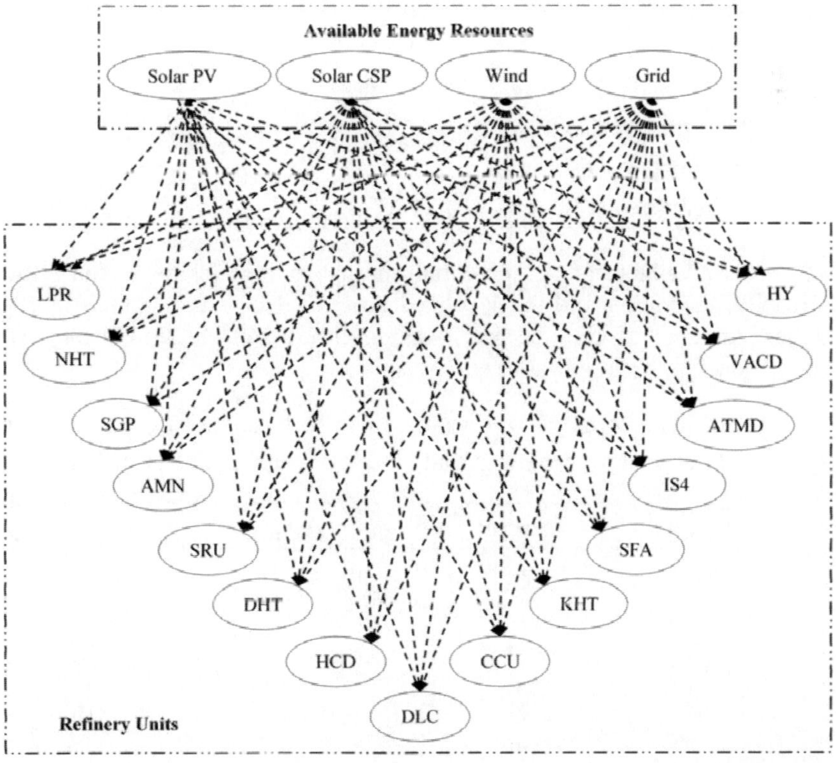

Figure 1. Superstructure diagram for the crude oil refinery units connected with all available energy resources.

2.2 Data collection

Each refinery unit was simulated in Aspen HYSYS and the amount of energy required by each unit was determined. Moreover, literature was surveyed in order to find the carbon emissions produced from each unit. Table 1 shows the energy demand and CO_2 emissions data, obtained from these two sources.

Table 1. Energy required by each unit and CO_2 emissions for each unit in the refinery.

Refinery unit	Abbreviation	MJ/year	g CO2/ MJ
Hydrogen plant	HYD	3.38 x107	0.362
Sulfur Recovery Unit	SRU	2.42 $x10^8$	0.056
Amine plant	AMN	2.53 $x10^7$	0.056
Saturated Gas Plant	SGP	1.00 $x10^8$	0.168
Naphtha Hydrotreater	NHT	1.63 $x10^7$	0.187
Reformer	LPR	1.34 $x10^8$	0.998
Kerosene Hydrotreater	KHT	7.07 $x10^6$	0.187
Diesel Hydrotreater	DHT	8.50 $x10^6$	0.187
Hydrocracker	HCD	1.80 $x10^8$	0.561
Delayed Coker	DLC	3.31 $x10^7$	0.312
Catalytic Cracking	CCU	3.29 $x10^8$	0.686
Sulfur Acid Alkylation	SFA	2.35 $x10^8$	0.000
C4 Isomerization	IS4	2.20 $x10^7$	0.062
Unsaturated Gas Plant	UGP	1.11 $x10^8$	0.168
Atmospheric Distillation	ATMD	2.06 $x10^5$	1.684
Vacuum Distillation	VACD	4.01 $x10^6$	0.561

Table 2 shows the available and/or potential energy sources in the region with CO2 emissions due to electricity generation and the levelized cost of electricity (LCOE) [8-10]. It is observed that the highest amount of CO2 emissions was produced from the grid energy source (i.e. natural gas) in comparison to other renewable energy sources.

Table 2. Potential energy sources in Abu Dhabi with CO2 emissions due to electricity generation and the levelized cost of electricity [11].

Source	gCO_2/MJ	LCOE $/kWh	Capacity (MJ/year)
Solar CSP	9.2	0.18	7.6 $x10^8$
Solar PV	36.8	0.27	6.3 $x10^7$
Wind	2.2	0.07-0.13	7.2 $x10^6$
Grid	119	0.05-0.07	$3.7x10^{11}$

2.3 Assumptions

The following assumptions were made whilst developing the model:

- Crude oil feed of 100,000 barrels of crude oil per day
- Cost of electricity generated from each source is independent of the unit it is consumed within.
- Intermittent energy is stored and thus, available to be used throughout the year

2.4 Mathematical model

The ε (epsilon) constraint method was used where the cost of energy was defined as the objective function and the amount of CO_2 emissions was posed as a constraint. Thus, mathematical expression of this problem statement consists of minimizing cost (objective function) while observing inequality constraints and equality denoting the limitations for demand and supply of energy, and amount of emissions, respectively. It is written in a general form as the following Linear Programming (LP) problem:

$$\min z = \sum_{p=1}^{6}\sum_{d=1}^{16} lcoe_{p,d}\, x_{p,d} \tag{1}$$

subject to

$$(1\text{-}\alpha)\, 7.92\times 10^{7} \leq g \leq \alpha(5.44\times 10^{8}) \quad \alpha \in [0,1] \tag{2}$$

$$\sum_{p=1}^{6} x_{p,d} - b(d) \geq 0 \tag{3}$$

$$\sum_{d=1}^{16} x_{p,d} - a(p) \leq 0 \tag{4}$$

$$g = \sum_{p=1}^{6}\sum_{d=1}^{16} gh\, g_{p,d}\, x_{p,d} \tag{5}$$

where:

z	total cost of producing electricity
$x(p,d)$	energy from energy supplier to energy demand
p	energy supplier (i.e. solar CSP, solar PV, grid, and wind)
d	energy demand (i.e. refinery units)
$a(p)$	production capacity of energy supplier (MJ / year)
$b(d)$	energy demand by each unit in the refinery (MJ)
$ghg(p,d)$	carbon dioxide emission by each energy supplier (CO2 g / MJ)
$lcoe(p,d)$	cost of energy production (USD / MJ)
α	weight varying between 0 and 1

3. Results and discussion

Results obtained from the simulated refinery unit as well as the optimization of the developed model are presented. The carbon dioxide emission was posed as a constraint with an assigned weight, α, that ranges between 0 and 1. A value of α=0 signifies a focus on minimizing carbon dioxide emissions with no regard to cost. Conversely, a value of α=1 signifies a focus on minimizing cost with no consideration of carbon dioxide emissions. Figure 2 shows the changes in the cost and carbon dioxide emissions as

IOP Conf. Series: Earth and Environmental Science **83** (2017) 012018 doi:10.1088/1755-1315/83/1/012018

alpha varies between 0 and 1. The cost is found to be minimum when emissions are maximum, and vice versa.

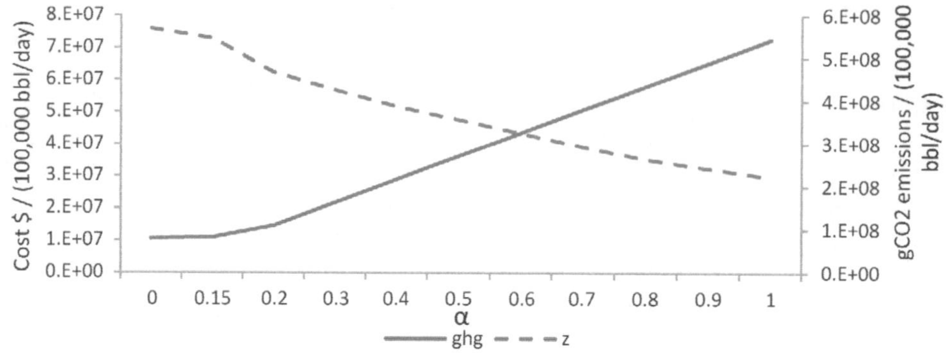

Figure 2. Cost and CO2 emissions with respect to α

Furthermore, a Pareto front was constructed, based on the results obtained from the developed model, as seen in Figure 3. This Pareto curve shows the optimal cost corresponding to the carbon dioxide emissions emitted by the refinery when renewable energy is integrated optimally.

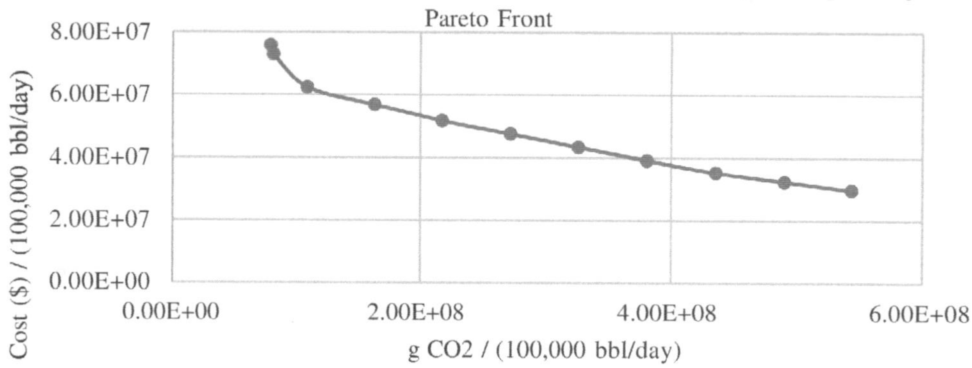

Figure 3. Cost and CO2 emissions optimum points.

Furthermore, figures 4 and 5 show the energy distribution between energy sources and the refinery units at α equal to 0 and 1, respectively:

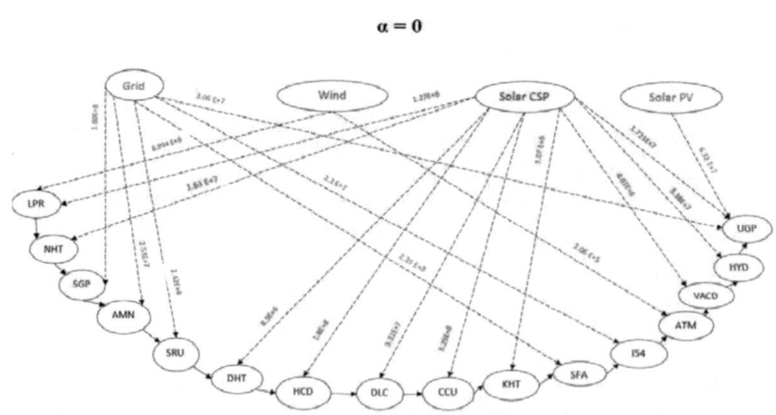

Figure 4. Optimal energy distribution to refinery units at α=0.

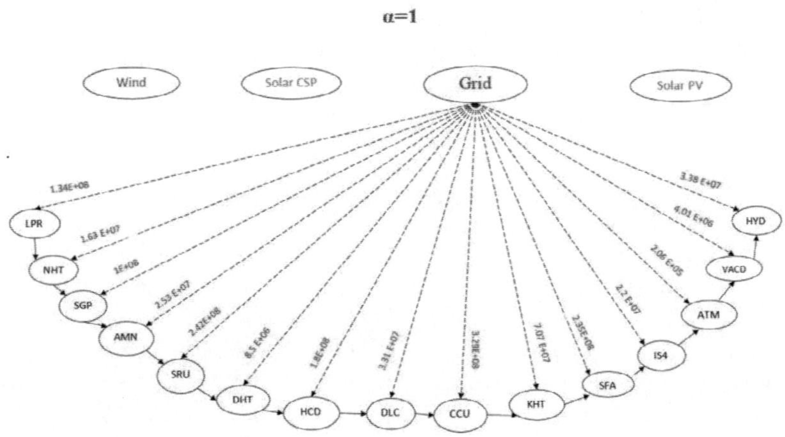

Figure 5. Optimal energy distribution to refinery units at α=1.

4. Economic Analysis

The economic feasibility for integrating renewable energy sources to a refinery in Abu Dhabi was examined. Solar PV, solar CSP, and wind energy sources are studied for high and low values of calculated Levelized Costs of Electricity (*LCOE*). However, these *LCOE*s are dynamically estimated for energy generation using the following mathematical formulae [11]:

$$CRF = \frac{D(1+D)^N}{(1+D)^N - 1} \tag{6}$$

$$LCOE = \frac{Capital\ Cost \times CRF \times (1 - TD_{PV})}{8760 \times Capcity\ Factor \times (1-T)} + \frac{fixed\ O\&M}{8760 \times Capacity\ Factor} + \frac{Variable\ O\&M}{1{,}000 \frac{KW}{MW}} \tag{7}$$

where

Capital cost	cost of plant
CRF	capital recovery factor
T	tax rate paid
DPV	present value of depreciation
8760	number of hours in a year
Capacity factor	yearly average percentage of power as a fraction of capacity
Fixed O&M	fixed operating and maintenance cost
Variable O&M	variable operating and maintenance cost

Table 3. Calculated economic and environmental parameters for available renewable energy resources at different *LCOE*.

Source	Solar PV (low *LCOE*)	Solar PV (high *LCOE*)	Solar CSP (low *LCOE*)	Solar CSP (high *LCOE*)	Wind (low *LCOE*)	Wind (high *LCOE*)
LCOE	0.05	0.56	0.04	0.23	0.04	0.12
Carbon credit value w/renewable ($/ton of CO_2)	8.68	8.68	8.68	8.68	8.68	8.68
CO_2 emissions w/grid (tons)	543.42	543.42	543.42	543.42	543.42	543.42
CO_2 w/source (tons)	167.45	167.45	41.704	41.704	10.11	10.11
Total capital cost ($MM)	70.5	392	86.0	517	56.4	187

Daily capacity (MWh)	47.0	47.0	47.0	47.0	47.0	47.0
Total fixed cost ($MM)	0.355	5.2	2.33	5.40	0.482	2.82
Fixed cost per year ($M)	11.8	172.2	77.5	180	16.1	94.0
Price per kWh in Abu Dhabi Industrial ($)	0.04	0.04	0.04	0.04	0.04	0.04
Annual Cost of 17564 MWh Grid Electricity（$MM）	16.5	16.5	16.5	16.5	16.5	16.5
Total Amortized Payments ($MM)	170	948	207	1,249	136	454
Total savings per year	16.5	16.3	16.4	16.3	16.4	16.4
Payback period (years)	7	83	10	77	6	40
Project lifetime (years)	30	30	30	30	30	30

As seen from Table 3, all sources of renewable energy are economically feasible at low *LCOE* with wind being most viable and solar CSP the least.

5. Sensitivity Analysis

A sensitivity analysis was conducted, on Solar CSP, to determine how critical parameters impact the payback period and LCOE with changes in capital cost and capacity. Other renewable technologies yielded similar results; hence, not presented in this paper.

Table 4. Sensitivity analysis for capital cost and capacity factor on *LCOE* and payback period.

Capital Cost	*LCOE*	Payback Period Years	Capacity Factor	*LCOE*
1830	0.04	10	25.3	0.29
2000	0.08	11	30	0.24
3000	0.1	29	40	0.18
4000	0.13	40	50	0.15
5000	0.15	50	60	0.12
6000	0.18	59	70	0.11
7000	0.21	69	80	0.09
8000	0.23	79		
9000	0.26	89		
10000	0.29	99		
11000	0.31	108		

As evident from the table above, as capital cost increases, the LCOE and payback period increases. In addition, since the assumed lifetime of each project is 30 years, the feasible capital cost is 3000. On the other hand, as capacity factor increases, the LCOE decreases significantly. Thus, indicating that technical improvements can help reduce *LCO*E substantially.

6. Conclusion

In this study, a model was developed to determine the optimal production planning for an oil refinery while reducing GHG emissions. The model incorporates the daily production, the supply and demand for energy, the supply and demand of each product as well as the CO2 constraint. A petroleum refinery with a set of different process units was simulated using Aspen HYSYS with a capacity of refining

100,000 bbl of crude oil blend is refined per day. From this refinery, the energy consumption by each unit was estimated. Also, a superstructure was designed to show the units within the refinery connected to available energy sources that could meet their energy demand. Furthermore, the CO2 emissions for each units within the refinery were estimated and the cost of the available energy sources. In addition,

the developed model was used to determine the optimal distribution of energy to the different units within the refinery using GAMs which were later expressed by a Pareto curve. This curve shows the optimal cost for the energy supplier versus CO2 emissions from different sources. Finally, economic feasibility studies and sensitivity analyses were conducted in this work for the integrated renewable energy sources in Abu Dhabi, based on different factors.
Based on this study, it is feasible for integrating renewable energy into the refinery. However, for future work, the following areas need to be incorporated within the scope:

- Energy hubs; only electricity through the grid and renewable sources were considered. A multi-energy hub network may be developed that involves additional energy input such as natural gas for on-site generators, heat streams, etc.
- Intermittent nature of renewable energy; an average annual potential of renewable energy sources such as solar and wind were considered. A more detailed study can be carried out that considers daily, monthly or seasonal changes in these sources of energy and determine the optimum conditions to operate at.
- Storage systems can be considered in future work that enhances reliability in integrating renewable energy systems with current energy systems.

7. References

[1] N. Kholod, M. Evans, and V. Roshchanka, "Energy Efficiency as a Resource: Energy Efficiency's Role in Meeting Ukraine's Energy Needs," 2015.
[2] L. Brennan and P. Owende, "Biofuels from microalgae :a review of technologies for production, processing, and extractions of biofuels and co-products," Renewable and Sustainable Energy Reviews, vol. 14, pp. 557-577, 2011.
[3] G. B. Leite, A. E. Abdelaziz, and P. C. Hallenbeck, "Algal biofuels: challenges and opportunities," Bioresource Technology, vol. 145, pp. 134-141, 2013.
[4] H. Khatib, "IEA World Energy Outlook 2011—A comment," Energy policy, vol. 48, pp. 737-743, 2012.
[5] M. K. Lam and K. T. Lee, "Microalgae biofuels: a critical review of issues, problems and the way forward," Biotechnology advances, vol. 30, pp. 673-690, 2012.
[6] A. Groysman, "Physico-Chemical Properties and Corrosiveness of Crude Oils and Petroleum Products," in Corrosion in Systems for Storage and Transportation of Petroleum Products and Biofuels, ed: Springer, 2014, pp. 1-21.
[7] X.-M. Li, B. Zhao, Z. Wang, M. Xie, J. Song, L. D. Nghiem, et al., "Water reclamation from shale gas drilling flow-back fluid using a novel forward osmosis–vacuum membrane distillation hybrid system," Water Science and Technology, vol. 69, pp. 1036-1044, 2014.
[8] E. watch, "Economy Watchhttp://www.economywatch.com/economic-statistics/United-Arab- Emirates/Electricity_Production/."
[9] "Shams Power Company Saudi Arabia, http://www.shamspower.ae/en/the-project/factsheets/overview/."
[10] M. C. Energy, "Masdar clean energy, http://www.masdar.ae/en/energy/detail/masdar-city- solar-pv plant."
[11] U. E. Information. http://en.openei.org/apps/TCDB/. Available:http://en.openei.org/apps/TCDB/, 2017

IOP Conf. Series: Earth and Environmental Science **83** (2017) 012019 doi:10.1088/1755-1315/83/1/012019

Assessing CO_2 Mitigation Options Utilizing Detailed Electricity Characteristics and Including Renewable Generation

K Bensaida[1,2], Colin Alie, A Elkamel[1,3] and A Almansoori[3]
1 Department of Chemical Engineering, University of Waterloo, 200 University Avenue W, Waterloo, ON N2L
2 Department of Mechanical Engineering, National Engineering School of Sfax, Tunisa
3 Department of Chemical Engineering, The Petroleum Institute, Khalifa University, UAE
E-mail: aelkamel@pi.ac.ae

Abstract. This paper presents a novel techno-economic optimization model for assessing the effectiveness of CO_2 mitigation options for the electricity generation sub-sector that includes renewable energy generation. The optimization problem was formulated as a MINLP model using the GAMS modeling system. The model seeks the minimization of the power generation costs under CO_2 emission constraints by dispatching power from low CO_2 emission–intensity units. The model considers the detailed operation of the electricity system to effectively assess the performance of GHG mitigation strategies and integrates load balancing, carbon capture and carbon taxes as methods for reducing CO_2 emissions. Two case studies are discussed to analyze the benefits and challenges of the CO_2 reduction methods in the electricity system. The proposed mitigations options would not only benefit the environment, but they will as well improve the marginal cost of producing energy which represents an advantage for stakeholders.

1. Introduction

The utilization of carbon-based fuels as energy sources significantly contributes to the production of pollutant gases. The accumulation of these gases in the atmosphere eventually leads to global warming. Global warming is defined as the rise of the earth's temperature due to anthropogenic greenhouse gas emissions (GHG) [1]. In the case of electricity systems, the energy sources varies amongst renewable (e.g., hydroelectric), nuclear, and fossil-based sources (e.g., oil, coal, natural gas) [2]. Each country or region has its own electricity system, whose operation is managed by a central authority with the objective of satisfying the seasonal electricity demand. According to the National Inventory Report 1990-2012 [3], the electricity sector accounted for 12% of Canada's GHG emissions in 2012.

Several strategies have been proposed in the literature to reduce pollutant gases emitted by the electricity sector. Since CO_2 emissions are one of the most harmful pollutants, the aforementioned strategies have focused on developing potential CO_2 mitigation methods and estimating the methods effectiveness. Accordingly, there are two methodologies used for such purpose: the techno-economic assessment of individual plants, and the medium-to long-term electricity system planning. The first methodology entails calculating a performance metric for each mitigation action; the better the value of the metric the better the mitigation strategy. This methodology includes two important techno-economic approaches. The first consists of calculating the associated cost of CO_2 capture (CCC); while the second calculates the cost of CO_2 avoided (CCA). The latter refers to the CO2 emissions that are actually mitigated as a result of the mitigation action. Based on Mariz et al. [4] and Paitoon et al. [5] works, CCA is the preferred mean for the evaluation of CO_2 mitigation strategies since it estimates the average unit cost of CO_2 reductions, while still affording one unit of electricity to consumers [6]. Moreover, Guillermo Ordorica-Garcia [7], Rao and Rubin [8], David Singh [9], and Akimoto et al. [10] have arrived at similar findings.

On the other hand, the second method identifies the investments that will best satisfy electricity demand and other system constraints over a given planning horizon. The models used for this purpose are extended with CO2 mitigation strategies and CO2 emission constraints (or, equivalently, a CO2 tax).The greater the activity of a mitigation technology in the optimal solution, the better the mitigation strategy. For instance, Turvey and Anderson [11], Johnson and Keith's [12], Elkamel et al. [13], and Sparrow and Bowen [14] have contributed with valuable information regarding CO_2 mitigation in electricity system planning models. In addition, the International Energy Agency (IEA) has developed an economic model for long-term analysis of national and international energy markets named MARKAL (Market Allocation). Over the years, a 'Canadian' version of MARKAL, Extended-MARKAL, has been further developed in order to enhance its accuracy and features. For instance, the Extended-MARKAL version consider multi-regional contributions [15]-[18], multi-pollutant emissions calculations [18], stochastic assessments [19], accommodation of price elasticity of demands [20], and accommodation of international trade in CO_2 emission permits [21].

Based on the above discussions, previous studies undertaking techno-economic assessments of CO_2 mitigation options have generally disregarded the detailed operating characteristics of the electricity system. The aforementioned works have made generic assumptions on the performance of the generating units with scare or no validation of parameters such as: capacity factor, unit heat rate, and fraction of CO2 captured. This study presents a novel approach to understand the impact of CO_2 mitigation strategies in the electricity generation sub-sector. The following CO_2 mitigation strategies are applied: load balancing with the 'top down' approach, carbon tax regulations, fixed and flexible carbon capture and storage (CCS) methods. The latter is defined as the reunion of technologies and techniques that enables the capture of carbon dioxide (CO_2) emissions produced from the combustion of fossil fuels in the industrial sector to reduce CO_2 emissions [22]. Accordingly, an electricity modeling framework is developed. The framework comprises detailed information of the electricity system's operation. The developed electricity system framework is based on deregulated power networks including markets for both real and reserve power. The consumers are price-insensitive, and its generators bid their units' power at the marginal generation costs. The electricity system operator provides hourly dispatch instructions seeking to maximize social welfare while respecting the physical constraints of the units and transmission system. There are three phases in the electricity system modeling: pre-dispatch, real time operation and market settlement. Each phase entails solving an optimization problem. The first phase involves a dynamic optimization problem whereas the remaining phases are fed by the first phase dynamic results. The optimization model proposed in this work uses the IEEE RTS '96 (IEEE One-Area Reliability Test System – 1996) [23] as test case. Therefore, the performance of the electricity system is benchmarked with GHG regulations in the form of a carbon tax at \$15, \$40 and \$100/tonne CO_2. Additionally, two different CO_2 capture methods are included in the electricity system (i.e., fixed and flexible CO_2 capture), and their impacts are assessed on the entire electric system. To the author knowledge, this approach has not yet been considered in any techno-economic or electricity system planning study.

2. Problem statement

In this work, an electricity system model is developed based on Ontario's electricity system operation. Parameters describing the technical and economic performance of Ontario's electricity system generating units were not readily available, whereby the IEEE RST '96 (IEEE One-Area Reliability Test System–96) was selected as test case. The IEEE RST '96 system features several appropriate characteristics for this case study. Such characteristics include: physical properties of the transmission system, electricity supply provided by large centralized and dispatchable generating units; and primary energy source (i.e., fossil fuels, uranium, or moving water). Additionally, the IEEE RST '96 have served as reference case for numerous electricity system studies; which is an advantage when validating information [24],[25]. Simulating an electricity system involves solving both a loadflow problem and an economic dispatch problem. The loadflow problem determines possible power flow losses within transmission and distribution lines. At this stage, the net power injected at each bus reflects the electricity demand for a single moment in time, and a particular response of the generating units in the system to that demand. The economic dispatch problem identifies the optimum output level of the generators that satisfy the electricity demand, while meeting technical and operational requirements. In this work, the electricity system is studied with reference to its three main phases: pre-dispatch, real-time operation, and market settlement. Additionally, the consumers are price *insensitive* and do not submit offers to buy electricity. In other words, the price depends only on the electricity demand which is, as per the price-insensitive assumption, referred as to *inelastic*.

3. Model formulation

This section presents a summary of an optimization model with the aim to build the three primary phases of an electricity system model. Figure 1 shows the relationship between the three model optimization phases (e.g., pre-dispatch, real-time operation and market settlement) considering the key inputs and outputs, and also the parameters used to estimate the benefits and costs of the evaluated mitigation options.

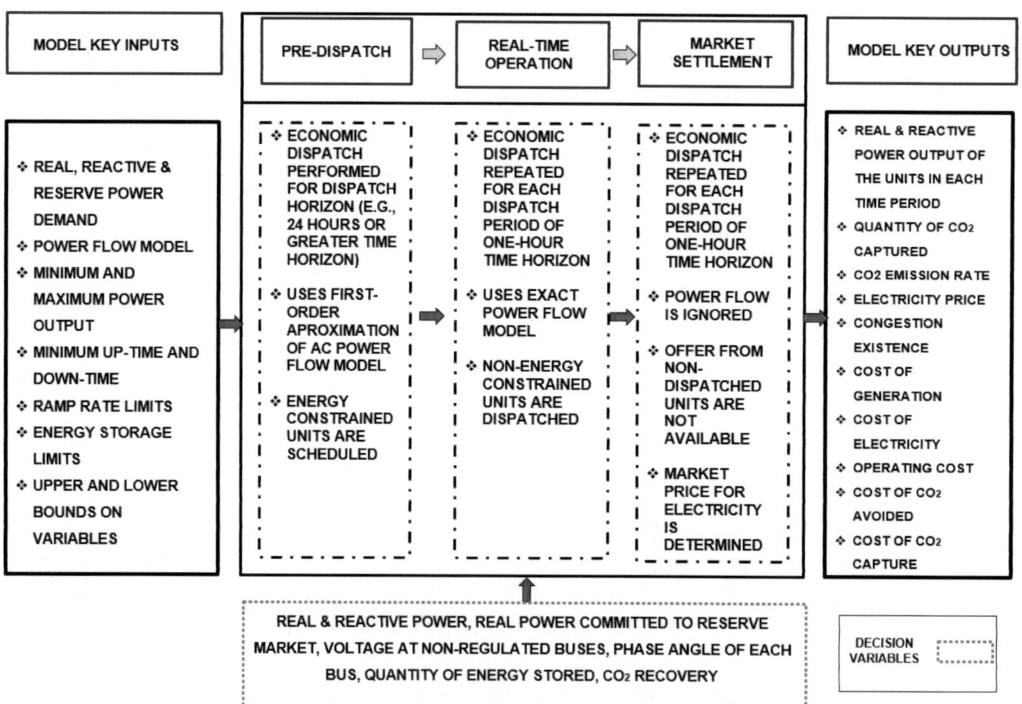

Figure1. General layout of the electricity system simulator optimization model

3.1. Phase 1: Pre-dispatch

The pre-dispatch phase targets to optimize the power availability and distribution in time within the electricity system model. This is accomplished by executing preliminary scheduling of the generating units well in advance. The electricity system operating and maintenance costs can be subdivided into two categories: fixed and variable. Given that fixed maintenance and operating costs do not vary as a function of the power output, the economic dispatch problem objective function (z) can be simplified to:

$$z = \sum_{n \in NG} \int_0^{P_n^S} \left[\left(\frac{dC_n^{VOM}}{dP_n^S} \right) \right] dP_n^S \tag{1}$$

Where the indices n and s represent the type of generating unit and the type of power supplied, respectively. The objective function is formulated as follows:

$$z = \sum_{n \in NG_D} \sum_{b=1}^{N_b} y_{bn} IHR_{bn} FC_n L_n + \sum_{r \in RM} C^{import} . RM_r^{slack} \tag{2}$$

where N_b is the number of bids per generating unit, L_n is the time duration (in this work L_t is considered to be equal to 1 hour), y_{bn} is the quantity of bids accepted by a particular generating unit, RM_r^{slack} is the reserve power pertaining to the slack bus, and C^{import} is the cost of the electricity imported from outside the grid. In this work, C^{import} is set at a ten percent premium to the most expensive bid of any generator in the system. Likewise, the last term in the objective function represents the cost required to facilitate reserve power from outside of the electricity system.

The economic dispatch problem has been formulated as an MINLP problem that consists of 420 equations, 33 binary variables, 501 continuous variables, and 165 nonlinear constraints. The optimization problem was solved using the MINLP solver DICOPT [26] in the General Algebraic Modeling System (GAMS). At this point the objective function has been formulated only to solve the economic dispatch problem. Therefore, in order to transform it into a proper pre-dispatch objective function; a time index is added to the variables and dynamic constraints are incorporated in the equation. Additionally, it is imperative to provide 'special' dynamic equations; otherwise, the implicit assumption is that the electricity system is undergoing a *black-start* (i.e., recovering from a state in which all generation is shut-down). Since dynamic constraints have been added to the pre-dispatch problem, considering a black-start in the system may lead to infeasible solutions in this phase. Consequently, the following equation illustrates the final pre-dispatch objective function:

$$\min_{\eta} z = \sum_{t=1}^{T} \sum_{n \in NG} u_{nt} HI_n FC_n + \sum_{t=1}^{T} \sum_{n \in NG_D} \sum_{b=1}^{N_b} y_{bnt} IHR_{bnt} FC_n L_t + \sum_{t=1}^{T} \sum_{r \in RM} C^{import} . RM_{rt}^{slack} \tag{3}$$

Where η represents the following set of decision variables

$$\eta = \left\{ \begin{array}{l} u_{n,t}, y_{bn,t}, P_{n,t}, P_{n,t}^S, P_{n,r,t}^R, Q_{n,t}^S, P_{k,t}, Q_{k,t}, I_{k,t}^{\mathrm{Re}}, I_{k,t}^{\mathrm{Im}}, \theta_{k,t}, |V_{k,t}|, x_{n,t}^{on}, x_{n,t}^{off}, w_{n,t}, \\ X_{n,t}^{on}, \psi_{n,t}^{on}, X_{n,t}^{off}, \psi_{n,t1}^{on}, P_{r,t}^R, \rho_{n,r,t}, E_{k,t}, RM_{r,t}^S, RM_{r,t}^{Slack} \end{array} \right\} \tag{4}$$

The main difference between the economic dispatch and the pre-dispatch objective function consists of the unit start-up (u_{nt}) which represents the energy input required by a thermal unit (that has been off) to initiate operations. Based on the previous definition, the unit's variable maintenance and operating cost is now expressed in terms of the start-up and the fuel costs. To a first approximation, the start-up cost is the energy expenses (fuel) required to start-up a particular power unit, where HI_n represents the heat input required to cold-start a unit. The binary variable representing the unit start-up takes the value of one if the unit started-up in the time period; or zero otherwise. This can be expressed as follows:

$$u_{n,t} \geq \omega_{n,t-1} - \omega_{nt} \qquad \forall n \in NG, t \in T \tag{5}$$

In the case of units experiencing black-start (e.g., units with non-zero start-up cost), the unit start-up (u_{nt}) is zero in the optimal solution, and for units whose start-up costs are zero the solution is indeterminate.

The capacity utilization is expressed in terms of the unit's availability and capacity. The unit's availability is the power that a generator can produce in a given time period; while the unit's capacity is the nominal power that the unit is designed to produce. The present constraint specifies that the capacity utilization $(P_{n,t})$ of a unit n in a given time period t is equal to the sum of the bids accepted in the time period $(y_{bn,t})$, i.e.

$$P_{n,t} = \sum_{b=1}^{Nb} y_{b\,n,t} \qquad \forall n \in NG, t \in T \tag{6}$$

The capacity utilization of each unit in each time period must equal the sum of the unit's contribution to the real and reserve market in a given time period t:

$$P_{n,t} = P_{n,t}^{S} + \sum_{r \in RM} P_{n,r,t}^{R} \qquad \forall n \in NG, t \in T \tag{7}$$

Where $P_{n,t}^{S}$ is the power supplied to the real market and $P_{n,r,t}^{R}$ is the power supplied to the reserve market.

Minimum and maximum real and reactive power outputs are represented by the following constraints:

$$(1-\omega_{n,t})P_n^{\min} \le P_{n,t}^{S} \le (1-\omega_{n,t})P_n^{\max} \qquad \forall n \in NG, t \in T \tag{8}$$

$$(1-\omega_{n,t})Q_n^{\min} \le Q_{n,t}^{S} \le (1-\omega_{n,t})Q_n^{\max} \qquad \forall n \in NG, t \in T \tag{9}$$

Where $Q_{n,t}^{S}$ is the reactive power supplied to a generating unit n at a particular time t, $Q_n^{\max}$ and $Q_n^{\min}$ are the maximum and minimum reactive power outputs, respectively. Additionally, ω_{nt} is a binary variable used to illustrate the state of a generating unit n in time period t. The binary variable assumes a value of one if the unit is "off", and zero otherwise.

There are generating units within the electricity system that are constrained not only in terms of power output, but also in terms of energy output. One example of energy constrained units are the hydroelectric generating units. These units cannot generate more power than that produced by the water contained in its reservoir. This is expressed as follows:

$$E_{k,t} = E_{k,t-1} + \left(\dot{E}_{k,t} - \sum_{n \in NG_k} P_{k,n,t}^{S} \right) L_t \qquad \forall k \in N^{ST}, t \in T \tag{10}$$

$$P_{k,t} L_t \le E_{k,t} \tag{11}$$

Where $E_{k,t}$ is the electric energy in a particular bus k at a particular time t, and $\dot{E}$ represents the energy inflow rate. Equation (10) defines the available energy, and outlines the net additions in each time period t. Also, it defines the energy availability at time period $t-1$. On the other hand, the unit's output limit can be calculated using Equation (11). At this point, the electricity system is started one day in advance (of the actual initial period of interest) to avoid any black-start that could turn the problem's solution into infeasible. To avoid anomalies in the results during the period of interest, the initial pre-dispatch period occurs over a 48-hour period.

The net real power injected at each bus $(P_{k,t})$ is the difference between the total output from a generating unit (P_{nt}^{S}) and the local demand (P_{kt}^{D}). The same situation applies to the reactive power (Q_{kt}), except for the buses with shunt admittance to ground (given the buses extra reactive power). The net power available at each bus can be calculated as:

$$P_{k,t} = \sum_{n \in NG_k} (P_{nt}^{S}) - (P_{kt}^{D}) \qquad \forall k \in N, t \in T \tag{12}$$

$$Q_{kt} = \begin{cases} \sum_{n\in N_k} Q_{nt}^S - Q_{kt}^D & \forall k \notin N^{shunt} \\ \sum_{n\in N_k} Q_{nt}^S - Q_{kt}^D + |V_{kt}|^2 & \forall k \notin N^{shunt} \end{cases} \quad \forall t \in T \tag{13}$$

In modern electricity systems reliability is important. Therefore, from the pool of available capacity, a portion is selected for a back-up role. The reserve requirements used in this study are based on the Ontario's electricity system operation; which adhered to the NERC (North American Electricity Reliability Corporation) [27]. Reserves are required to preserve the generation/load balance, as well as to compensate for the variability and uncertainty of load (e.g., regulation, load following, and forecast uncertainty). Also, reserves respond to forced outages of conventional generation (contingency reserves). The reserve power requirements $\left(P_{n,r}^R\right)$ are represented by the load of operating capacity exclusively committed to the reserve market. In the present work, three reserve markets are considered and the total power committed to each is expressed as follows:

- Ten-minute spinning reserve:

$$RM_{10^{sp}}^S = \sum_{n\in NG} P_{n,10^{sp}}^R (1-\omega_n) \tag{14}$$

- Ten-minute non-spinning reserves:

$$RM_{10^{ns}}^S = RM_{10^{SP}}^S + \sum_{n\in NG,\tau_n^{up}=0} \omega_n P_{n,10^{ns}}^R \tag{15}$$

- 30-minute non-spinning reserve:

$$RM_{30}^S = RM_{10^{ns}}^S + \sum_{n\in NG} P_{n,30}^R (1-\omega_n) + \sum_{n\in NG,\tau_n^{up}=0} \omega_n P_{n,30}^R \tag{16}$$

Where RM_r^S represents the power supplied to the reserve market.

3.2. Phase 2: Real-time operation

In an active power network, the system operator is constantly updating the demand forecast since this can change at any given time. Therefore, generating units require constant changes in their power output to regulate voltage, and respond to contingencies in an economical and optimal way. The real-time operation phase can be described as a simplified pre-dispatch phase problem. Though, there are small differences between the two phases in terms of the actual demand, generator outputs, and power flows. For instance, real-time operation is no longer a dynamic problem; nevertheless, time dependency is preserved. The state of time-dependent variables is specified using parameters whose values are obtained from the solution of the problem for the previous time period. The problem used in the real-time operation phase considers the economic dispatch problem for a single time period. As a result, the objective function is formulated as follows:

$$\min_{\eta} z = \sum_{n\in NG} u_n HI_n FC_n + \sum_{n\in NG_D} \sum_{b=1}^{N_b} y_{b,n} IHR_{b,n} FC_n L_t \frac{1}{10^3} + \sum_{r\in RM} C^{import} . RM_r^{slack} \tag{17}$$

$$\eta = \left\{u_n, y_{b,n}, P_n, P_n^S, P_{n,r}^R, Q_n^S, P_k, Q_k, I_k^{\mathrm{Re}}, I_k^{\mathrm{Im}}, \theta_k, |V_k|, x_n^{on}, x_n^{off}, w_n, RM_r^S, RM_r^{Slack}\right\}$$

Since time has been removed from the objective function, the real-time operation phase is now a deterministic problem; which requires less computational effort. Additionally, the definitions that were previously discussed in the pre-dispatch phase regarding the generating unit constraints, minimum and maximum power output, and reserve power constraints are also used for the real-time operation.
Furthermore, the real-time operation phase requires no power flow model simplification. During this phase, the actual performance of the electricity system is described, which means it is crucial to use the complete power flow model as shown below:

$$P_k^S = I_k^{Re} |V_k| \cos\theta_k + I_k^{Im} |V_k| \sin\theta_k \qquad \forall k \in N \tag{18}$$

$$Q_k^S = I_k^{Re} |V_k| \sin\theta_k - I_k^{Im} |V_k| \cos\theta_k \qquad \forall k \in N \tag{19}$$

In the present electricity system, one of the main goals of the pre-dispatch phase is to determine a plan for using energy constrained units (e.g., hydroelectric generating units). Unlike other generating units in the IEEE RTS' 96 system, the hydroelectric units have a minimum real power output of zero. Thus, it is possible to find hydroelectric units that have zero real power output and non-zero reactive power output. This situation is tolerated during the pre-dispatch phase, and such results are used to initialize the real and reactive power in the real-time operation phase. As a result, Q_n^S is fixed at zero for any hydroelectric unit where $P_n^S = 0$.

3.3. Phase 3: Market settlement

The power flow model is removed from the market settlement. This leads to the assumption that the generating units and loads are connected to the same bus. Since the power flow model is ignored, the references (i.e., variables and constraints) related to this model such as: $I_k^{Re}, I_k^{Im}, \theta_k, |V_k|$ are also ignored. Moreover, references related to the reactive power such as Q_n^S and Q_k are eliminated, as well as the minimum and maximum reactive power output. Accordingly, the market settlement objective function is formulated using the same structure applied in the real-time operation. However, since the power flow is ignored at this stage, the decision variables set is rewritten as follows:

$$\eta = \{u_n, y_{b,n}, P_n, P_n^S, P_{n,r}^R, x_n^{on}, x_n^{off}, RM_r^S, RM_r^{Slack}\} \tag{20}$$

Correspondingly, some of the definitions previously discussed on capacity utilization (6), power disaggregation between real and reserve market (7), and reserve power in the electricity system (14)-(16) are calculated following the real-time operation phase structure. Additionally, based on the assumption that all the generating units and loads are connected to a single bus, the net power available at each bus switch into the supply/demand balance. This can be formulated as follows:

$$\sum_{n \in NG} P_n^S \geq \sum_{k \in N} P_k^D \tag{21}$$

In order to ensure the availability of reactive power in the system, the following constraint is added:

$$\sum_{n \in NG} Q_n^{max}(1-\omega_n) \geq \sum_{k \in N} Q_k^D \tag{22}$$

4. CO_2 mitigation options assessment techniques

In order to evaluate the effectiveness of a particular mitigation strategy, it is important to account for the CO_2 emissions that are in fact mitigated as a result of a particular mitigation action. In this work, calculating the cost of CO_2 avoided (CCA) is considered the most reliable method for this matter. This is formulated as follows:

$$CCA = \frac{(CoE)-(CoE)_{ref}}{(CEI)_{ref}-(CEI)} \tag{23}$$

where CoE represents the cost of electricity, and CEI is the CO_2 emission intensity. Accordingly, by definition CCA is the ratio of the incremental cost of the CO_2 mitigation action to the incremental change in CO_2 emissions. The values for CoE and CEI can be calculated as follows:

$$CoE = \frac{\sum_{n \in NG} \dot{C}_n^{FOM} P_n^{max}}{HPY \sum_{n \in NG} CF_n P_n^{max}} + \frac{\sum_{n \in NG} FC_n HR_n CF_n P_n^{max} L}{\sum_{n \in NG} CF_n P_n^{max}} \tag{24}$$

$$CEI = \frac{\sum_{n\in NG} HR_n EI_n^{CO2} CF_n P_n^{max} L}{\sum_{n\in NG} CF_n P_n^{max}} \quad (25)$$

Where $\dot{C}_n^{FOM}$ represents the annual fixed operating and maintenance costs, P_n^{max} indicates the maximum real power value, FC_n represents the fuel cost, HR_n is the heat rate, HPY denotes the annual operating time, and EI_n^{CO2} represents the fuel emission intensity.
In this work, setting carbon tax prices is considered an accurate method for the assessment of CO_2 mitigation strategies. The emission cost can be expressed in terms of the heat input to boilers as follows:

$$CO_{n,t}^{CO_2} = u_{n,t} HI_n EI_n^{CO_2} TAX^{CO_2} + \dot{q}_{n,t} EI_n^{CO_2} TAX^{CO_2} L_t \quad (26)$$

Where TAX^{CO_2} is the CO_2 emission's tax and $\dot{q}_{n,t}$ is the boiler's heat input. The first term of the equation accounts for the fuel consumed during the start-up, whereas the second term accounts for fuel consumed during normal operation. For that reason, it is convenient to express the permit cost in terms of the incremental heat rate. Consequently, the unit's variable operating and maintenance cost of a generating unit n at a time period t can be formulated as follows:

$$C_{n,t}^{VOM} = C_{n,t}^{start-up} + C_{n,t}^{fuel} + C_{n,t}^{CO_2} \quad (27)$$

Where $C_{nt}^{CO_2}$ is the cost per unit of CO_2 that a generating unit emits.

The electricity system simulator is used to assess the effectiveness of the load balancing approach to reduce GHG emissions. Accordingly, the marginal emission cost is calculated taking the first derivative of the first term in Equation (26) with respect to $P_{n,t}^S$. Furthermore, in order to capture the relationship between carbon tax (TAX^{CO2}) and CO_2 emissions, a surrogate model is developed. Such surrogate model is a reduced order model that attempt to represent the solution space of the models they are based upon but with fewer variables. The development of such reduced-order mathematical model is used to represent the coal-fired electricity generating unit with CO_2 capture. Aspen Plus® was used as simulation tool to evaluate the coal-fired generating unit. This unit is modelled after the 500 MWe units at the Ontario power generation's (OPG) Nanticoke station in Ontario, Canada. These subcritical units are designed to burn subbituminous coal and to generate 1500 tonne per hour of steam at 538 °C and 165 bar with a single, 538 °C reheater. Therefore, the reduced-order model only needs to represent the Pareto optimal frontier of the power plant [28].The unit's power plant process model can be found in Colin's [29].
On the other hand, approaches to capture CO_2 in coal-fired power units fall into one of three different categories: pre-combustion, oxy-fuel combustion, or post-combustion capture. Post-combustion capture (PCC) methods using amine (i.e., MEA) solvents and are regarded as the best near-term CCS option. Therefore, this approach is selected to develop an integrated model for a power plant with CO_2 capture [29]. A generating unit that captures CO_2 does not need to obtain permits for the fraction of CO_2 that has been captured assuming that it is all permanently stored. Hence, a new cost component is required to represent the rebate that generating units receive for the quantity of CO_2 captured. Also, it is assumed that the solvent consumption rate is proportional to the CO_2 capture rate.

5. Case studies

In this section, two case studies are analyzed considering two GHG mitigation strategies: load balancing, and carbon tax regulations. These case studies aim to evaluate the effectiveness of different mitigation strategies in the electricity generation sub-sector.

5.1. Case study 1: reducing GHG emissions through load balancing

The first case study discuses load balancing as GHG mitigation strategy. This strategy differs from others since it does not require new capital investment. Accordingly, the IEEE RTS '96 system was used as test platform for this approach assessment. Moreover, detailed information on the power plants within the IEEE RTS '96 system is summarized in Table 1. The parameters used in this analysis are compiled in Table 2. The capacity factors are taken from the base-case simulation of the IEEE RTS 96', the annual cost ([[C˙]]_n^FOM) and the rest of the parameters (e.g., fuel costs, net heat rates, incremental heat rates, cold start unit heat input) are taken from Grigg et al23

Table 1. Summary of generating units power output

Unit type					$\overline{HR}_n$		
Bus	Fuel	Capacity MWe	Number	CF	Time-weighted Btu/kWh	Energy-weighted Btu/kWh	$N^{start-up}$
Abel	#2 Fuel Oil	20	2	0.02	14821	14607	7
Abel	Coal	76	2	0.65	12475	12080	0
Adams	#2 Fuel Oil	20	2	0.05	14673	14592	10
Adams	Coal	76	2	0.7	12408	12064	0
Alder	#6 Fuel Oil	100	3	0.39	11465	10535	3
Arne	#6 Fuel Oil	197	3	0.28	9816	9696	16
Arthur	#6 Fuel Oil	12	5	0.02	16017	16017	25
Arthur	Coal	155	1	0.28	10951	10680	0
Asser	Coal	155	1	0.48	10428	9965	0
Astor	Nuclear	400	1	1	10000	10000	0
Attlee	Nuclear	400	1	1	10000	10000	0
Aubrey	Hydro	50	6	0.64	N/A	N/A	N/A
Austen	Coal	155	2	0.53	10197	9931	0
Austen	Coal	350	1	0.83	9508	9505	0

Table 2. Parameters of units at Austen, Arne and Alder in the reference case

Parameter	Reference-case values		
	Austen	Arne	Alder
CF	0.826	0.278	0.393
$\mathbf{HR}_n$ (Btu/kWh)	9500	9600	10000
$\dot{C}_n^{FOM}$ $(\$/MW/year)$	25000	7500	7500
P_n^{max} (MWe)	350	591	300
EI_n^{CO2} $(lb\ CO_2/MMBtu)$	210	170	170
FC_n $(\$/MMBtu)$	1.2	2.3	2.3

The final approach considers load shifting from Austen (350 MWe) to the units at Arne and Alder. In scenario 1, Arne load balancing is limited by the capacity of the 350 MWe unit at Austen. In this scenario, at maximum load balancing, $CF_{Austen} = 0$ and $CF_{Arne} = 0.767$. On the other hand, in scenario 2, Alder load balancing is limited by the capacity of the 100 MWe units at Alder. In this scenario, at maximum load balancing, $CF_{Austen} = 0.306$ and $CF_{Arne} = 1.0$. The estimated CoE, CEI, and CCA for the two scenarios of interest are shown in Table 3. Concurrently, it is worth to recall that CCA is defined as the carbon price at which the mitigation action 'breaks even' with the reference case. Therefore, since the carbon price obtained exceeded the \$65/tonne CO2, it would be economical to transfer load from Austen to Arne. This action would reduce CoE and achieve CO2 emission reductions of up to 48 tonne CO2/h. As shown in Table 3, when load balancing is applied between

Austen and Alder, a carbon price exceeding $87/tonne CO2 is required to make this approach economically feasible. Thus, CO2 emissions can be reduced up to 24 tonne CO2/h in relation to the reference case. It is important to keep in mind, that the overall rate of CO2 emissions from the system is approximately 1000 tonne CO2/h.

Table 3. Cost of CO_2 avoided for load balancing scenarios

Parameter	Scenario 1		Scenario 2	
	Initial	Final	Initial	Final
CoE ($\$/MWh_e$)	18.59	25.4	17.85	23.04
CEI ($t\,CO_2/MWh_e$)	0.845	0.74	0.866	0.806
CCA ($\$/t\,CO_2$)	65		87	
$(\Delta CO_2)^{max}$ ($t\,CO_2/h$)	48		24	

These results indicate that load balancing could immediately trigger emission reductions. The basis used in the present analysis is representative of the basis employed in many published studies [13]; therefore, it is worth to consider its validity. For instance, the basis includes heat rate (HR) values corresponding to those of the generating units at base loads. Implicit in the above analysis is that the location of the units in relation to the reference case, and the loads in the system is unimportant. In other words, a unit of power injected at Alder or Arne is undifferentiated from a unit of power injected at Austen. This is further reinforced by the observation that there is limited unused capacity along the transmission lines that connects Alder to the rest of the system. Therefore, the transmission system may have implications on the effectiveness of load balancing that the previous analysis fails to capture. Other factors that denotes the validity of the basis, involves the 350 MWe unit at Austen (see Figure 2), and the units at Alder and Arne; which have an important role in satisfying the requirements for reserve power in the IEEE RST '96. This may limits the extent to which the load can be shifted from the 350 MWe unit at Austen to the units at Alder or Arne.

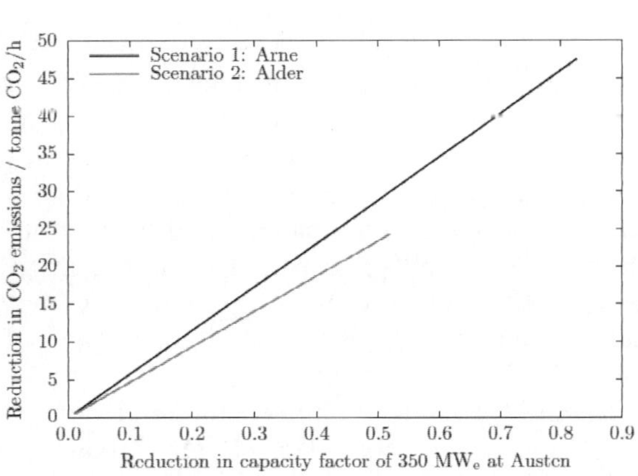

Figure2. Effect of load balancing on CO_2 emission reductions.

5.2. Case study 2: Adding GHG regulation to the electricity system

This second case study discusses GHG regulations as the proposed mitigation strategy. Accordingly, generators are required to pay for every unit of CO_2 emitted to the atmosphere. This means that, in addition to the unit's variable operating and maintenance cost, there is now a contribution based upon the quantity of CO_2 that the unit emits. The unit's variable operating and maintenance cost is shown in (32). Figure 3 shows the composite supply curve for the IEEE RTS '96 for different levels of carbon prices. In the vertical axis, it is shown the electricity offer price ($/MWhe); while the horizontal axis comprises the electric energy (MWhe). As can be seen in the figure, as the carbon price increases so does the marginal cost of each bid and this increase takes place in a manner that is proportional with respect to the carbon price and unit's incremental heat rate. Consequently, bids from coal units are at the higher end of the composite supply curve and vice versa for bids from oil-fired units. Accordingly, for carbon prices greater than the maximum analyzed in this work ($100/tonne CO2), the relative position of the units matches the one based merely on the CO2 emission intensity.

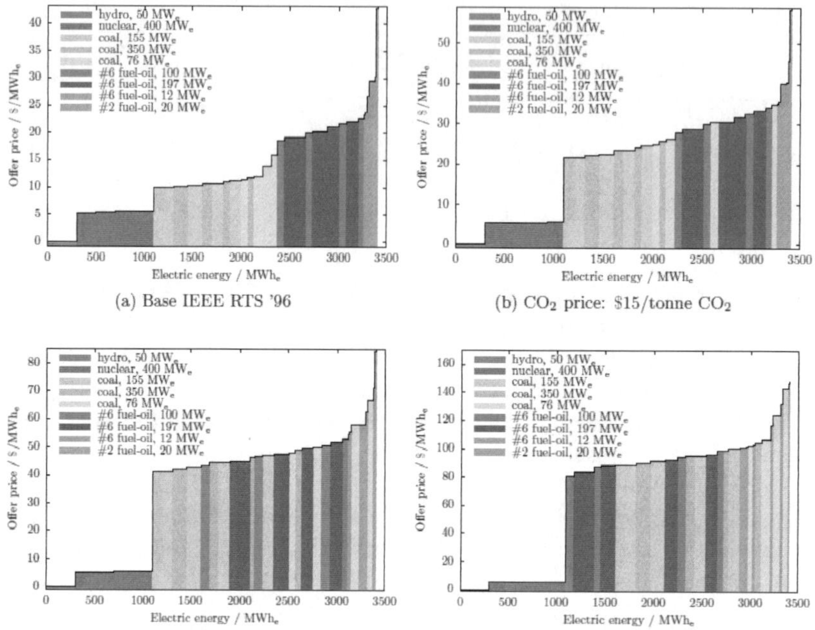

(a) Base IEEE RTS '96

(b) CO_2 price: $15/tonne CO_2

(c) CO_2 price: $40/tonne CO_2

(d) CO_2 price: $100/tonne CO_2

Figure 3. Composite supply curves for electricity system for different levels ocarbon prices [30]

For the present case study, the power system is simulated for one full week under three different carbon prices: $15/tonne CO_2, $40/tonne CO_2, and $100/tonne CO_2. In fact, $15/tonne CO_2 is the permit price proposed by the Canadian federal government. Such a permit price level is perceived as sufficient to stimulate CCS where CO_2 is an input to the production of a saleable commodity. On the other hand, a $40/tonne CO_2 is comparable to the most optimistic cost of CO_2 avoided reported for CCS. Accordingly, a $40/tonne CO_2 permit price would be sufficient to make CCS economic in various sectors [31], whereas a $100/tonne CO_2 is approximately the permit price considered necessary for the widespread adoption of CCS [32]. These three permit prices comprise the range of values expected if the adoption of serious regulations of GHG emissions takes place.

Based on the above, understanding the impact of incorporating carbon prices within the electricity system is necessary in the present work. Therefore, the selected carbon prices are applied to different types of generating units (within the power network); whereas, its impact is evaluated in terms of the generating units' capacity factor. Figure 8 illustrates that the average capacity utilization of the nuclear plants (at Astor and Attlee) and hydroelectric units (at Aubrey) remain unchanged by the GHG regulation. These units are non-emitting, and consider lower marginal operating costs than the fossil-fired units. Thus, in practice they are fully utilized in the base case, and remain so after carbon prices are imposed. Additionally, it is observed that as the carbon price increases, the capacity utilization increases for oil-fired units, and decreases for coal-fired units (with the exception of the 376 MWe generating unit with 85% CO2 capture installed at Austen). This indicates that the higher the carbon price, the lower the utilization of high-emitting generating units. Another estimation of the carbon price impact in the electricity system is given by the change in average power output. Figure 4 reveals that the coal-fired units (e.g., the 76 MWe units at Abel and Adams, the 155 MWe units at Arthur, and the units at Asser and Austen) present a significant reduction in their average power output and corresponding emissions. Accordingly, as the stringency in GHG emissions regulation increases, the effect on the units' utilization also increases. For example, higher CO2 permit prices results in power supply shifts from higher (i.e., coal-fired plants) to lower (i.e., natural gas and oil fired plants) emission intensity units.

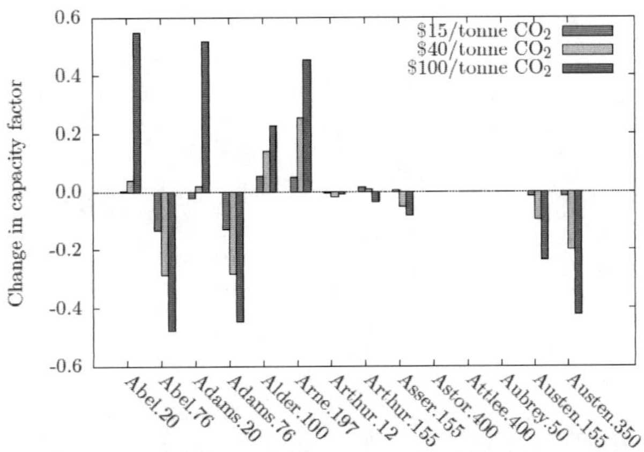

Figure 4.Capacity factor for different CO2 permit prices for the electricity system buses[30]

Accordingly, the CO_2 emissions are lower when a carbon price is incorporated in the system, and thus the greater the carbon price the lower the emissions. Table 4 summarizes the results in terms of CO_2 emissions for the base case and different stringencies of GHG regulation. Therefore, the higher the GHG permit prices the lower the CO_2 emission intensity (CEI) as shown in Table 4.

Table4. Summary of CO2 emissions and reductions

Scenario	$\dot{m}^{CO_2}$ $(t\ CO_2/h)$	$\Delta\ CO_2$ $(t\ CO_2/h)$	CEI $(t\ CO_2/MWh_e)$
Base Case	995	-	0.483
\$15/tonne CO_2	980	14.9	0.476
\$40/tonne CO_2	959	36.5	0.466
\$100/tonne CO_2	920	75.0	0.447

Although, the increase in fuel costs is significant, the cost of acquiring CO_2 permits is the main cause for most of the generation cost increase. As a result, the higher the carbon tax price, the higher the *CoE* to produce power from high emission-intensity generating units as the coal-fired ones. In Table 5, the first column displays the *CAA* calculated using values of *CoE* that do not include the cost of acquiring CO_2 emission permits; while for the second column the cost of CO_2 permits is included. In terms of electricity price, the CCA results demonstrates that the greater the permit price, the greater the electricity price.

Table5. Cost of CO_2 avoided for load balancing scenarios

Scenario	$CCA\ w/o\ permits$ $(\$/tonne\ CO_2)$	$CCA\ w\ permits$ $(\$/tonne\ CO_2)$
\$15/tonne CO_2	29	1049
\$40/tonne CO_2	61	1156
\$100/tonne CO_2	64	1306

The energy benefit is also important in the present analysis, as it represents the revenue earned by the generators from selling power into the electricity market. Table 6 illustrates the change in net energy benefit obtained by generators at different GHG regulation levels. The most interesting observation is that most of the generators are more profitable with GHG regulations than without it, which means that the net energy benefit increases along with the carbon price.

Table 6. Change in net energy benefit due to GHG regulation

Scenario	$Net\ energy\ benefit$ $(\$/MWh_e)$	$\Delta\ Net\ energy\ benefit$ $(\$/MWh_e)$
Base case	13.31	-
\$15/tonne CO_2	16.30	2.99
\$40/tonne CO_2	23.41	10.10
\$100/tonne CO_2	47.54	34.23

6. Conclusion

In this work, a mixed-integer nonlinear programing (MINLP) model for the techno-economic assessment of CO2 mitigation options in the electricity generation sub-sector was presented. Two scenarios were considered: load balancing and GHG regulations. In order to assess the validity of the proposed GHG mitigation strategies, an electricity system was modelled in GAMS for the IEEE RST '96 test case. The key element of this work includes the development of a short-term generation scheduling model. The model considers detailed operation of both the electricity system, and each one of the presented mitigation options. The present optimization model aims to reduce the electricity generation sub-sector GHG emissions while enhancing the units' utilization, maintaining suitable production costs and electricity prices.

References

[1] Cooper CD, Alley FC. Air Pollution Control: A Design Approach. USA: Waveland Press Inc.; 1994.

[2] The Canadian Electricity Association. Electricity Generation Choices: Canada's Electricity Supply. 2014. http://powerforthefuture.ca/electricity-411/electricity-generation-choices/canadas-electricity-supply/.

[3] MacDonald D, MacDonald L, Matin A, et al. National Inventory Report 1990-2011: Green House Gas Sources and Sinks in Canada.; 2013.

[4] Mariz C, Ward L, Ganong G, Hargrave R. Cost of CO2 recovery and transmission for EOR from boiler stack gas. In: Riemer P, Wokaun A, eds. In Greenhouse Gas Control Technologies: Proceedings of the 4th International Conference on Greenhouse Gas Control Technologies. Elsevier Science Ltd (Vol. 1).; 1999.

[5] Tontiwachwuthikul P, Chan C, Kritpiphat W, et al. Large scale carbon dioxide production from coal-fired power stations for enhanced oil recovery: a new economic feasibility study. J Can Pet Technol. 1998;37(11):48-45.

[6] Rubin ES, Rao AB, Chen C. Understanding the cost of CO2 capture and storage for fossil fuel power plants. In: 28th International Technical Conference on Coal Utilization and Fuel Systems. Clearwater, FL: Carnegie Mellon University; 2003.

[7] Ordorica-Garcia JG. Evaluation of combined-cycle power plants for CO2 avoidance. 2003.

[8] Rao AB, Rubin ES. A technical, economic, and environmental assessment of amine-based CO2 capture technology for power plant greehouse gas control. Environ Sci Technol. 2002;36(20):4467-4475.

[9] Singh DJ. Simulation of CO2 capture strategies for an existing coal fired power plant - MEA scrubbing versus O2/CO2 recycle combustion. 2001.

[10] Akimoto K, Kotsubo H, Asami T, et al. Evaluation of carbon sequestrations in Japan with a mathematical model. In: Gale J, Kaya Y, eds. Greenhouse Gas Control Technologies:

Proceedings of the 6th International Conference on Greenhouse Gas Control Technologies. Elsevier Science Ltd (Vol. 1).; 2002:115-119.

[11] Anderson D, Turvey R. Electricity Development in Turkey: A Case Study Using Linear Programming. In: Electricity Economics: Essays and Case Studies. Baltimore: The Johns Hopkin University Press; 1977:184-200.

[12] Johnson TL, Keith D. Fossil electricity and CO2 sequestration: how natural gas prices, initial conditions and retrofits determine the cost of controlling CO2 emissions. Energy Policy. 2004;32(3):367-382.

[13] Hashim H. An optimal fleet-wide CO2 mitigation strategy for a network of power plants. 2006.

[14] Sparrow FT, Bowen BH. Modelling Electricity Trade in Southern Africa: User Manual for the Long-Term Model. West Lafayette, Indiana; 2001.

[15] Loulou R, Kanudia A. Advanced bottom-up modelling for national and regional energy planning in response to climate change. Int J Environ Pollut. 1999;12(2/3):191-216.

[16] Loulou R, Kanudia A. Extended MARKAL: A brief User manual for its stochastic programming and multi-region features. Groupe d'etudes Rech en Anal des d´ecicions. 1997.

[17] Loulou R, Kanudia A. Joint mitigation under the Kyoto protocol: A Canada-USA-India case study. Energy Policy. 1998.

[18] Loulou R, Kanudia A, Lavigne D. GHG abatement in central Canada with inter-provincial cooperation. Energy Stud Rev. 1996;8(2):120-129.

[19] Loulou R, Kanudia A. Minimax regret strategies for greenhouse gas abatement: methodology and application. Oper Res Lett. 1999;25(5):219-230.

[20] Kanudia A, Shakula PR. Modelling of uncertainties and price elastic demands in energy-environment planning for India. Omega. 1998;26(3):409-423.

[21] Loulou R, Kanudia A. The Kyoto protocol, inter-provincial cooperation, and energy trading: A systems analysis with integrated MARKAL models. Energy Stud Rev. 1999;9(1):1-23.

[22] What is CCS? Carbon Capture Storage Assoc. 2015. http://www.ccsassociation.org/what-is-ccs/. Accessed March 2, 2015.

[23] Albrecht P, Allan R, Billinton R, et al. The IEEE reliability test system — 1996. IEEE Trans Power Syst. 1999;14(3):1010-1021.

[24] Chowdhury AA, Koval DO. Generation reliability impacts of industry-ownded distributed generation sources. Proc 38th IAS Annu Meet. 2003.

[25] Billinton. RFG and R. Economic costs of power interruptions: a consistent model and methodology. Electr Power Energy Syst. 2006;28:29-35.

[26] Drud A. CONOPT - A large-scale GRG code. ORSA J Comput. 1992;(6):207-216.

[27] The North American Electric Reliability Corporation (NERC). NERC IVGTF Task 2.4 Report: Operating Practices, Procedures and Tools.; 2011. http://www.nerc.com/files/ivgtf2-4.pdf.

[28] Nuchitprasittichai A, Cremaschi S. Optimization of CO2 capture process with aqueous amines using response surface methodology. Comput Chem Eng. 2011;(35):1521-1531.

[29] Alie C. CO2 capture with MEA: integrating the absorption process and steam cycle of an existing coal-fired power plant. 2004.

[30] Alie C, Elkamel A, Croiset E, and Douglas P L. using short-term resource scheduling for assessing effectiveness of CCS within electricity generation subsector. AICHEE Journal. 2015;62(12):4210-4234.

[31] Dissou Y. Cost-effectiveness of the performance standard system to reduce CO2 emissions in Canada: a general equilibrium analysis. Energy. 2005;27(3):187-207.

[32] International Energy Agency (IEA). World Energy Outlook. Paris, France; 2008.

From Smart-Eco Building to High-Performance Architecture: Optimization of Energy Consumption in Architecture of Developing Countries

M Mahdavinejad[1] and N Bitaab [2]

1 Associate Professor, Department of Architecture, Tarbiat Modares University, Tehran, Iran
2 M. Sc. Student, Department of Architecture, Tarbiat Modares University, Tehran, Iran
E-mail: mahdavinejad@modares.ac.ir

Abstract. Search for high-performance architecture and dreams of future architecture resulted in attempts towards meeting energy efficient architecture and planning in different aspects. Recent trends as a mean to meet future legacy in architecture are based on the idea of innovative technologies for resource efficient buildings, performative design, bio-inspired technologies etc. while there are meaningful differences between architecture of developed and developing countries. Significance of issue might be understood when the emerging cities are found interested in Dubaization and other related booming development doctrines. This paper is to analyze the level of developing countries' success to achieve smart-eco buildings' goals and objectives. Emerging cities of West of Asia are selected as case studies of the paper. The results of the paper show that the concept of high-performance architecture and smart-eco buildings are different in developing countries in comparison with developed countries. The paper is to mention five essential issues in order to improve future architecture of developing countries: 1- Integrated Strategies for Energy Efficiency, 2- Contextual Solutions, 3- Embedded and Initial Energy Assessment, 4- Staff and Occupancy Wellbeing, 5- Life-Cycle Monitoring.

1. Introduction

Why energy conservation is important? It is one of the most challenging questions regarding to contemporary architecture of developing countries such as Iran. [1] Cultural issues have a lot to do with architecture [2] not only in form and shape but also in architectural design process and strategies. [3] In recent trends in contemporary architecture of developed countries, more than ever, new technologies have dominated and future architecture might be seen from the viewpoint of high-tech developments. It is very important to explain that future legacy in architecture is in need of smart contribution towards technology. [4] Developing a new paradigm for performance of recent theories in emerging cities is depends on eco-friendly development as well [5] especially in advanced mega-cities. Future architecture and new technology are two inseparable phenomena. Looking back to traditional architecture oriental countries shows that ecology has a lot to do with architecture as well. [6] Therefore in order to meet future legacy in architecture, it is essential to find a biophilic trends towards technology. [7] Shifting to parametric program for development and contemporization of indigenous and intelligent pattern might be seen as a way to sustainable architecture. Smart materials, recombinant nano-structured smart materials, recently developed hybrid technologies, biophilic recycled materials etc. [8] have made a meaningful changes in our perspective regarding to the future

[9]; sometimes promising and sometimes threatening [10]. It might be concluded that there are no chance for future architecture unless via interaction of nature and technology.

2. Literature review

Energy efficient architecture and planning is going to a dominant paradigm for future architecture. There are considerable numbers of researches regarding to significance of understand historic buildings and traditional architecture oriental countries through energy efficiency perspective [11]. Literature review of the paper show that meaningful number of researches concentrating on searching for sustainable building technology and urban development, energy technology and policy [12] and its essential criteria such as shape and form, air pressure, natural ventilation wind flow, thermal comfort and indoor temperature [13]. Optimisation of building shape and orientation for better energy efficient architecture [14] in one hand; and searching for flexible approach to architectural design as a tool for achievement eco-friendly buildings [15] in other hand, brought a new atmosphere to state of the art of the issue.

The concept of sustainability in contemporary architecture and its significant relationship with vernacular architecture [16] seems to be a secure way to sustainable development. Traditional architecture oriental countries is a scene to find invaluable samples of natural ventilation efficiency [17] Architectural application of new materials and flexible formworks [18] for architectural design process, besides reunderstanding the role of vernacular architecture in sustainable design [19] while adopting recombinant materials in contemporary energy efficient architecture [20] and search for smart solutions for innovative architecture [21] might be seen as a way towards regenerative environments for future. Recent trends concentrated on passive and low energy architecture, zero energy development for cities, buildings and societies.

3. Theoretical framework

From raising idea of "innovative technologies for resource efficient buildings" to latest developments in computational CFD approaches towards zero-energy architecture; there are considerable number of researches regarding to meet high-performance architecture and planning. It is a controversial issue to imagine characteristics of future architecture while there are meaningful differences between architecture of developed and developing countries. In emerging cities like Dubai and other Dubaization followers in developing countries, there is influential shadow of technology-oriented concepts for contemporary architecture of developing countries. Formal document of "Smart-ECO buildings towards 2020/2030: innovative technologies for resource efficient buildings" [22] might be as basis to understand concept of smart-eco building. In other words, "Sustainable Smart-ECO Buildings" might be understand as an integrated energy and architecture design (IEAD) [23] as a holistic design applying innovative technologies. [24] It seems that green eco architecture as mentioned by JOSE'SEBASTIAN RICCIARDI [25] might be seen as a goal for green strategies for development of smart-eco buildings. Although sustainability of buildings has been studied for a long time, there is no universally accepted definition of a "sustainable", "ecological", or whatever definition may be used, buildings. According to the resulting vision, a Smart-Eco building in 20 years should:

- Be designed from a lifecycle point of view
- Be constructed with limited resources and minimized energy consumption and waste production
- Have minimized operational complexity while allowing easy monitoring of technical and environmental performances
- Be adaptable to changes in capacity, type of users and performance requirements
- Include local issues in all aspects of design construction use and dismantling
- Facilitate ease of dismantling – reuse, recycle, restore. [22]

Other related issues to sustainable development such as economic sustainability has a lot to do with energy [26] and efficiency in energy audit in architectural performance.

Recent researches are focused on efforts to make buildings smarter in order to overcome interoperability problems to assert control over our buildings such as offices, hotels and airports while cutting costs by streamlining building operations like air conditioning and lighting. [27] Theoretical framework summarized based on holistic viewpoint are ranked as 1- Technical Performance, 2- Economic, 3- Sustainable, 4- Regional Pre-conditions and 5- Social.

4. Methodology

This paper is to analyze the level of developing countries' success to achieve smart-eco buildings' goals and objectives in case of emerging cities of West of Asia those are selected as case studies of the research.

Ten buildings selected randomly from among recently acclaimed buildings in West of Asia. To analyze the cases, research team set focus group discussion (FGD) in order to analyze the cases. FGD includes five expert regarding to the different aspects of the issue such as 1- architect (designer), 2- architect (engineer), 3- sociocultural critic, 4- project manager and 5- An expert for life-cycle assessment (LCA) who knows about life-cycle cost (LCC) and life-cycle cost analysis (LCCA).

5. Case studies

5.1. Aspire Tower, Doha

Aspire Tower, also known as The Torch Doha, is a 300-metre-tall (980 ft) skyscraper hotel located in the Aspire Zone complex in Doha, Qatar. Designed by architect Hadi Simaan and AREP and engineer Ove Arup and Partners. The tower served as the focal point for the 15th Asian Games hosted by Qatar in December 2006. [28] Aspire Tower is also known as Al Aziziyah, or Sports City Tower which built to house the Symbolic Flame of the 2006 Asian Games. The design symbolizes a hand grasping the torch that sits at the tower's top. (Figure 1) [29]

5.2. Financial Harbour, Bahrain

Bahrain Financial Harbour (BFH) is a large-scale commercial development project, mainly completed in 2009, in Manama, the capital of Bahrain. The majority of the project is being constructed on reclaimed land. The two tallest twin-towers (Commercial East and Commercial West) are currently listed as the tallest completed towers in Bahrain, with a height of 260 m. [30] This was raised as an attempt towards new generation of Dubaization in West of Asia. (Figure 2)

5.3. World Trade Center, Bahrain

The Bahrain World Trade Center (also called Bahrain WTC or BWTC) is a 240-metre-high, 50-floor, twin tower complex located in Manama, Bahrain. The towers were built in 2008 by the multi-national architectural firm Atkins. The Bahrain World Trade Center integrates large-scale wind turbines into its design; and together with numerous energy reducing and recovery systems, this development shows an unequivocal commitment to raising global awareness for sustainable design. (Figure 3) [31]

Figure 1. Aspire Tower, Qatar. [29]

Figure 2. Bahrain Financial Harbour. [30]

Figure 3. World Trade Center, Bahrain. [31]

IOP Conf. Series: Earth and Environmental Science **83** (2017) 012020 doi:10.1088/1755-1315/83/1/012020

5.4. Chelsea Tower, Dubai
Chelsea Tower is a 250 m skyscraper located on Sheikh Zayed Road in Dubai, United Arab Emirates. The 49 storey building is occupied by a hotel and residential apartments. Chelsea Tower is the 17th-tallest building in Dubai, and one of the tallest residential buildings in the world. When completed in 2005, Chelsea Tower was the fifth tallest building in the city. [32] The building as a monumental statue could easily be seen on Sheikh Zayed Road in Dubai. (Figure 4)

5.5. Jumeirah Emirates Towers
Jumeirah Emirates Hotel Tower, also known as Emirates Tower Two, is a 56-storey hotel in the city of Dubai, United Arab Emirates. The hotel includes 40 luxury suites and is operated by the Jumeirah International Group. At a structural height of 309 m, Emirates Towers Hotel is the smaller of the two sister towers. It ranks as the 48th-tallest building in the world. It is the world's third-tallest all-hotel building. Construction was completed on 15 April 2000. (Figure 5) [33]

5.6. Liberation Tower, Kuwait
Liberation Tower is the symbol of Kuwaiti liberation, the representation of country's resurgence, second tallest tower in Kuwait, and the fifth tallest telecommunication tower in the world. Officially unveiled by the late Kuwaiti Amir, Sheikh Jaber Al-Ahmad Al-Jaber Al-Sabah on 10th March 1996, this 372meter tall tower is 40 meters taller than the Eiffel Tower. The tower is so-named following the multinational coalition that led to liberation of the nation from seven months of Iraqi occupation during the Gulf war. (Figure 6) [34]

Figure 4. Chelsea Tower, Dubai. [32]

Figure 5. Jumeirah Emirates Towers, Dubai. [33]

Figure 6. Liberation Tower, Kuwait. [34]

5.7. Doha Tower, Qatar
Doha Tower, also known as Burj Doha and previously named as Burj Qatar and Doha High Rise Office Building, is an iconic high rise tower located in West Bay, Doha, Qatar. Doha Tower comprises 46 floors above ground, 3 floors below ground and a total gross floor area of approximately 110,000 m². It has no central core, leaving more internal space available to its occupants. [35] The building inspired from Persian mashrabiya and diagrid (diagonal grid) structure. (Figure 7)

5.8. Kingdom Centre, Riyadh, Saudi Arabia
Kingdom Centre is a 99-storey, 302.3 m skyscraper in Riyadh, Saudi Arabia. It is the third tallest skyscraper in the country after the Abraj Al Bait Towers and the Burj Rafal, and is the world's third tallest building with a hole after the Shanghai World Financial Center and Tuntex Sky Tower. (Figure 8) [36]

5.9. Burj al-Arab Tower, Dubai
The Burj al-Arab is a luxury hotel located in Dubai, United Arab Emirates. It is the third tallest hotel in the world; however, 39% of its total height is made up of non-occupiable space. Burj Al Arab stands on an artificial island 280 m from Jumeirah beach and is connected to the mainland by a private curving bridge. The shape of the structure is designed to mimic the sail of a ship. (Figure 9) [37]

5.10. Capital Gate Tower, Abu Dhabi

Capital Gate is a skyscraper in Abu Dhabi adjacent to the Abu Dhabi National Exhibition Centre designed with a striking lean. At 160 m and 35 stories, it is one of the tallest buildings in the city and inclines 18° to the west. The building has a diagrid especially designed to absorb and channel the forces created by wind and seismic loading, as well as the gradient of Capital Gate. (**Figure 10**) [38]

Figure 7. Doha Tower, Qatar [35]

Figure 8. Kingdom Center, Saudi Arabia. [36]

Figure 9. Burj al-Arab Tower, Dubai [37]

Figure 10. Capital Gate, Abu Dhabi [38]

6. Discussion

The FGD was set in order to analyze ten buildings selected randomly from among recently acclaimed buildings in West of Asia. The results might be seen in table 1.

Table 1. A slightly more complex table with a narrow caption.

		Technical Performance	Economic	Sustainable	Regional Pre-conditions	Social	Total
1	Aspire Tower	2	4	1	0	4	2.2
2	Financial Harbour	1	0	2	1	5	1.8
3	World trade center	4	4	4	4	3	3.8
4	Chelsea Tower	2	3	0	0	2	1.4
5	Jumeirah Emirates Towers	1	2	3	2	3	2.2
6	Liberation Tower	1	2	1	0	0	0.8
7	Doha Tower	5	5	3	2	1	3.2
8	Kingdom Centre	3	1	2	1	1	1.6
9	Burj al-Arab Tower	4	1	4	5	2	3.2
10	Capital Gate	5	3	3	3	3	3.4
	Total	2.8	2.5	2.3	1.8	2.4	-

Quantitative as well as quantitative approach towards the main objectives of the research shows that "Technical Performance" is the only criterion in which the result is over 2.5. It might be concluded that the most essential issues in this test, like "Economic Sustainable", "Sustainable", "Regional Pre-conditions" and "Social" are not in satisfactory condition. In other words, the projects could not easily be called smart-eco building even sustainable although they might be seen sustainable individually.

The statistical outcomes show that there is not enough evidence to name recently acclaimed architecture of West of Asia as "smart' while there are lots of building information modeling schedules in levels of architecture and planning adopted to enhance the level of energy efficiency in these buildings.

7. Conclusion

Based on research achievements, smart-eco buildings have some parameters to distinguish and introduction as follows: 1- Technical Performance, 2- Economic, 3- Sustainable, 4- Regional Pre-conditions and 5- Social. Therefore, those buildings which have not all of these parameters despite of their particular advantages and disadvantages cannot stay on Smart-eco buildings. Due to the analyzes most of the high rise buildings in west Asia technical performance while their costs of build dramatically are high and are not economic. At the next stage, has been considered to sustainability and regional pre-conditions whereas attention to social item is very slightly; in some ways we can deduct that this item has sacrificed for technics and other aspects.

A comprehensive overview on selected case studies shows upcoming dilemma regarding to sustainability despite the fact that those buildings enjoy latest developed technologies: 1- Imported Materials: meaningful amount of the construction materials imported from other countries which imposed shipping costs and glass-house gas (GHG) emission. 2- Predesigned and Borrowed plans: Some essential architectural plans borrowed from foreign-based consultants and there are lack of meaningful data to make architectural critical decisions. 3- Foreign Workforce: major part of workforce especially in skilled-worker side, invited from among other countries which endanger staff wellbeing more than other criteria. 4- Over-Design: The buildings dissipation in adoption of smart and recombinant materials and high technologies. 5- Over-Energy Consumption: Considering embedded energy for manufacturing the building materials as well as shipping and other related issues shows high level of energy consumption.

Therefore the concept of high-performance architecture and the idea of performative building might be different in developing countries. The research summarized the key issues for optimization of energy consumption in contemporary architecture of developing countries: 1- Integrated Strategies for Energy Efficiency: The building industry is in need of an integrated strategies for optimization of energy consumption developing countries especially based on regional pre-conditions, 2- Contextual Solutions: Concentration on contextual design process, inspiration from context and integration of building and site, 3- Embedded and Initial Energy Assessment: It is essential to consider embedded and initial energy consumption as well as running issues, 4- Staff and Occupancy Wellbeing: high-rise buildings and great-scale redevelopment plans are in need of humanistic approaches towards design and performance, 5- Life-Cycle Monitoring: The buildings should be devoted to eco-friendly strategies in performance period as well.

The results of the paper suggests that it is a necessity to transfer to high-performance architecture concept in order to optimize the level of energy consumption in architecture of developing countries. Although the concept of smart-eco building was an influential step towards sustainability in architectural design process, there is immediate need for some more comprehensive concepts based on other essential issues in contemporary architecture of developing countries such as West of Asia, in order to meet sociocultural sustainability, economic sustainability and environmental sustainability. The results of the research emphasize that it is essential to integrate concept of smart-eco building and the concept of biophilic architecture in order to develop meaningful energy efficient building in contemporary architecture of developing countries.

8. Acknowledgments

This paper is result of a M. Sc. Dissertation of Ms. Najme Bitaab in TMU (Tarbiat Modares University) under supervision of Prof. Mohammadjavad Mahdavinejad.

9. References

[1] Mahdavinejad M, Zia A, Larki A.N., Ghanavati S. and Elmi N 2014 Dilemma of green and pseudo green architecture based on LEED norms in case of developing countries *International Journal of Sustainable Built Environment* **3:2** 235-46
[2] Pourjafar M, Amini M, Hatami Varzaneh E and Mahdavinejad M 2014 Role of bazaars as a unifying factor in traditional cities of Iran: the Isfahan bazaar *Front. Arch. Res.* **3:1** 10-9
[3] Mohtashami N, Mahdavinejad M and Bemanian M 2016 Contribution of city prosperity to decisions on healthy building design: a case study of Tehran *Front. Arch. Res.* **5:3** 319-31

[4] Kasraei H, Nourian Y and Mahdavinejad M 2016 Girih for domes: analysis of three Iranian domes *Nexus Netw* **18:1** 311–21.

[5] Mahdavinejad M, Amini M, Bemanian M and Hatami Varzaneh E 2014 Developing a new paradigm for performance of educating city theory in advanced technology mega-cities, case: Tehran, Iran *Journal of Architecture and Urbanism* **38:2** 130-41

[6] Ghasempourabadi M, Mahmoudabadi Arani V I, Bahar O and Mahdavinejad M 2011 Assessment of behavior of two-shelled domes in Iranian traditional architecture: the Charbaq School, Isfahan, Iran *WIT Transactions on Ecology and the Environment* **155** 1223-33

[7] Mahdavinejad M, Siyahrood S A, Ghasempourabadi M and Poulad M 2012 Development of intelligent pattern for modeling a parametric program for public space (case study: Isfahan, Mosalla, Iran) *Applied Mechanics and Materials* **220-3** 2930-5

[8] Mahdavinejad M, Rafsanjani L H, Rasoolzadeh M and Nazari M 2014 Challenges regarding to usage of nanostructured materials in contemporary building construction *Advanced Materials Research* **829** 426-30

[9] Pakdehi S G, Rasoolzadeh M and Moghadam A S 2016 Barium oxide as a modifier to stabilize the γ-Al2O3 structure *Pol. J. Chem. Technol.* **18:4** 1-4

[10] Mahdavinejad M, Nazari M, Khazforoosh S 2014 Commercialization strategies for industrial applications of nanomaterials in building construction *Advanced Materials Research* **829** 879-83

[11] Mahdavinejad M, Javanroodi K and Rafsanjani L H 2013 investigating condensation role in defects and moisture problems in historic buildings, case study Varamin Friday Mosque in Iran *World Journal of Science, Technology and Sustainable Development* **10:4** 308-24.

[12] Mahdavinejad M and Javanroodi K 2016 Impact of roof shape on air pressure, wind flow and indoor temperature of residential buildings *International Journal of Sustainable Building Technology and Urban Development* **7:2** 87-103.

[13] Mahdavinejad M and Javanroodi K 2014 Natural ventilation performance of ancient wind catchers, an experimental and analytical study – case studies: one-sided, two-sided and four-sided wind catchers *Int. J. Energy Technology and Policy* 10:1 36-60.

[14] Mahdavinejad M and Fallahtafti R 2015 Optimisation of building shape and orientation for better energy efficient architecture, *Int. J. Energy Sect. Manage.* 2015, 9(4) 593 – 618.

[15] Mahdavinejad M, Ashtiani S R, Ebrahimi M and Shamshirband M 2013 Proposing a flexible approach to architectural design as a tool for achievement eco-friendly multi-purpose buildings *Advanced Materials Research* **622** 1856-9

[16] Loo L D and Mahdavinejad M 2017 The concept of sustainability in contemporary architecture and its significant relationship with vernacular architecture of Iran *Journal of Sustainable Development* **10:1** 132 – 41

[17] Najaf khosravi S, Saadatjoo P, Mahdavinejad M and Amindeldar S 2016 The effect of roof details on natural ventilation efficiency in isolated single building *Proc. Int. Conf. on Passive and Low Energy Architecture - Cities, Buildings, People: Towards Regenerative Environments (Los Angeles)* (USA: Passive and Low Energy Architecture) pp 1301-8

[18] Bemanian M, Mahdavinejad M, Karam A and Ashtiani S R 2012 Architectural application of smart materials for non-flexible structures made by flexible formworks *Applied Mechanics and Materials* **232** 132-6

[19] Mahdavinejad M, Ghaedi A, Ghasempourabadi M and Ghaedi H 2013 The role of vernacular architecture in design of green sidewalk (case study: Iran, Shushtar) *Applied Mechanics and Materials* **261-2** 65-68

[20] Mahdavinejad M, Ghanavati S, Elmi N, Larki A N and Zia A 2014 Recombinant materials and contemporary energy efficient architecture Advanced Materials Research **936** 1423-7

[21] Sauchelli M, Lobaccaro G, Masera G and Fiorito F 2013 Smart Solutions for Solar Adaptive Façade Preliminary studies for an innovative shading device. *InXIX IAHS World Congress,* Milan, Italy 2013 Sep.

[22] Iannaccone G, Imperadori M and Masera G. Smart-ECO buildings towards 2020/2030: innovative technologies for resource efficient buildings. Springer; 2014 Jul 14.

[23] Zambelli E, Frontini F, Masera G and Salvalai G Sustainable Smart-ECO Buildings: an integrated energy and architecture design (IEAD) process to optimize the design of the new Politecnico di Milano campus in Lecco, Italy.

[24] Iannaccone G, Imperadori M and Masera G Holistic Design Applying Innovative Technologies. In Smart-ECO Buildings towards 2020/2030 2014 (pp. 13-36). Springer International Publishing.

[25] RICCIARDI JS Green eco architecture. Proposta per una fattoria didattica a Berzo San Fermo inserita all'interno del progetto SEAP 20.20. 20.

[26] De Angelis E, Pansa G and Serra E 2014 Research of economic sustainability of different energy refurbishment strategies for an apartment block building *Energy Procedia* **48** 1449-58.

[27] Snoonian D. Smart buildings. IEEE spectrum. 2003 Aug **40:8** 18-23.

[28] Wikipedia (2017). (https://en.wikipedia.org/wiki/Aspire_Tower) (2017-05-20)

[29] Aspire Tower, 2017 (https://skyscrapercenter.com/building/aspire-tower/535) (2017-05-20)

[30] Bahrain Financial Harbour, 2017 (http://www.marcopolis.net/bahrain-financial-harbour-business-location-in-the-middle-east.htm) (2017-05-22)

[31] Bahrain World Trade Center, 2017 (https://skyscrapercenter.com/building/bahrain-world-trade-center/) (2017-05-22)

[32] Chelsea Tower, 2017 (http://legacy.skyscrapercenter.com/images/) (2017-05-22)

[33] Jumeirah Emirates Towers, 2017 (http://www.skyscrapercenter.com/building/emirates-tower-two/456) (2017-05-22)

[34] Liberation Tower, 2017 (https://www.visit-kuwait.com/attractions/liberation-tower-kuwait.aspx) (2017-05-22)

[35] Doha Tower, 2017 (https://skyscrapercenter.com/building/doha-tower/1083) (2017-05-22)

[36] kingdom Center, 2017 (https://www.salini-impregilo.com/en/projects/completed/civil-industrial-buildings/) (2017-05-22)

[37] Burj al-Arab Tower, 2017 (https://www.skyscrapercenter.com/building/burj-al-arab/) (2017-05-22)

[38] Capital Gate, 2017 (http://www.ctbuh.org/TallBuildings/F/CapitalGateTowerAbuDhabi/ /3380/) (2017-05-22)

High-performance Sonitopia (Sonic Utopia): Hyper intelligent Material-based Architectural Systems for Acoustic Energy Harvesting

F Heidari [1] and M Mahdavinejad [2]

1 Senior Production Editor, M. Sc. Student, Department of Architecture
Tarbiat Modares University, Tehran, Iran
2 Senior Production Editor, Associate Professor, Department of Architecture
Tarbiat Modares University, Tehran, Iran
E-mail: mahdavinejad@modares.ac.ir; f-heidari@modares.ac.ir

Abstract. The rate of energy consumption in all over the world, based on reliable statistics of international institutions such as the International Energy Agency (IEA) shows significant increase in energy demand in recent years. Periodical recorded data shows a continuous increasing trend in energy consumption especially in developed countries as well as recently emerged developing economies such as China and India. While air pollution and water contamination as results of high consumption of fossil energy resources might be consider as menace to civic ideals such as livability, conviviality and people-oriented cities. In other hand, automobile dependency, cars oriented design and other noisy activities in urban spaces consider as threats to urban life. Thus contemporary urban design and planning concentrates on rethinking about ecology of sound, reorganizing the soundscape of neighborhoods, redesigning the sonic order of urban space. It seems that contemporary architecture and planning trends through soundscape mapping look for sonitopia (Sonic + Utopia) This paper is to propose some interactive hyper intelligent material-based architectural systems for acoustic energy harvesting. The proposed architectural design system may be result in high-performance architecture and planning strategies for future cities. The ultimate aim of research is to develop a comprehensive system for acoustic energy harvesting which cover the aim of noise reduction as well as being in harmony with architectural design. The research methodology is based on a literature review as well as experimental and quasi-experimental strategies according the paradigm of designedly ways of doing and knowing. While architectural design has solution-focused essence in problem-solving process, the proposed systems had better be hyper intelligent rather than predefined procedures. Therefore, the steps of the inference mechanism of the research include: 1- understanding sonic energy and noise potentials as energy resources, 2- recognition of transductor and other similar mechanisms, 3- developing an integrated, hyper intelligent and material-based system, 4- examining the productivity, performance and efficiency of proposed systems in commercial buildings and office departments of Tehran as case study.The results of the research show that high-performance Sonitopia concept might be helpful for adoption in contemporary architecture of developing countries such as Iran in order to better energy efficiency. It is intelligent energy systems (IES) enjoy electromechanical energy converters based on performance-oriented design in over-crowded architectural spaces. The results indicated significance of concentrating on smart, intelligent and recombinant materials in order to achieve higher performance and productivity.

IOP Conf. Series: Earth and Environmental Science **83** (2017) 012021 doi:10.1088/1755-1315/83/1/012021

1. Introduction

1.1 Energy Processing

1.1.1 Environmental pollution. Restrictions on energy and environmental issues is one of the most important concerns in the energy sector, because of the importance of this issue, different policies in order to secure energy resources and reduce environmental pollution have been adopted. Some of these policies include plans to support renewable energy sources and intelligent systems with high programmability to reduce energy consumption. For consumption There is also the potential to produce as much pollution . nowadays pollution has become one of the main challenges for management So that Countries in addition to their domestic policies, follow this issue in the international realm. Europe Union Approved the 20-20-20 plan in 2007. The 2020 package is a set of binding legislation to ensure the EU meets its climate and energy targets for the year 2020. The package sets three key targets: 20% cut in greenhouse gas emissions, 20% of EU energy from renewables and 20% improvement in energy efficiency. [1] Construction and housing with more than 40% of energy consumption is the largest energy consumer in Iran. The energy consumption of buildings in Iran is more than 2.5 times the average global consumption and big cities like Tehran have high levels of air pollution which is caused by Fossil energy consumption. While more than 98 percent of the energy consumption of buildings in Iran are provided oil and gas products. shows that housing is one of the main reasons of Creating pollution. This section contains about 26.4% of carbon dioxide emissions. The following must be said that the buildings in Tehran Produces more than 40 percent of the carbon dioxide in the province. reducing energy consumption in the building and housing sector will have a significant impact on the total energy consumption of the country. Reducing energy consumption of buildings in terms of economic and environmental values is imperative. [2] Considering environmental issues began in the 1960s and its main focus was on industrial pollution due to the rapid growth of industrial economy. [3] in the late 1970s, trade issues and environmental issues peaked and Environmentalists protested the deplorable environmental situations of the development of trade, opposition and wide meetings were held around the world Which they say because of trade liberalization and increased exports, economic activity and also pollutive activities has spread and Inappropriate use of resources and energy increased, Pollution and environmental issues in crowded cities are not limited to a particular region And its effects will have a large-scale of impact . This pollution will impact Environment directly or indirectly, the direct impact includes the single major source of emissions and up to two to three stages However, in an indirect emission secondary sources, which are the primary sources of consumer, Create pollution. This pollution is mainly caused by man-made environments. Among the indirect pollution noise pollution in large cities can be noted, noise pollution can be infected with multiple sources. The sources in the city include noise pollution caused by traffic congestion, noise generated by unwanted noise pollution caused by mechanical machinery and equipment. Noise pollution can be divided into three parts: 1. Spherical or point 2. Linear cylinder or 3. Surface. Point sources of pollution such as presses, woodworking and turning, drilling or workshops, alone, cause noise pollution and increased load ambient sound environment Alone causes noise pollution and increased load of ambient sound in the environment. the human nervous system reacts in different ways to an increase in noise level of the environment. for example, if a person regularly uses Loud noises, it will lose its hearing capillary cells and does cells will not replace and reproduce. Volume reduction in heart rate, breathing and pulse rhythm, drop in electrical resistance of the skin, reduction in skin temperature, impaired gastric peristalsis, the production of juice and saliva, dilated pupils, increased blood pressure and narrowing of the blood vessels are the different typed of respond that the nervous system gives to the increase in the ambient noise level. sound is a vibration that propagates as a typically audible mechanical wave of pressure and displacement, through a transmission medium such as air or water. Humans can hear sound waves with frequencies between about 20 Hz and 20 kHz. Sound above 20 kHz is ultrasound and below 20 Hz is infrasound. Other animals have different hearing ranges. Soundproofing relates to the overall ability of a building element or building structure to reduce the sound transmission through it. Two types of sound insulation might be referred to – airborne sound insulation and impact sound insulation.

Airborne noises are transmitted by air and atmosphere. The radio, or people carrying on a conversation, are good examples of airborne noises. Impact noises propagate through solids and result from a shock on them. The footsteps of a person and the sound of an object falling on the floor are examples of impact noises. Sound is what we hear. Noise is unwanted sound. The difference between sound and noise depends upon the listener and the circumstances. Noise does not specify the type of sound and there are only vaguely heard audio source, in this case, due to unknown sources it will cause a noise traffic in certain areas. This may occur in residential buildings in dense urban context and in more important cases in office and commercial buildings which are very busy and productive in hours of noise traffic Monitoring and analyzing sound waves is investigated from several fronts which The three main branches can be noted as "Noise pollution control, sound source (energy production and other approaches), transmission and Transmission and separators ". Noise pollution may be intentionally or unintentionally form from one source of energy that will have devastating effects than other states. Sound waves produced by sources of energy for conversion into electrical energy have a direct relation The distance between the supply and sound absorbers. For example, for designing a sound into electrical energy converter in commercial construction to begin, attention should be payed to the potential in the building itself and then for the development and even saving energy Through facade to expand its relationship with the exterior transducers. Similar systems can be used as combined with electromagnetic or mechanical systems using a combination of smart materials (e.g., using the properties of polycrystalline). Recent advances in the field of micro-electronic processing and transmission of wireless sensor (WSN) has led to the use of this technology to be extremely common. In any case usage in industry, chambers of monitoring and control, urban traffic control systems and even get information on the health of a building structure are the cases of their often usage. [4] Configuration of the system consists of an external energy source (solar energy, heating energy, wind energy, sound energy and other resources), absorbing and converters, and in the final stage the design goal of using the system (production + save energy) that as a whole by an electronic system alone or in combination, guided and controlled. Expansion of the usage of this type of system is subjected to the potentials of points of use. noise Traffic in dense urban areas Both in terms of traffic and in terms of human will enhance the potentials in sound energy efficiency. The use of this potential may be provided at the same time and leads to energy savings and therefore its transferred to the municipal power. Among the strengths in the direction of sound energy efficiency in the urban context the following situations can be noted :1. Sound energy of urban traffic and local industry2. Sound energy from the crowd in the dense areas .3. Sound energy from the sound of the percussion movement of people on the surfaces (inside and outside)4. The reflected sound energy in the vicinity of dense tissue5. Artificial sound produced by enhancers.(fig1)

2. Research history

systems designed in Previous researches are often designed in a combination of several sources of energy and its main objective is the analysis of transmissions and the potential impact of the percussion sound of the user's movements and other types of vibration. such as: Sound absorption coefficient of the collisions, the sound transmission coefficient. In such systems, which are mostly used as a therapeutic strategy for non-insulated buildings, non-isolated surface with Low sound absorption coefficient have the highest efficacy. This means that energy-absorbing and sound amplifiers beneath the surface collisions. with the increase in amount of energy transfer, from the Output Level, would have saved more energy in the ending point of the Converters and batteries. Examples of this type of system can be a combination of poly-crystal systems with oscillator and amplifier. These systems consist of terminals electrochemical (battery) that is charged by electric energy absorbers The type of electricity and voltage change from AC to DC by the oscillator device (using the piezoelectric properties). and in the end a Rectifier device will flow transmitters between the piezoelectric converter and the batteries. And will offer a uniform flow as electrical output either directly or stored in resource. Later, using nonlinear electronic circuits and energy-induced HS (SSHI) the efficiency in the production and conversion of sound energy into electrical energy increased. [5] In a nutshell this system, by shifting the voltage across the piezoelectric energy reduces the absorption from the absorber and its conversion speed. And this generates more energy in less time. Before The main point was the usage of a feature of polycrystalline materials which its Property was used in the most of the Smart Sonic Systems as an absorber Which, by virtue of their special Property will produce energy. This type of material may have natural features in itself in order to produce energy or indirectly with a trigger or catalyst to achieve reaction stage.(fig 2,3)

The potentials of acoustic energy	acoustic energy resulted from traffic		acoustic energy resulted from overcrowding			acoustic energy resulted from percussion sound	acoustic energy resulted from recurrence and refraction	Producing synthetic sound (resonators)
	acoustic energy source					Main road	auxiliary road	
	traffic	population	Percussion	vicinity	installation	Location (separate/ compound)	plan	perspective
Utilizing population potential+ traffic+ user's percussion + vicinity	✸	✸	✸	✸	✸	Commercial building in crowded urban fabric exposed to very dense vicinity Main road + auxiliary road Very high number of user population	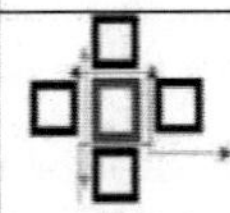	
Utilizing population potential I+ traffic + user's percussion	✸	✸	✸	✸	✸	Commercial building in semi-crowded urban fabric exposed to dense vicinity Main road + auxiliary road high number of user population the building does not have vicinity from one front		
Utilizing population potential + user's percussion	✸	✸	✸		✸	Commercial building in an urban fabric with average density and with less dense vicinity Main road + auxiliary road Fair number of user population the building does not have vicinity from two fronts	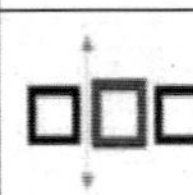	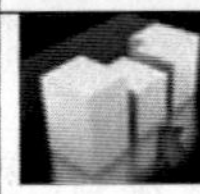
Utilizing percussion potential + installations			✸		✸	Commercial building in an urban fabric with low density and minimum vicinity Main road The building in an urban fabric with minimum noise pollution and with vicinity to one front		
Utilizing percussion potential + installations			✸		✸	Commercial building in a non-condensing urban fabric no vicinity no noise pollution from external source local access	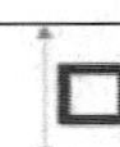	

Figure 1. City potentials for the use of Acoustic energy sources [author]

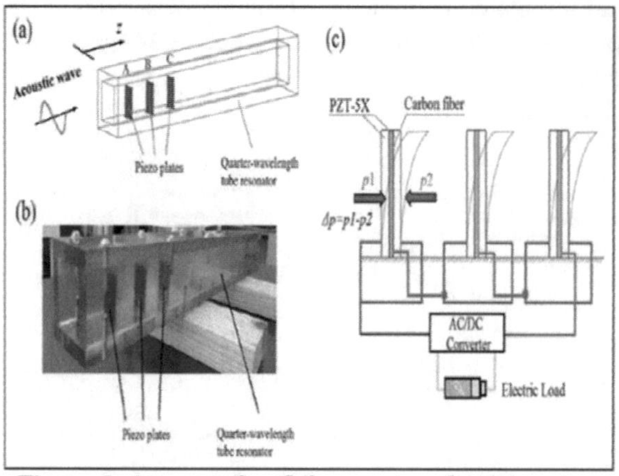

Figure 2. An example of the smart sonic system to produce electrical energy Components: the trajectory of the wavelength of sound-absorbing panels piezoelectric cavity resonance wavelength of sound, batteries and converters[6]

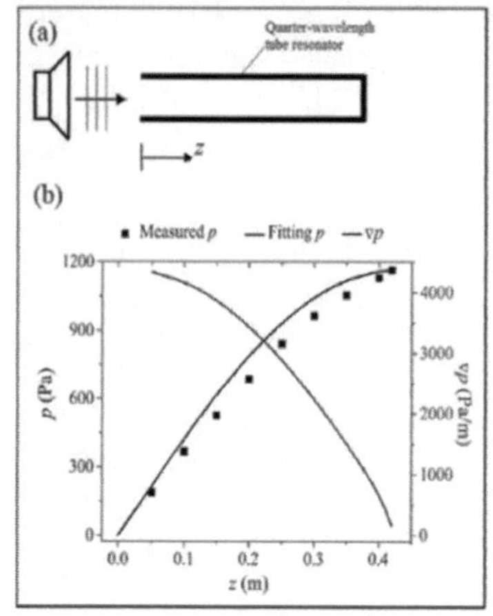

Figure 3. resonance wavelength Cavity generated by sound source intensity of sound pressure and pressure gradient graph [6]

3. Sonic energy

3.1 Intelligent systems

3.1.1 Energy generation. The quality of sound is determined by frequency. Measured in units of hertz (Hz), frequency is the number of complete cycles of vibration above and below the static pressure in a unit of time. The human ear detects sensitive frequency is approximately 1,000 hertz. Vibration is a periodic back-and-forth motion of the particles of an elastic body or medium, commonly resulting when almost any physical system is displaced from its equilibrium condition and allowed to respond to the forces that tend to restore equilibrium. This physical phenomenon is so-called the Wave. To get a rough picture of how we create sound waves we consider Pendulum as an example. A pendulum is defined as a mass, or bob, connected to a rod or rope, that experiences simple harmonic motion as it swings back and forth without friction. The equilibrium position of the pendulum is the position when the mass is hanging directly downward. Pendulums weight in its movement, pushes forward a thin layer of air molecules and This causes one a side of the weight to be a molecular density and in the other hand to be a molecular dilution. Dilution means the incrassation of the distance between molecules and density means the reduction of their distance. If we pull a rubber tire from its ends the length of it increases, the reason for this is that the distance between rubber molecules are increased in the mid portion and the molecules are Gathered in the ends of the rubber tire and so The distance between the molecules of the two ends decreases. Thus, in the middle molecular density and at both ends molecular dilution are caused [7] .Now, if we let go of the rubber molecules revert back to their original location. Perhaps one of the simplest solutions to the conversation of sound in to electric energy lies in a device called microphone. Microphones receive sound waves and convert them into electrical signals. Among the most important Microphones, dynamic microphones, tape and capacitive can be noted. In dynamic microphones a coil is glued to the rear of a membrane, and there is a strong magnet surrounding this coil. [8] When sound waves hit the microphone, the membrane moves to the rhythm of the sound waves, and the coil on its back moves along with it. The relative movement of the coil within its (stationary) magnetic gap induces a small signal voltage in this coil. There's your microphone, a device that converts sound into an electrical signal. Among other examples used for the production of energy by sound speakers can be noted. a considerable point in energy generation using sonic is the direct conversion of sonic energy too electrical energy. [9] Among all the examples that even people have lived with the through ages, without considering the magnificent structure of sound and its potentials, energy transmission was from sonic form to Mechanical vibration. thus can be noted that the main way of converting sonic energy to electrical energy was through a mechanical interface, in conventional methods the goal was to achieve a mechanical vibration, In the piezoelectric material by the vibrations of the material by the percussion type sounds and the properties of polycrystalline materials converts sound by a mechanical interface into pure electrical energy. later on using convertors and Transformers eased the way of converting sonic energy into other typed of energy and Vice versa. Among these systems sonic engines can be noted. this engine hast the ability to convert the wind energy to the sonic energy through a pipe resonator. sonic engine means that by the aerodynamic effects converts wind power to sonic energy. In this way, instead of using mechanical movements as energy converters, systems such as Converters, transformers and heat refrigerators were used, which the main point in the process of energy conversion was the absence of mechanical interfaces in the reverse mechanism of the sonic system. [10](fig 4,5,6,7)

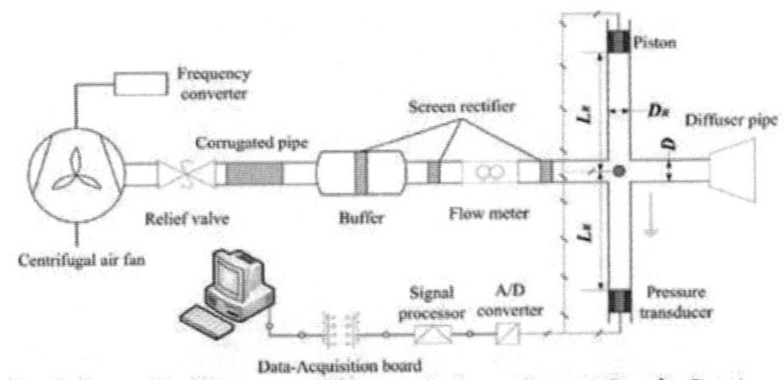

Figure 4. Schematic diagram of the prototype Smart Sonic System (sonic energy converter without mechanical inference)[10]

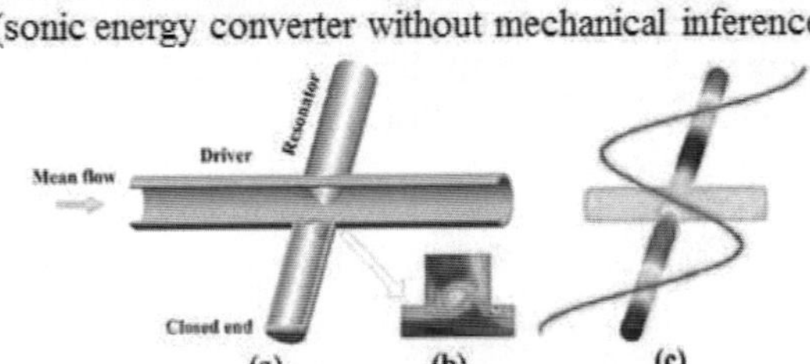

Figure 5. a. engine that starts audio stream b. Eddy current c. Distribution of sound pressure in the pipe[10]

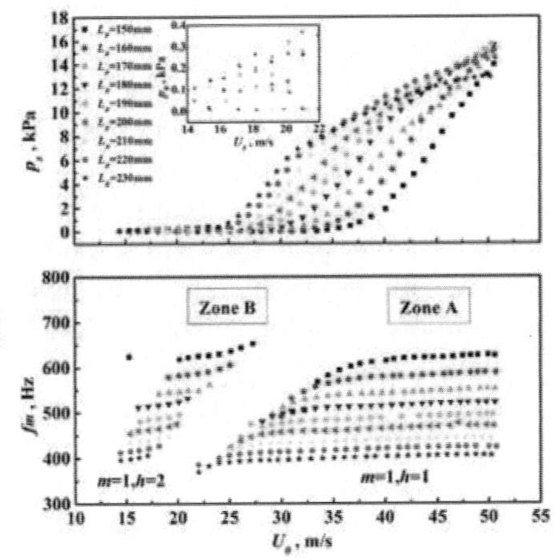

Figure 6.A variety of sound pressure at different points in the audio amplifier tube [10]

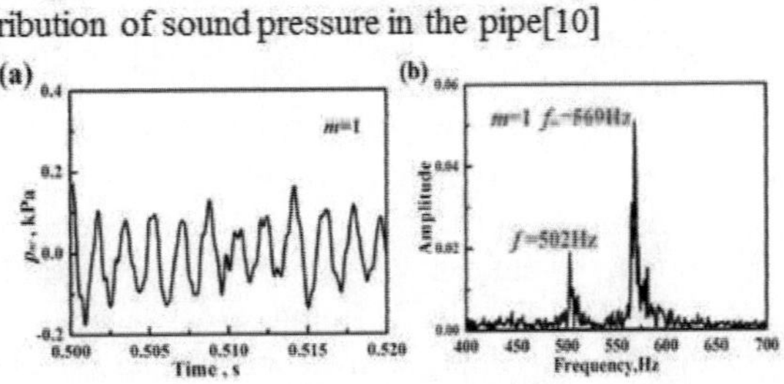

Figure 7.a.The pressure releases wavelength in different time periods b. Spectral analysis of frequencies[10]

4. Principles and approaches to reduction in noise pollution and energy production in architecture

4.1 Designing smart sonic systems (noise absorption and electrical energy generation)

Among all the different types of energy in the universe, from Non-renewable to renewable energy sources, Wind energy, solar, geothermal, wave and fossil fuels were used more often in architecture and other fields. In the meantime, sonic energy and its potentials in optimizing energy usage in architecture and constructions were less used. When an object vibrates creates an Unstable waveform state, when the waves are more distanced from its source of production, they are gradually destroyed and absorbed by the environment. The phenomenon that Stimulates the ear (with mechanical vibration) is called sound. And the space that this phenomenon happens is called and acoustic field. The main goal in design the following system is to aim to reduce noise pollution by absorbing Obtrusive sounds and using its potentials to optimize energy consumption in buildings. The following design has to main resources. 1. internal sound energy sources in buildings itself like the noise from Uninsulated

building installations and the movement of the users in the building and also the sound itself when the users are talking (In office and commercial buildings that have a lot of users at certain hours) 2. External sound energy sources which this sources are the main in the buildings energy supply. because of large facades these resources have a wide surface impact level. And the façade will be the most effective factor in absorbing sound energy by the designed system. Smart Sonic Facades will have a different from the traditional optimized smart thermal facades. Optimized sonic building forms are one of the factors which can interact with the building itself to multiply the sound Absorptions and can be used to optimize the process of generating energy. The Functional mechanism of the designed system Is as follows :1. Guided noise in crowded urban areas to building facades.2. absorption of the noise pollution which are guided to the facades by the optimized sonic forms and the smart sonic system placed in the 2nd layer of the system .3. and in the end absorbed Noise pollution is converted into electrical energy by the mechanical movement as a model for creating a magnetic field to produce electrical energy.
In short we have 3 types of Energy Conversion in the designed mechanism Sound energy into mechanical energy, mechanical energy into electromagnetic energy and electromagnetic energy into electrical energy, absorption and the guiding noise by the optimal sonic form (Such as conical forms part of the cone, the cone of hyperbolic paraboloid) will lead to the movement of the elastic membrane of the system and by that it will move the hydraulic jack , which the hydraulic jack itself has magnets in its end which by its forth and back movement in the main axel of the rotated coils will create an magnetic field which leads to generating electrical energy . the elastic membranes can be wet in order to improve the movement of sound. Because sound moisture has a direct relation with each other and the moisture in the membranes can help to guide the sound through the cone. The strength of the magnets and the number of turns of the coil is also an important factor in the design.(fig8,9)

5. Conclusion

Now days with the high potentials of sound energy in architectural design, it is a wise move to design architectural elements with the approach of using sound energy in it. absorption of noise pollution by the Smart Sonic System that is used in the exterior facades and also in interior parts will help us to achieve a more sound optimized design for our buildings. Using this system in double skin facades (exterior skin as an element for guiding the noise through the system and the internal skin for the absorption of the noise pollution) can help us archive the maximum amount of efficiency specially in the denser and Populated areas. Guided, absorption and energy production are among the principals of designing with the approach of using sound energy. Using this system with its different types and forms in the walls of the urban highways and using the potential of the sound energy which is often found in the modern and developing cities, can help us generate electrical energy by optimizing and its maximum efficacy levels. This can lead to the point that the building which are equipped with this design can harvest energy not only for themselves but in a larger scale. can produce so much electrical energy that can save and transfer them to the cities energy plant. Due to the increasing development of technology and industry in developing and developed countries, noise pollution will also increase with the Urban sprawl and this shows us this using this system is extremely Vital and important for creating an ideal Sonitopia.(fig10,11)

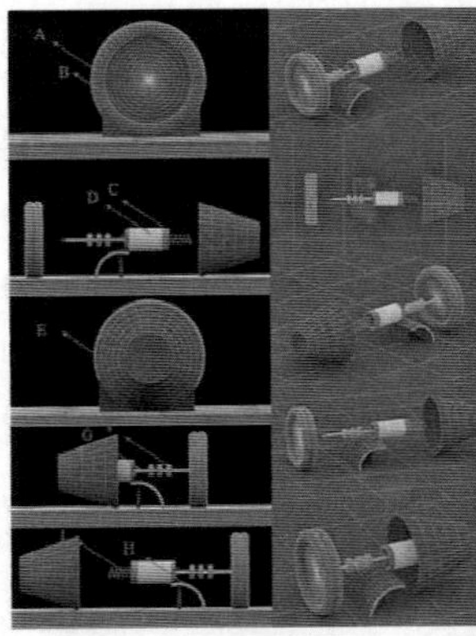

Figure8. A. Elastic tube to keep elastic membrane
B. Elastic Membrane vibrated by the energy of the absorbed sound
C. Pipe guiding hydraulic jack
D. coil Rotated around the pipe to generate a magnetic field
E. Conic form and the support for the tail part
F. Magnets to create a magnetic field (by moving in the middle of the coil pipe)
G. Pipes attached to the elastic membrane to move the hydraulic jack
H. end Spring for the back and Forth Movement of the jack and energy generation by reciprocating motion
I. The main support base [Author]

Figure10. Analysis of the optimal form for the Sonic façade
A. Using different optimal sonic forms for guiding the noise
B. The second layer is added and it includes the smart sonic system in itself
C. Multiple optimized sonic forms for guiding the noise through the system and generating energy
D. Sonic forming with the aim to maximize the speed of the Sound waves to produce more energy
[Author]

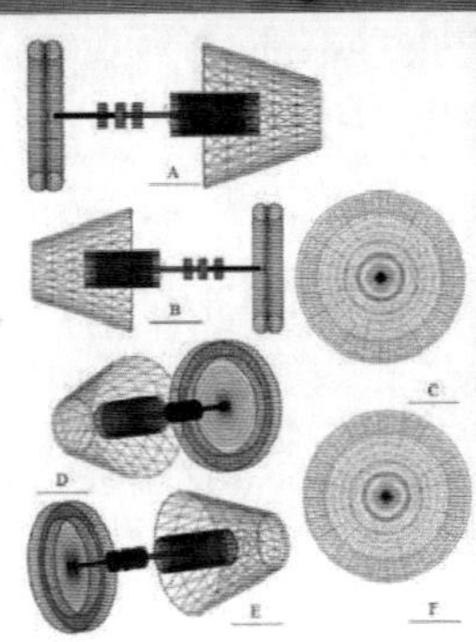

Figure9. Schematic design of the high-performance Sonic system[Author]

Figure11. Fabricated Prototype and a mechanism Analysis mechanism of the Smart Sonic Systems[Author]

6. References

[1] Stankeviciute, L. and Criqui P. (2008) "Energy and Climate Policies to 2020" : The impact of European"20/20/20"approach," International journal of Energy sector Management, vol2, pp.252-273.

[2] Hollinger, Keith H. (2008), Trade Liberalization and The Environmental Study of Natta's Impact in Elpaso, Texas and Juarez, Mexico.Virginia Polytechnic Institute and state university,1-79.

[3] Khalil, s.Inam,z.(2006),Is Trade Good for Environment ? A Unit Root Co-Integration Analysis. Journal of Pakistan Development Review,45(4) ,1187-1196

[4] Ullahkhan,Fan Izhar(2016)"Hybrid Acoustic Energy harvesting using combined electromagnetic and Piezoelectric conversion ,Review of Scientific Instrument ,Vol 87.

[5] Guyomar D,Badel A,Lefeuvre E, et al.(2005)"Toward energy harvesting using active materials and conversion improvement by nonlinear processing",IEEE Transactions on Ultrasonics,Ferroelectrics,and Frequency Control 52(4):584-595.

[6] Li,B. and Hoyou ,J.(2012)"Experimental study on self-powered synchronized switch harvesting on Inductor circuits for multiple Piezoelectric plates in Acoustic Energy harvesting'', Journal of Intelligent material system and structure ,1-10.

[7] Mahdavinejad M, Ghasempourabadi M, Ghaedi H. The role of form compositions in energy consumption of high-rise buildings (case study: Iran, Tehran), Adv Mat Res 2012; 488-489: 175 – 181.

[8] Mohtashami N, Mahdavinejad M, Bemanian M. Contribution of City Prosperity to Decisions on Healthy Building Design: A case study of Tehran, Front. Arch. Res. 2016: 5: 319–331.

[9] Ghiabaklou,Zahra.(2013)"Fundamentals of Building Physics 1: Acoustic ",Tehran: Publication of Jihad Amirkabir University

[10] Sun,D.Xu,Y.Chen,H.Shen,Q.Zhang,X.Qiu,L.(2013) ''Acoustic Characteristics of a mean flow acoustic engine capable of wind energy harvesting : Effect of resonator tube length'', Energy procedia Elsevier ,361-368.

IOP Conf. Series: Earth and Environmental Science **83** (2017) 012022 doi:10.1088/1755-1315/83/1/012022

New Trends on Green Buildings: Investigation of the Feasibility of Using Plastic Members in RC Buildings with SWs

M H Arslan[1] and H D Arslan[2]

1 Civil Engineering Department, Selcuk University, Konya, Turkey
2 Architecture Department, Necmettin ErbakanUniveristy, Konya, Turkey
E-mail: mharslan@selcuk.edu.tr

Abstract. Shear walled (SW) reinforced concrete (RC) buildings are considered to be a type of high seismic safety building. Although this structural system has an important seismic advantage, it also has some disadvantages, especially in acoustic and thermal comfort. In this study, experimental studies have been conducted on RC members produced with plastic material having circular sections to determine structural performance. RC members have been produced with and without 6 cm diameter balls to analyze the structural behaviour under loading and to investigate the thermal performance and sound absorption behaviour of the members. In the study, structural parameters have been determined for RC members such as slabs and SWs produced with and without balls to discover the feasibility of the research and discuss the findings comparatively. The results obtained from the experimental studies show that PB used in RC with suitable positions do not significantly decrease strength but improve the thermal and acoustic features. It has been also seen that using plastic balls reduce the total concrete materials.

1. Introduction

Concrete is a man-made stone composed of natural resources aggregate (sand and gravel combination), a binding agent (cement) and water. The need for new resources in concrete manufacturing is apparent because the aggregate making up 75 % of the concrete's weight is extracted from natural resources and increased building construction increases the concrete demand and decreases natural resources.

In recent years, recycled aggregates made from recyclable plastic, similar solid wastes or building wastes, and waste car tires have been preferred for reducing concrete cost and restricting the use of natural resources. The material used in these studies is mixed in the concrete at a definite ratio, and discussions about how the mechanical characteristics of the concrete changes are made by observing the compressive strength of the manufactured concrete [1], [2]. Materials such as plastic, tires, and recycled aggregates that are substitutes for the aggregate very negatively affect the compressive strength of the concrete if their usage ratios exceed specific ratios.

Producing structural members using plastic space formers, which is the newest method for structural members, was developed in Europe and found an application area in Europe and the USA. The Bubbledeck and Cobiax firms first discovered this innovation and started to use hollow PB with diameters of 18-45 cm inside RC floor slabs [3]. These new designs were first applied by these two companies. The weight of the building decreased significantly because of the decrease in concrete consumption with this invention. Experimental studies in accordance with Eurocodes were performed in research centers of several European countries. In the literature [4], research on this concept has

focused on the shear load carrying capacity and the punching strength of the RC slab section because the space former affects the shear capacity strength. According to these studies, due to these hollow balls, the weight of the floor decreases by 35 %. Moreover, the thermal strength decreases by 17-39%, meaning that the hollows balls significant contribute to heat insulation. In the literature [5], using the balls in a flat slab with thicknesses ranging between 20 and 50 cm is common.
The main goal of using plastic material is based on the stress distribution and the deformation shape of a typical RC section under bending moment. Under bending, the neutral axis (approximately the middle of the section) and its surroundings have no effect on the bending moment capacity or the loading capacity of a RC member. Therefore, the related zero stress region can be removed from the RC section.
Shear walled (SW) reinforced concrete (RC) systems are the primary monolithic housing systems in many countries due to their fast construction technique in which in-situ concrete is poured into shear walls and floor slabs simultaneously. The economic gains from the faster construction than conventional systems increased the demand for this type of construction. Moreover, the standardized dimensions and precast elements that are added to the system increase the quality of the carrying system. In these buildings, all of the vertical carrying members are made from shear walls, and the slab system is flat plate; therefore, the lateral rigidity of the system is greater than in classic frame and shear wall–frame structures. In all of the studies based on lessons from earthquakes [6], researchers mention the importance of larger cross-sectional areas such as shear walls in a structural system. For this reason, shear walled buildings are known as a high seismic safety building type in earthquake prone regions such as Turkey, Greece and Italy. After the earthquakes in Turkey in 1999, more than 100,000 RC buildings were heavily damaged or totally collapsed and more than 250,000 people needed houses .Turkey is located on a significant seismic belt, and the main reason for the damage in classic RC systems that happened after the last earthquakes is the carrying system and less use of the shear walls than needed. (A substantial number of the demolished and heavily damaged buildings had no shear walls). For this reason, the earthquake performance of the SW buildings carrying systems that consist of shear walls and additional architectural elements is considerably better compared to the buildings with RC frame systems; the carrying system configuration makes the buildings safe.
In SWRC buildings, where slabs are mainly subjected to the bending moment, shear walls are subjected to especially high bending moments with shear forces and low axial loads. The effects of the shear force and axial force on the section are usually negligible, especially in RC sections such as slabs under the effect of bending moments. Therefore, such a contraction in the neutral axis will not substantially change the behaviours of elements such as floors. However, the behavior of the shear walls, which are the most critical members under earthquake loading, are not as basic as slabs. The main objective is similar to the other studies [7], [8], and the initial aim of this study is to explore the shear wall behaviour produced with PB under lateral loading.
Beyond the seismic advantage of the SW building, this type of system also has some disadvantages, especially in acoustic and thermal comfort. For instance, non-structural components such as facade walls, stairs, landings and partition walls are produced as precast members to expedite the construction phase and can cause some comfort problems in SW buildings. In addition to the acoustic and thermal disadvantages of the SWsystem, several weaknesses with regard to its use can be classified as lack of architectural solutions from being a modular system and having standard formworks, insufficient numbers of heat and sound absorbing elements in the building, details that are omitted to reduce cost by the manufacturing firms and noise and heat problems [9]. In the buildings where SW system are applied, finer sections are created using concrete with low porosity and a high modulus of elasticity, and the sound absorption capacity remains insufficient due to using hollow masonry units inadequately. Joint defects emerge during the suspension process of the precast elements that are applied on the external walls of the SW buildings and suspended on the wall; the quality of the grouting materials and high tolerances of the components considerably exceed the ideal values in manufacturing lead to substantial losses in heat and sound. Various researches about both consumer satisfaction and perceptual evaluations on heat, soundetc. in SW systems were performed. Several studies also question the quality concept based on some visual effect criteria within the scope of the conducted studies according to the inner climate. Additionally, some of the studies emphasized the

quality loss caused by the wrong combination details of the precast wall panels used in SW buildings [10].

2. Research significance and scope

The significance of this study is to produce RC structural members using plastic space formers (PB) that find application areas, especially in SW buildings due to their weak acoustic and thermal comfort. The PB used in the RC section do not negatively affect the structural behaviour under static and dynamic loads (earthquake, wind etc.) and positively change the thermal and acoustic parameters; therefore, the use of plastic ball materials in members of buildings with RC SW systems will be a new technique in civil engineering practice. In this study, experimental studies have been conducted on RC members produced with plastic material with circular sections (PB). The main aim of the study is to eliminate the thermal and acoustics insulation deficiencies in SW buildings, lighten the structural system, reduce concrete consumption and maintain the lateral and horizontal load carrying capacities of the structural members using PB in the RC members. In this scope, research has been performed on the feasibility of using PB with 6 cm diameter in RC systems. RC members have been produced with and without balls to analyse the structural behaviour under static and quasi-static loading and to investigate the thermal performance and sound absorption behaviour of the members. In the study, the structural parameters of the RC members such as slab and shear wall produced with and without balls have been determined, the feasibility of the research has been explained and the findings have been discussed comparatively.

3. Experimental program

In the experimental study, two different test groups with different targets and their corresponding results were analyzed. The targets of the experiments are a) Studying the changes in the characteristics of horizontal and vertical load carrying capacities that are observed due to the use of PB in the shear walls and slabs, b) Determining the extent to which the PB used in the shear walls and slabs improved the heat and sound insulation in the system section compared to the RC sections without balls; and c) Analyzing the pros and cons of using PB in RC members.The ball diameter was selected as 6 cm, and the cross section of a typical section is shown in Figure 1. The balls replaced the neutral axis of the sections. Because the main function of the balls is to create a hollow inside the RC section, the criteria of having a hollow inside and an adequate wall thickness to avoid any deformation during concrete casting are regarded as sufficient. The balls were purchased from toy stores, and no specific dimensions or material manufacturing were requested because of the size of the experimental study. The balls used in the shear walls were kept within the neutral axis during concrete casting.

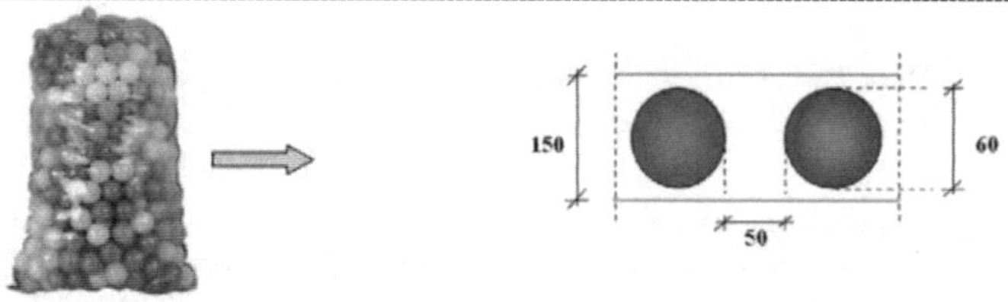

Figure 1. Typical Concrete and Ball Section (mm)

3.1. Shear wall tests

To conduct a behavioral analysis of the shear walls under cyclic lateral loading or quasi-static lateral loading resembling the earthquake effect, two specimens at full scale were tested. The shear walls with 150 mm thickness were produced with and without balls, and these specimens were called S_1 and S_2, respectively. The concrete used in the shear walls had a compressive strength of approximately 29 MPa. According to TEC-2007 (Turkish Earthquake Code, 2007) [11], the minimum comprehensive strength of concrete must be 20 MPa, and the minimum tensile strength of the reinforcement bars must be 420 MPa. The RC shear wall was 150×1050 mm in section. Longitudinal deformed bars of 16ϕ8 mm and a mesh reinforcement horizontal web bars of ϕ8/150 mm were used in the shear wall. The test samples produced with and without balls and the reinforcement layouts of the SW (dimensions in mm) are given in Figure2. The shear walls were cast horizontally and then lifted and placed. A wooden mold kept at room temperature for nearly 28 days was used to cast the concrete of the RC shear wall in a vertical position where the reinforcement was placed on both faces of the wall. In the tests carried out on the plane frame model, the out-of-plane behaviors were prevented or ignored. The earthquake-simulating reversed-cyclic loading was applied either from the top story or, at different rates (triangular load distribution), from the story levels representing real earthquake behavior of the building. In the tests, particularly those that used the quasi-static loadings, the load was applied as load-controlled until the yield displacement of the frame system and was then displacement-controlled. The lateral load-top displacement graphs obtained during the tests of the two test samples are presented in Figure3.

3.2. Slab tests

To conduct a behavioral analysis of the slabs under static vertical loading, two specimens at full scale were tested. The slabs with 150 mm thickness were produced with and without balls, and these specimens were called F_1 and F_2, respectively. The length and width of all of the slabs were 2100 mm and 800 mm, respectively, with a clear span of 1950 mm, and these dimensions were kept constant throughout the study. The thickness of the slabs was 150 mm according to the requirements of the TBC-500-2000 [12]. A concrete cover of 30 mm was used for the longitudinal reinforcements. Initially, the slabs were designed as under-balanced. The longitudinal reinforcement ratio was 0,0035. The slabs had same longitudinal reinforcement in the upper and lower sections. In the slab samples, the same concrete and reinforcement steel bars were used as with the shear walls. The balls used in the slabs were kept within the neutral axis during concrete casting. For this purpose, the wires passing through the balls' axes were fixed on the formwork heads to suspend the balls.

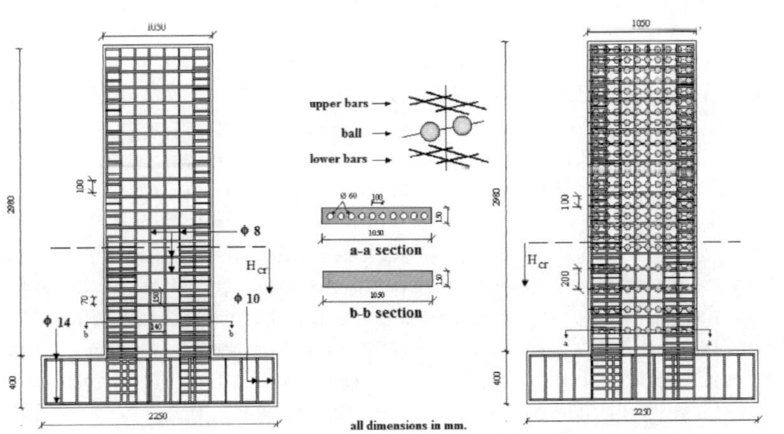

Figure 2. Reinforcement Layouts of RC Frames and Shear Walls (S_1 and S_2) (dimensions in mm)

The simply supported F_1 and F_2 slabs were tested under monotonically increased loads, using a 500-kN-capacity rigid steel loading frame. The slabs were instrumented to measure the applied load and the midspan deflections. The loads were applied at the mid-point of the slab. The test samples F_1 and F_2 produced with and without balls and the vertical load-displacement graphs obtained during the tests of the two test samples F_1 and F_2 are presented in Figure 4.

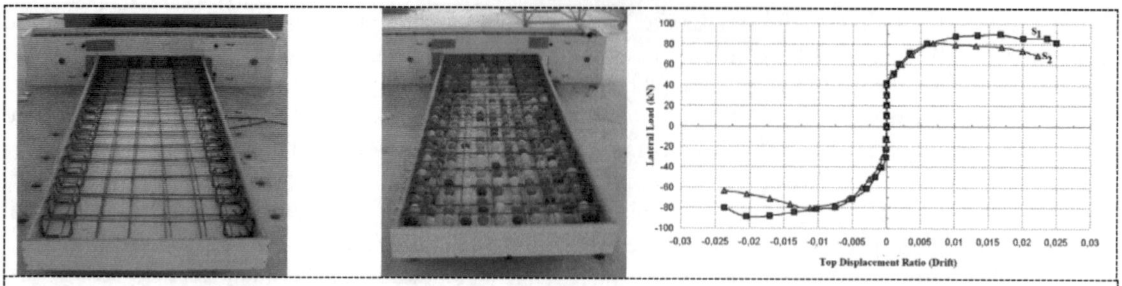

Figure 3. Test Samples (S_1 and S_2) and Base Shear-Top Story Displacement Envelope Curves (S_1- S_2)

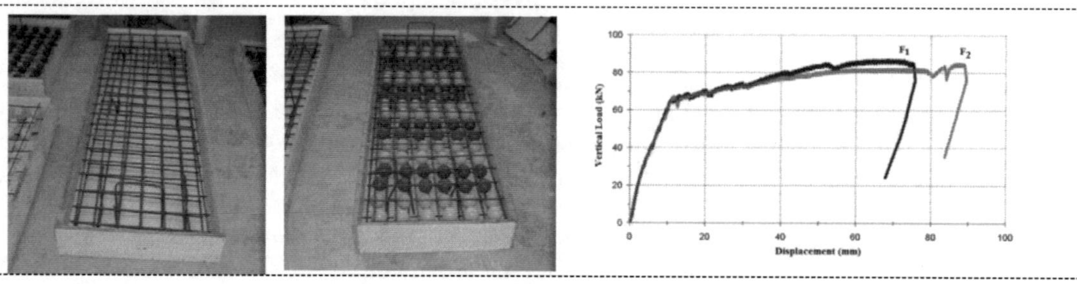

Figure 4. Test Samples (F_1 and F_2) and Load-Displacement Capacity Curves of the Slabs (F_1 and F_2)

3.3. Thermal test

After the results of the loading experiments for the first target were deemed satisfactory, the second phase of the experimental study began to obtain the acoustic and thermal characteristics of the specimens. The members produced with and without balls are shown in Figure 1. The main object of the thermal tests presented here is to investigate the effect of the balls on the thermal transmittance of the RC members. For this aim, six different specimen walls with 150 mm thickness were manufactured both with and without balls. The walls were 1200 x 1200 mm in plan. The concrete used in the sample (M_1 and M_2) walls had a compressive strength of approximately 30 MPa. The average data that were calculated in the experiments and the behaviors of the six samples under the influence of heat were compared and interpreted. The data showed that the thermal conductivity value of the section changed depending on the inner structure of the sample and the quantity of hollows. Figure 5 shows the M_1 and M_2 sample performances under the influence of heat. The thermal conductivity (k) is the main parameter of thermal performance. According to the experimental results in the literature [13], the aggregate volume fraction, the moisture condition, the age of the concrete, the mortar conductivity, the W/C ratio and the cement paste are the main factors that affect the conductivity of concrete. Because the samples were produced, stored and tested under equal conditions, all of the parameters of the samples were regarded as equal. The mass per unit volume of the sample without balls, M_1, was 2257 kg/m3, and the mass per unit volume of the sample with balls, M_2, was calculated as 1024 kg/m3. The density of the section and the ratio of the hollows and/or porosity in the section affect the heat conductivity value. The balls used in the RC section may be expected to ensure effective heat insulation because of their distribution in the section and smaller diameter. In other words, a high number of pores is more appropriate than a low number of pores. However, the properties and locations of the balls should be constant due to the reasons mentioned previously. The heat storage property of the construction component is related to the heat transmission coefficient of the material. Thus, the thermal properties of all types of materials used as construction components should be analyzed in detail when performing thermal comfort studies. When the values obtained from the heat tests were analyzed, a decrease in the thermal conductivity value from 1,974 to 0,626 can be seen as an increase in the RC section to the desired levels of thermal comfort. According to the table, the thermal conductivity variations of the samples showed a similar trend as their density variations. According to Turkish Standards TS 825 [14], the thermal conductivity (k) is given as 2,5

W/mK and 1,1 W/mK for normal and lightweight RC materials, respectively. In this study, M_2 had a lower thermal conductivity (k) than the value for lightweight concrete given in the related code.

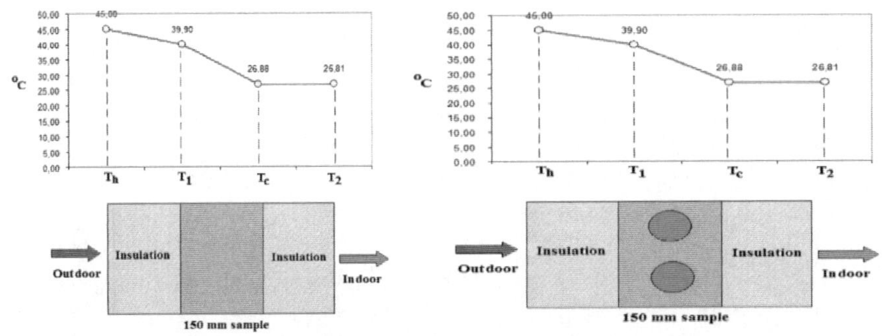

Figure 5. Thermal Transmittance Through M_1 and M_2

3.4. Acoustic test

Acoustic measurements of the sound insulation in the RC members with and without balls were performed [15]. The difference in the sound (noise) levels from one side of a wall to the other indicates the sound transmission loss through the wall. In this experiment, the sound transmissions of the samples were obtained, and the sound absorption coefficients of the samples were measured under the same humidity conditions. The relative humidity was 50 %, and the frequency of the sound was between 100 and 3150 Hz. In the literature [16], the same method has been used in acoustic measurements of sound insulation.Six specimen walls with 150 mm thickness were manufactured both with and without balls. The walls were 120 x 120 cm in plan. The concrete used in the samples walls had a compressive strength of approximately 30 MPa. In Table 1, the sound absorption values are given for A_1 and A_2, respectively. For example, if the sound generated inside a room is 90 decibels (dB) and 40 db is measured on the other side of the wall (adjoining room), then a reduction of 50 dB is achieved. Acoustic tests relate the sound loss through a wall at various frequencies. The results are averaged to provide a single absolute value number. This rating system is necessary to compare other wall systems to a specific wall design. This absolute value is known as the Sound Transmission Class (STC). STC is a rating system used to estimate the sound insulation properties of a wall. According to the test, the average values of A_1 and A_2 are 52,75 dB and 73,25 dB, respectively. In Europe, some countries have detailed regulations [17] for STC values with four classes: A, B, C and D. The A and B classes (approximately 63 and 58, respectively) represent higher levels of acoustic comfort. For instance, the Ontario Building Code [18] requires an STC rating of 50 as a minimum acceptable value and STC of 55 in specific areas. Due to changing lifestyles, i.e., condominium living, many builders prefer to design for STC of 55 or more if end users are demanding and willing to pay for higher quality.In addition, STC findings are based on laboratory results under ideal working conditions, but on-site construction conditions are not the same. Therefore, wall assemblies constructed in the field have significantly lower STC than the laboratory ratings.

4. Results and conclusion

In this study, different experimental studies were conducted on RC members produced with PB. First, RC members produced with and without balls were analyzed under loading. Then, the thermal performance and the sound absorption behavior of the members were explored experimentally. The following are the research findings from this study.

Table 1. Measure Sound Levels for Different Frequencies in A_1 and A_2

Frequencies (Hz)	Sound levels inside of the wall (dB) A_1 and A_2		Sound levels outside of the wall (dB) A_1 and A_2		Sound Absorption (dB)	
100	81.77	81.81	34.87	14.41	46.9	67.4
125	83.47	83.51	34.97	14.51	48.5	69.0
160	84.87	84.91	33.97	13.51	50.9	71.4
200	86.27	86.31	34.37	13.91	51.9	72.4
250	86.77	86.81	35.07	14.61	51.7	72.2
315	87.47	87.51	35.37	14.91	52.1	72.6
400	88.67	88.71	35.57	15.11	53.1	73.6
500	89.67	89.71	35.87	15.41	53.8	74.3
630	90.27	90.31	35.37	14.91	54.9	75.4
800	92.37	92.41	34.97	14.51	57.4	77.9
1000	92.97	93.01	34.27	13.81	58.7	79.2
1250	93.17	93.21	35.17	14.71	58.0	78.5
1600	92.77	92.81	36.87	16.41	55.9	76.4
2000	93.07	93.11	39.07	18.61	54.0	74.5
2500	92.77	92.81	44.07	23.61	48.7	69.2
3150	92.67	92.71	45.17	24.71	47.5	68.0

- Using 6 cm diameter PB in RC shear walls with 15 cm thickness, the lateral load carrying capacity of the specimens decreased by approximately 1-7 % at the failure loading level. At the yielding and maximum loading levels, the specimen behavior was very similar. The addition of balls in the shear walls decreased the initial stiffness but had no effect on the final stiffness. The ductility of the walls was approximately equal. The load-displacement curves are used to determine the general behaviors and strengths of the specimens. The addition of balls in the specimen did not significantly change the strength or ductility. Therefore, the PB did not considerably affect the SW behavior.
- The vertical load carrying capacity did not change with the presence of the balls in the slab members. The inertia rigidity and the seismic energy absorption of the two sections was the same in the slabs. The cracks appeared at the tension region of the slab when the concrete stress at the extreme tensile fiber reached the flexural tensile strength of concrete. The slab specimens showed the same crack development up to failure.
- The thermal conductivity coefficient (k) in the reference section was calculated as 1,974 W/m^2K; the same coefficient in section with balls was 0,626 W/m^2K. According to the tests, the thermal conductivity variations of the samples showed a similar trend as their density variations, and decreasing the density improved the sound and thermal insulation performance. The experiments showed that PB in RC increased the absorption coefficient of these walls.
- Acoustic tests measured sound loss through a wall at various frequencies. For all of the frequency levels, the sound absorption level was very different in the two members. For instance, for the reference section produced without balls, the STC coefficient was approximately 52,75 decibels (dB). However, the section with balls had a coefficient of 73,25dB. The RC wall produced with balls had a higher level of acoustical comfort.
- As a result of the study, the thermal and acoustic comfort conditions will be improved for the residents and a positive contribution to a sustainable environment will be ensured due to the use of the balls inside the slabs and walls in SWwork systems.

In this study, structural and comfort parameters have been investigated in RC sections such as slabs and shear walls produced with and without balls. The results obtained from the experimental studies show that 6 cm diameter PB used in RCSW and slabs with suitable positions do not significantly decrease the strength but show important improvements in thermal and acoustic features. In future work and applications, manufacturing PB with a diameter of 6 cm located inside the formwork members in recycling facilities will reduce the production cost of construction. Moreover, using plastic waste, which has the longest degradation period, as a raw material significantly contributes to the protection of our natural resources, the prevention of environmental pollution, affordability and energy conservation.From the environmental aspect of the study, reducing the harmful effects of concrete and cement production on nature and creating an eco-friendly carrying system over the long term by manufacturing the PB using plastic waste are indirect outcomes of this method. At this point, comparing the cost of the PB with the concrete cost will not be as meaningful; likewise, the cost of raw materials will be minimized by manufacturing PB by recycling methods, particularly for the purpose of recycling plastic waste.In addition, economic solutions can be found in foundation systems by decreasing the concrete consumption and decreasing the weight of the building. Other than offering greater safety to the resident, the weight of the carrying system will be decreased by reducing the concrete inside the slabs and walls. Being a construction manufacturing system with a broad application area, the development described here will make a significant contribution to the construction industry and thereby the national economy.

5. References

[1] Giacomo M and Tarun RN 2016 Structural Concrete Made with Recycled Aggregates for Sustainable Concrete Design, Fourth International Conference on Sustainable Construction Materials and Technologies http://www.claisse.info/Proceedings.htm,USA
[2] Kleijer AL Lasvaux S Citherlef S and Vivani M 2017 Product-specific Life Cycle Assessment of ready mix concrete: Comparison between a recycled and an ordinary concrete, Resources, Conservation and Recycling, 122,210-218
[3] Joseph AV 2016, Structural Behaviour of Bubble Deck Slab, M-Tech Seminar Report
[4] Chung JH, Choi HK and Lee SC 2011 Shear Capacity of Biaxial Hollow Slab with Donut Type Hollow Sphere", Procedia Engineering, 14, 2219 -2222.
[5] A Churakov 2014 Biaxial hollow slab with innovative types of voids, Construction of Unique Buildings ad structures, 6, 21,70-88.
[6] Arslan M H and Korkmaz HH 2007 What is to be Learned from Damage and Failure of RC Structures During Recent Earthquakes in Turkey? *Engineering Failure Analysis*, 14, 1-22.
[7] Polak M A 2005 Punching Shear in RC Slabs. American Concrete Institute, Farmington Hills, MI 302
[8] Terec LR Terec MA 2013 The bubbledeck floor system: A brief presentation", Constructii, 2, 33-40.
[9] Topsun M Dincer K 2011 Modelling of a thermal insulation system based on the coldest temperature conditions by using artificial neural networks to determine performance of building for wall types in Turkey, *International Journal of Refrigereton*,34,1,362-377.
[10] Wojdyga K 2009 An investigation into the heat consumption in a low-energy building *Renewable Energy*, 34, 2935–2939.
[11] Turkish Earthquake Code (TEC) 2007 Regulations on Structures constructed in Disaster Regions, Ministry Of Public Works And Settlement, Ankara.
[12] TBC-500 2000 Requirements For Design and Construction of RC Structures, Ankara
[13] Kook-Han K Sang-Eun J Jin-Keun K and Sungchul Y 2003 An experimental study on thermal conductivity of concrete, *Cement and Concrete Research*, 33,363-371
[14] TS-825 2009 Turkish Code of Thermal Insulation Requirements for Buildings, 2009 Ankara.
[15] EN TS ISO 140-4, EN TS 2382. Acoustics-measurement of sound insulation on buildings and of building elements part 4: field measurements of air borne sound insulation between rooms.
[16] Binici HAksogan O Bakbak D Kaplan H and Isik B 2009 Sound insulation of fibre reinforced mud brick walls, *Construction and Building Materials*, 23,1035-1041.
[17] Garg N Sharma O and Marj S2011 Design Consideration of Building Elements for Traffic and Aircraft Noise Abatement, *Indian Journal of Pure and Applied Physics*, 2011, 49,437-450.
[18] Ontoria Building Code, http://ccmpa.ca/wp-content/uploads/2012/02/7-SndProp.pdf

Application of Waste Heat Recovery Energy Saving Technology in Reform of UHP-EAF

J H Zhao[1], S X Zhang[1], W Yang[1] and T Yu[1]

1 School of chemical engineering & energy, Zhengzhou University, 100 Science Avenue, High-tech zone, CN
E-mail: 35860031@qq.com

Abstract. The furnace waste heat of a company's existing 4 × 100t ultra-high-power electric arc furnaces is not used and discharged directly of the situation has been unable to meet the national energy-saving emission reduction requirements, and also affected their own competitiveness and sustainable development. In order to make full use of the waste heat of the electric arc furnace, this paper presents an the energy-saving transformation program of using the new heat pipe boiler on the existing ultra-high-power electric arc furnaces for recovering the waste heat of flue gas. The results show that after the implementation of the project can save energy equivalent to 42,349 tons of standard coal.The flue gas waste heat is fully utilized and dust emission concentration is accorded with the standard of Chinese invironmental protection, which have achieved good results.

1. Introduction

Iron and steel industry as an important pillar industry of the national economy. It provides basic raw materials also consume a lot of energy at the same time. According to statistics, China's annual output of steel is more than 600 million and energy consumption accounts for about 15% of the total energy consumption [1]-[5], but the energy efficiency is only 30%-50% [6]. The effective use of energy has become the key to China's economic development. The Iron and steel enterprises have a wealth of waste heat resources. A lot of waste heat generated in the production process is an important secondary energy. With the waste heat resources gradually being taken seriously, the waste heat recovery technology has been applied to steel industry. It is of great significance for conserving resources, improving environment quality, enhancing economic efficiency, realizing the circulation and optimal allocation of resources and sustainable development [7]. This article takes the example of a company intends to build an energy-saving technological transformation project of waste heat for four 100-ton steelmaking ultra-high-power electric arc furnaces to analyze and introduce.

2. Process status and problems

Iron and steel company has reached an annual output of 2.33 million tons of steel, steel plate 1.86 million tons of production scale, rolling system of industrial kiln, the annual consumption of about

150,000 tons of heavy oil. There are four VD (vacuum degassing) furnaces, supporting the installation of three fast gas boilers, fuel for natural gas. According to the statistics, a ton of steel's consumption of steam is 0.19GJ. According to an annual output of 4 million tons of steel production, consumption of steam is 80GJ. Figure 1 is no waste heat utilization of the furnace dust removal system process, the waste heat produced by waste production in turn through the settlement room, water-cooled flue, cooler, booster fan into the mixing chamber and from the roof catchment and the lid of the flue gas Mixed int→o the dust collector, through the fan discharge.

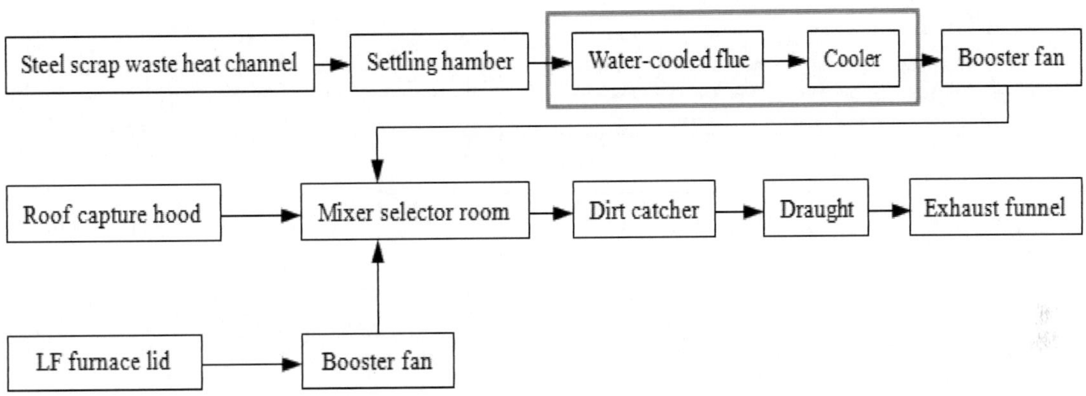

Figure 1. No waste heat utilization of the furnace dust removal system process

In recent years, although some of the enterprises in the fuel heating furnace for some transformations, because the fuel industry furnace process is simple, heavy oil burning directly in the kiln, resulting in heavy oil consumption (burning is not complete, low thermal efficiency), combustion products pollution of the environment, the degree of automation is not high enough, the labor intensity and the quality of heat treatment is not stable and some drawbacks difficult to overcome. At present, four 100t ultra-high power electric furnace flue gas waste heat is not carried out directly and discharged, electric furnace smelting generated in a large number of flue gas heat can not be recycled. At the same time, with the increase of the electric furnace steelmaking equipment production capacity, molten iron into the amount and oxygen intensity, the dust removal capacity of electric dust removal system is also weakened.

3. Transformation program

3.1. Overview of the program

In view of the current four 100t ultra-high-power electric arc furnaces generated flue gas waste heat is not used and discharged directly of the situation, we decided to add a waste heat recovery system to make full use of electric furnace flue gas waste heat. Because the medium in heat pipe through the phase change heat transfer, the heat pipe has good heat transfer performance, high efficiency and it can maximize the efficient use of waste heat. Based on the above advantages, we decided to use heat pipe waste heat boiler for the four sets of electric furnaces.After analysis and research, replacing the water-cooled flue and cooler with heat pipe heat exchanger, then the high temperature flue gas can be used by the waste heat utilization system into high temperature and high pressure steam to meet the daily life and production needs.

3.2. Waste heat utilization system process

According to 100t electric furnace waste heat process parameters and the use of requirements, there are waste heat utilization of the furnace dust removal system and furnace waste heat utilization process shown in Figure 2, Figure 3.The heat pipe heat exchanger in Figure 2 replaces the water-cooled flue and cooler of Figure 1. The industrial water in Figure 3 after the softening of water into the deaerator, and then pressurized by the pressurized water pump into the hot water preheater, after preheating into the heat pipe steam generator, in addition to the absorption of heat to vaporize the steam into steam, And finally through the heat collector outward delivery.

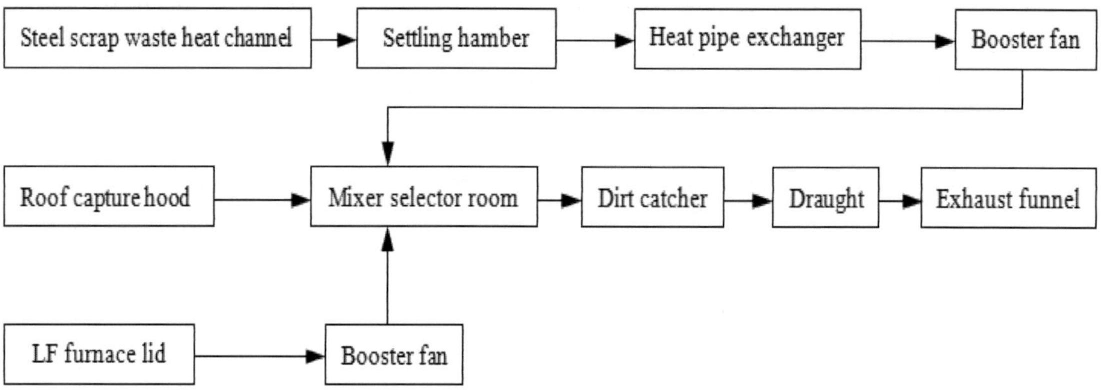

Figure 2. Waste heat utilization of the furnace dust removal system process

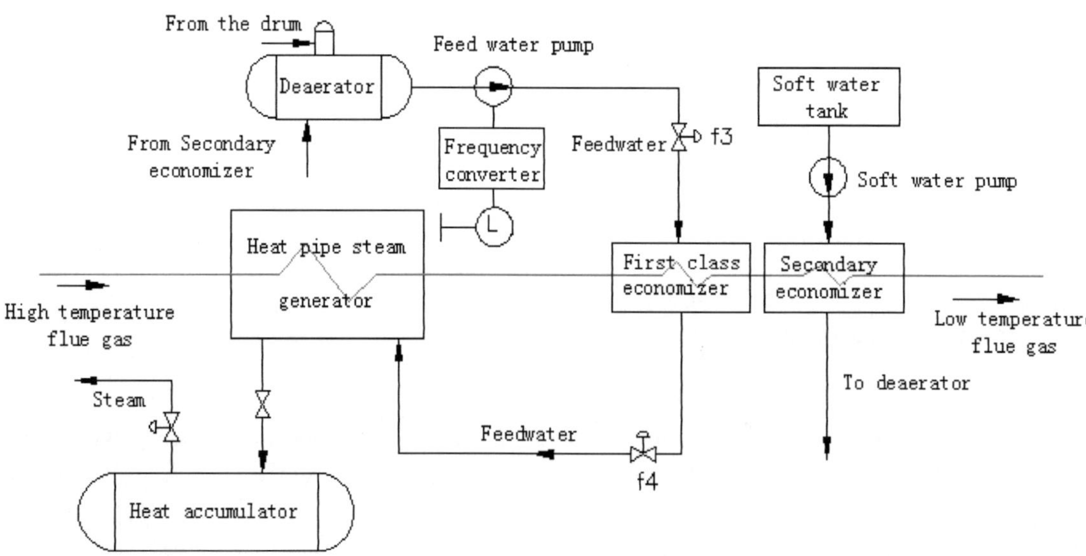

Figure 3. Main diagram of waste heat utilization process of electric furnace

3.3. Program feasibility analysis

3.3.1. System features

- The use of heat pipe will be fully separated from the hot water circulation system and the flue gas heat fluid, optimizing the working environment of the heat transfer element.

- The heat pipe elements in the equipment are independent of each other, and the hot and cold medium is subjected to secondary wall heat transfer. The fluid is not interlocked and does not affect the continuous and stable operation of the system when one or more heat pipes are damaged. At the same time, water, steam will not be due to heat pipe damage into the hot fluid.
- Designed to adjust the heat transfer area at both ends of the heat pipe can effectively adjust and control the wall temperature, to prevent low temperature acid dew point corrosion.
- The heat transfer process of the system belongs to the natural circulation, does not need any external power drive, the operation is simple, the maintenance is convenient, the work is reliable, the failure rate is low.

3.3.2. Advantages of heat pipe economizer

Table 1. Comparison between heat pipe economizer and cast iron economizer

Project	Unit	Heat pipe type	Cast iron type
Flue gas inlet temperature	℃	250	250
Flue gas outlet temperature	℃	190	190
Feed water inlet temperature	℃	20	20
Water supply outlet temperature	℃	50	50
Flue gas volume flow	$N \cdot m^3/h$	2988	2988
Gas volume flow	t/h	2.2	2.2
Minimum wall temperature	℃	136	35
Heat recovery	kW	78	78
Flue gas pressure drop	Pa	38	98
Unit volume of heated area	m^2/m^3	106.12	35.89
Dimensions	mm	1670×450×250	1350×600×700
Total capacity	m^3	0.188	0.608
Total quality	kg	320	720

As can be seen from Table 1, the same conditions of the heat pipe economizer and cast iron economizer in the following areas are quite different:

- Heat transfer system. Heat pipe economizer heat transfer strength of cast iron economizer 7.45 times, K heat pipe: K cast iron = 145.3: 19.5 = 7.45: 1;

IOP Conf. Series: Earth and Environmental Science **83** (2017) 012023 doi:10.1088/1755-1315/83/1/012023

- Minimum wall temperature. Heat pipe economizer for the 135 ℃, while the cast iron economizer only 35 ℃, a difference of 100 ℃;
- Smoke side pressure drop. Heat pipe economizer for the 38Pa, cast iron economizer for 98Pa;
- The volume of heat transfer per unit volume. Heat pipe economizer for the cast iron economizer 29.6 times;
- Total capacity. Heat pipe economizer compact structure, the total volume of cast iron economizer 1/3;
- Total quality. Heat pipe economizer less metal consumption, the total weight less than cast iron economizer 1/2.

In summary, the use of heat pipe economizer for boiler flue gas waste heat recovery, whether it is advanced technology or work reliability, service life and other aspects are significantly better than the traditional cast iron economizer.

3.3.3. Heat pipe waste heat boiler

Heat pipe waste heat boiler application heat pipe as a heat transfer element, absorb the higher temperature of the flue gas waste heat used to produce steam, the steam generated can enter the steam pipe network. The kind of waste heat utilization form, has the following performance characteristics:

- Has a super heat characteristics: heat transfer efficiency, strength, energy saving effect is remarkable;
- Heat pipe completely isolated heat and cold source, will not produce hot and cold fluid blending;
- Flue gas side for the tube outside the heat transfer, ash easy;
- Start the fast, the adaptability of the load fluctuations, generators or thermal oil furnace load in the range of 30% to 100% change, can produce steam;
- Stable operation, long service life;
- Easy to install, not limited by the installation location, without changing the original process system, structural design and location layout is very flexible;
- Short investment payback period.

Heat pipe heat exchanger as a new and efficient energy-saving equipment, in the steel, petrochemical and other industries have been widely used in waste heat recovery. Which can effectively recover the use of production or equipment in the process of running a variety of sensible exhaust gas, both to enable enterprises to obtain significant economic benefits, but also reduce pollutant emissions, to achieve energy saving,the dual purpose of protecting the environment.Therefore, the project selected advanced heat pipe waste heat recovery equipment to recover ultra-high power electric furnace emissions of waste heat.

4. Transformation effect

4.1. Energy saving evaluation

After the implementation of the project to save energy for the four ultra-high-power electric furnaces configuration of the four waste heat boilers recovery of steam, each waste heat boiler steam production is 14 t / h, steam pressure 1Mpa. After the implementation of the project to increase the energy consumption for electricity consumption, the new power facilities, the average effective power of 2094kw. According to the actual operation of steel-making electric furnace, it takes the actual operation of the project for 330 days, 24 hours a day running. In this calculation, the standard coal coefficient of steam is 0.10857 tce / t steam, and the standard coal coefficient is 0.350 tce / k.kwh.

Energy saving accounting (yearly):

- 4 sets of waste heat boiler recovery steam folding coal = 4 × 14 × 0.10857 × 24 × 330 = 48153 tce.
- the new power equipment consumption discounted standard coal = 2094 × 0.35 × 24 × 330 × 10–3 = 5804 tce.
- after the implementation of the project the actual energy savings of coal:

Year Actual Energy Saving = New Recycled Energy– New Energy Consumption = 48153–5804 = 42349 tce.

After the implementation of this project, the annual savings of energy equals to 42,300 tons of standard coal, waste heat boiler has replaced the company's original three natural gas boilers. After the system has been run, the heat has reached the use of high-temperature flue gas to produce steam for VD vacuum furnace production and other domestic use of steam targets, and after using the new technology of utilizing flue gas waste heat, the original VD vacuum furnace using natural gas boiler as a spare to save a lot of fuel costs, reducing production costs.

4.2. Economic feasibility analysis

The total investment of the project is 62.6458 million yuan. After the implementation of the project, the annual energy savings will be 4.2349 million tons, being calculated at 570 yuan/ton, it can save 2413.89 yuan, the payback period is 5.93 years. The profitability of the project will meet the requirements. The set of financing options can be accepted.

From the financial point of view of the technical program evaluation, the project is feasible. Projects can produce significant economic results, suggesting owners to take a positive investment strategy.

4.3. Environmental evaluation

The dust of the project first uses waste heat boiler to remove dust, and then uses the bag to do dust collection, the received dust, ash are all back to the company's internal production system to recycle iron, etc. It will not cause the environmental pollution.

Under normal production conditions, the amount of industrial waste water produced is less, standard discharge or recycling make a little impact on the environment: the neutralizing treatment of acid and alkali wastewater of chemical water treatment station is being done in the neutralization pond, discharging standards.

5. Conclusion

Through the 4 × 100t ultra-high power furnaces flue gas waste heat utilization of energy-saving technological transformation to waste heat boiler to replace the company's original natural gas boiler, while after using advanced heat pipe waste heat boiler technology, energy-saving technological transformation the exhaust heat is fully utilized. The new 4 sets of waste heat boilers can save a total of 4.2349 million tons per year. At the same time, the implementation of the project makes smoke (powder) dust and CO_2 emissions decreasing significantly, not only significantly reduce energy consumption, improve enterprise product quality, increase business income, promote the healthy and sustainable development of enterprises, but also in line with national industrial policy, helping to ease the government energy supply and energy project construction pressure. It also has a more realistic significance to reduce the waste gas pollution protection environment.

6. References

[1] Chen T, Chen Y, Chen L and Hu Q 2012 *J. The implementation of contract energy management problems and related recommendations. Construction Economy* **1** 67-69

[2] He W D and Zhang K 2013 *J. Decomposition Analysis of Influencing Factors of Carbon Emission in China 's Iron and Steel Industry. Journal of Industrial Technological Economics* **1** 3-10

[3] Dong H Z, Xue H F, Song L H and Zhang Q 2009 *J. Analysis on the variable factors of Energy Consumption Intensity in Iron and Steel Industry. Science Research Management* **3** 132-138

[4] Lin B Q and Liu J H 2001 *J. Estimating Coal Production Peak and Trends of Coal Imports in China. Energy Policy* **1** 512-519

[5] Zhang X and Feng J X 2013 *J. Latest advancement of energy performance contracting (EPC) and discussion on its development in industrial sector. Energy for Metallurgical Industry* **6** 3-6

[6] Zhang M 2015 *J. Waste heat recovery technology in the steel industry application and energy saving potential analysis. Resource conservation and environmental protection* **8** 8-9

[7] Yang W B, Chen Z L and Kong L B 2001 *J. Recycling technical discussion and effect analysis for the waste heat of the flue gas from electrical furnace. Energy forMetallurgical Industry* **1** 47-48

IOP Conf. Series: Earth and Environmental Science **83** (2017) 012024 doi:10.1088/1755-1315/83/1/012024

Carbon Capture and Sequestration- A Review

Akash Sood[1] and Savita Vyas[2]

1 Research Scholar, SoEEM, Rajiv Gandhi Proudyogiki Vishwavidyalaya, Bhopal
2 Assistant Professor, SoEEM, Rajiv Gandhi Proudyogiki Vishwavidyalaya, Bhopal
E-mail: akku.sood@live.com, savita_vyas@hotmail.com

Abstract. The Drastic increase of CO_2 emission in the last 30 years is due to the combustion of fossil fuels and it causes a major change in the environment such as global warming. In India, the emission of fossil fuels is developed in the recent years. The alternate energy sources are not sufficient to meet the values of this emission reduction and the framework of climate change demands the emission reduction, the CCS technology can be used as a mitigation tool which evaluates the feasibility for implementation of this technology in India. CCS is a process to capture the carbon dioxide from large sources like fossil fuel station to avoid the entrance of CO_2 in the atmosphere. IPCC accredited this technology and its path for mitigation for the developing countries. In this paper, we present the technologies of CCS with its development and external factors. The main goal of this process is to avoid the release the CO_2 into the atmosphere and also investigates the sequestration and mitigation technologies of carbon.

1. Introduction

The climate variability and changes are one of the evidence for global warming, it increases the average temperature of the globe. It is due to the increase of greenhouse gas in the atmosphere. CO_2 is the main component in greenhouse gas. The processes in industries are the main reason for the emission of CO_2. It is mostly emitted from the burnings of carbonaceous fuels. Greenhouse gas is also emitted by the natural phenomenon like agriculture and live stocks. The mechanism of nature is to absorb the CO_2 to maintain the biosphere balance in the atmosphere. The emission of greenhouse gas is increased when comparing with the initial revolution of the industries. It is due to the usage of fossil fuels, Thermal power generation, Logistics, and transport. Coal is mainly used in the power sector and it consumes 70% of Indian economy. The Economic growth of India is accelerated to add more 600000 MW by 2030 approximately [1]. The emission of CO_2 must be reduced to minimize the global warming. Nuclear energy, hydro energy, fossil fuels, coal power have to produce power in large amount to accommodate the need of power but clean coal technologies and coal combustion in the efficient and clean methods are required to be developed to reduce the emission level of CO_2. Capture and Storage Technologies are used to reduce the emission of Greenhouse gasses by capturing the CO_2 gas from the viable surface.

Carbon capture and sequestration is a physical process which involves in the capturing of CO_2 and its storage. CCS technology is used to reduce the emission of CO_2 in the atmosphere. The integrated system of CCS has the following process, the CO_2 will be captured and separated from the other gasses. Then it will be purified, compressed and transported to the sequestration site. CO_2 will be injected into the geological surface of the reservoir or it will be stored in the ocean. This review is focused to analyze, study and evaluate the importance of CCS technology to reduce the GHG emission to avoid global warming with the implications of economic. The main objective is to research about the Technology of CCS with power sector to understand the feasibility of the technologies. The

IOP Conf. Series: Earth and Environmental Science **83** (2017) 012024 doi:10.1088/1755-1315/83/1/012024

applicability of the power sector is applied to the demonstration of the requirements to establish the technology. The aim of CCS technology is listed as follows, it enhances the power plant efficiency with the latest technologies to reduce the emission of Carbon dioxide by using the capture technology. CO_2 is captured and it is separated from the gas streams which is emitted from the mixture of combustion gasses. The captured CO_2 will be transported to the underground storage. CO_2 will be stored in underground sedimentary basins, saline aquifers, and coal reservoirs. CO_2 will be stored in the potential areas of the country. Trapping mechanism will be used for the storage process. It is important to ensure the safety of the storage area. The main goal of the CCS technology is to reduce the amount of CO_2 in the environment. CO_2 emitted by the fossil fuels will be stored in secured places for hundreds of years with affordable prices.

2. CCS Scenario in India

Nearly 53.7% of electricity is generated through coal which means direct carbon emission to atmosphere. Around 40% of CO_2 emission is for power generation and further emission by the industrial side and transport will be 15% and 32% respectively. Then 11% of carbon released by the residential and commercial sectors [1]–[3]. These overall percentage of carbon dioxide are not flexible to CO_2 Capture and Sequestration. The CCS is well suited for large industry, power plant and stationary sources of carbon dioxide emission due to economic cost and high capital cost. But the oil refining company, cement production industry and, iron and steel manufacturing combust the fuels and release the carbon. These are all the intensive industries. The cost based on the flue gas cost properties, basically down with high carbon dioxide concentrations and less temperature. The sequestration of CO_2 accumulation in earth is due to unacceptable combustion of fuels such as fossil. Even though the cost invested in fuel acquisition, domestic infrastructure and reserves, the maintenance, and the alternate energy source combustion. The alternate sources are the wind, solar and biomass energy. This CCS technology provides the solution to the dilemma. The solution provided by the carbon emission reduction costs due to the storage of acquired CO_2 and transportation which will increase in helping to control the evaded cost of capture [4].

According to 2007-2008 in India, a particular care had been taken to carbon capture and sequestration field. During the financial year, a various number of researchers sponsored nearly thirty projects depends upon storage or sequestration. CCS technology is one of the interesting processes. By comparing with those years, after 2013 only some researcher's standby in open market. In India, there is no particular CCS policy needs to be expressed. Due to this policy, India lag on R&D dynamic approach. The absence of CCS will cause the cost about 70% increased at 2050 due to the ejection of carbon. In India, CCS is at an initial stage only, because the government does not consider the CCS technology is the key factor for climate and national energy policies [5]. Instead of involving in coordination CCS development process and coordination process, the DST (Department of Science and Technology) started the (ICOSAR) "Indian CO_2 sequestration Applied Research. These CCS projects are always executed in three steps given as absorbing, pre-combustion and post-combustion. The power sector does not yet have demonstration -scale. These levels provide the projects for carbon dioxide capture in 1988 [6]. In India, the four initiating key points are as follows:

- **Capture of pure carbon dioxide streams in Industry**
  The process is very easy to produce a pure stream of carbon dioxide. The examples are fertilizer and gas industries. The shortage of CO_2 at fertilizer plants in India because of the CO_2 usage produced from the production of ammonia. The excess of Carbon dioxide is also used for the urea manufacture. The waste supply of carbon dioxide replaces the generation of this excess CO_2 [7].
- **EOR (Enhanced Oil Recovery) offshore and onshore**
  The CO_2 developing plans are already initiated from the offshore sour gas by the Indian government at Gujarat, Hazira. This recovery is done at the onshore site about 70 km far away. Nearly 1200 tons of received carbon dioxide transportation takes place every day to the oil field. It maintains the pressure and reduces the oil viscosity.
- **CCS technology with power sector and coal**

This process starts with technology generation such as advanced supercritical steam conditions. It includes new power plant design in India, such as UMPP's ("unimpaired performance and less upfront costs").

- **Export CO_2 for foreign EOR activities**
  The UMPP project design is mainly for imported coal such as Mundra. The project integration formed by the shipping terminals [8].

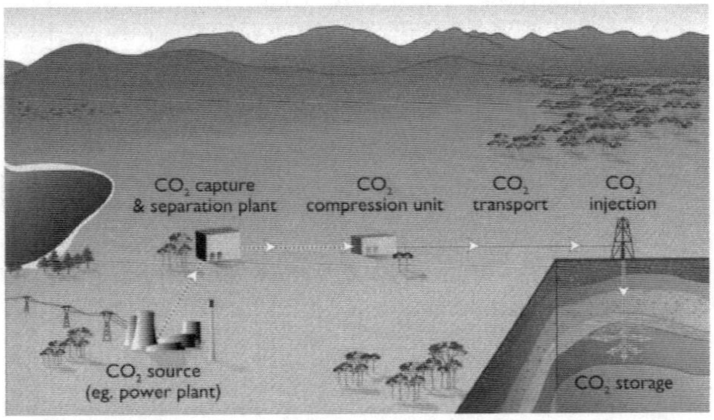

Fig. 2.1 Process flow of geo-sequestration of CO_2 [9]

CCS process has three main steps (as seen in fig. 2.1):

1. **Capture of CO_2**- CO_2 is captured from the industrial sources, power plants and natural gas as they have high content of CO_2.
2. **Transportation**- Transportation is the process of transferring the CO_2 to the storage site.
3. **Geological storage**: In the last step, the transferred CO_2 is saved in geological storage such as in deep saline formations, un-mine-able coal seams, depleted oil/gas fields or gas recovery sites.

The combustion processes used to capture the CO_2 gas are in three modes. They are pre-combustion mode, post-combustion mode and oxy-fuel combustion [10].

3. CO_2 Capture

There are three main sources available for emission of CO_2 gasses. They are from power plants, natural gas, and industrial gas. The most important CO_2 gas emission takes place in power plants. This CCS technology is used to reduce the CO_2 emissions. The pre-combustion technique is used to collect the gas from the coal combustion streams and depose it geographically. So, this method is used to protect the gas to enters into the atmosphere.

3.1. Carbon transportation

Transport the connection between the carbon captured and storage sites. This uses the pipeline process to transfer the carbon. Transport the connection between the carbon captured and storage sites. This uses the pipeline process to transfer the carbon. For transferring gas long distance, this pipeline technique is used. Liquid, gas and solid are the three states of carbon transformation available. The liquid and gaseous carbon are transported by using the commercial scale transport like pipelines, ships, and tanks. The use of ships in transferring gas is the most usable method. There are four types of ships that are used for this transportation. These four types of ships are used to transport the liquefied carbon from the large type of sources such as ammonia plants. The ships then transfer these liquefied carbons to the coastal distribution terminals [11]–[13].

3.2. Carbon Sequestration

This stage is the critical stage in this technology. The emissions of the carbon in the atmosphere are protected by this stage. In this sequestration stage, the captured carbon is stored in the geological site. Based on the economic point of view, the famous sites for storing the carbon is un-mineable coal seams, deep saline formations, enhanced oil or gas recovery (EOR or EGR) and depleted oil/gas fields. There are many techniques that are used for storing the carbon into the storage reservoirs [14].

3.3. Potential sources of GHG emissions

According to the ministry of power, India, there are five division available in India: Western, Eastern, Northern, Southern and North-Eastern. Coal is found in India in the central regions, eastern, and southern region. Lignite is found in big hydro-power and the southern region. But, for more power generation the lignite required to develop in the north-eastern and northern region. Efficient technologies are used in the growth of the power generation. At this time, there is also a need to maintain the GHG emission to protect the environment [10].

The power sector is one of the largest sources for emission of GHG. In the power sector, the fossil fuel-fired power plants are supplied the base load electricity. This is the major part that affects the environment by GHG emissions. There are some solutions to reduce the emission of GHG as listed below [15],

- Waste reduction and timely maintenance are efficient way to increase the efficiency of power plants.
- Recycling the old power plants and application of new technologies.
- Change the base load generation using the fossil fuel power plants to power plants as renewable energy.
- Should use CCS technology for upcoming power plants.

4. CCS Technologies

4.1. Oxyfuel combustion

This technology is used to remove the fuel in the air with oxygen. The air in atmosphere contains 78.08% nitrogen, 20.95% oxygen and the remaining percentage of air contains gasses like Helium Xenon, Neon etc. (inert gases). In this technology, the fuel is added to the existing atmospheric air for combustion. Inert gases in the atmospheric air do not reach the high temperature then the combustion is incomplete. But instead of air if oxygen is used in the process, a higher temperature can be attained. The process of Oxy- Fuel Combustion technique therefore is burning the fuel with pure oxygen. This process acts in the air as primary oxidant air [16]. The process flow of plant is shown in fig. 4.1 the process shows oxy-fuel cement plant, which runs on the oxyfuel technology, where the amount of CO_2 was raised in a rotary kiln by combusting oxygen and fuel to get high concentration of CO_2 flue gas which was sent to CO_2 purification unit.

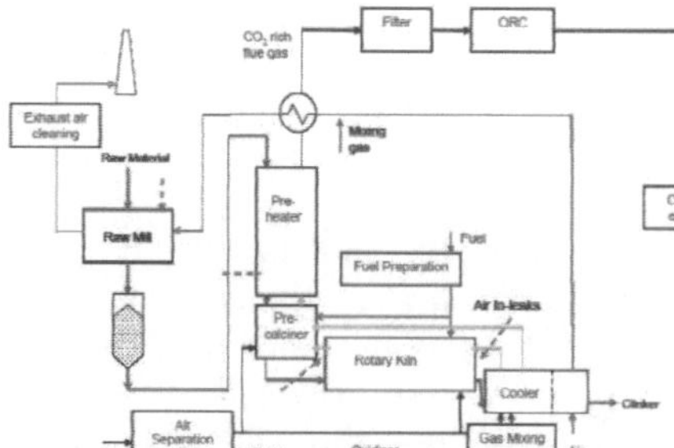

Fig. 4.1 Oxy-fuel cement plant configuration [15]

4.2. Pre-combustion

Pre-combustion is the process to obtain the fossil fuel without carbon. This process is done before the combustion process is completed. First, the coal is oxidized in steam. Then the coal is mixed with the air. This mixture forms the synthesis gas at high temperature. This gas is also referred as 'syngas' and it is the mixture of carbon mono-oxide, hydrogen, CO_2 and a smaller amount of methane. This process is called as gasification. The next level is water-gas shift reaction. In this level, the carbon monoxide and the water are converted to hydrogen H_2 and carbon dioxide (CO_2). In this mixture, the range of the CO_2 is 15-50%. After this pre-combustion method, the CO_2 is captured, transfers and then sequestrated, the fig. 4.2 shows the process of pre-combustion explain the process steps. As compared to the post-combustion method, pre-combustion is used to remove dilute CO_2 at low pressure. At high pressure, this combustion removes rich CO_2 gas before the hydrogen is combusted on the water-gas shift reaction. Pre-combustion is the more efficient technique at the time of availability of more concentrated CO_2. But, more expensive for the base gasification process [9], [15].

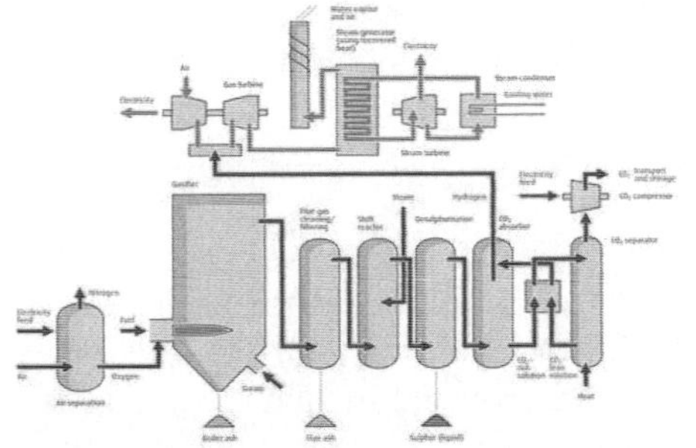

Fig. 4.2 Pre-combustion process [17]

4.3. Post-combustion

The post-combustion method is used to remove the CO_2 from the flue gasses. The post-combustion method in power plants uses the chemical absorption processes with solvents of monoethanolamine (MEA). This process is called as amine separation process. The Fig 4.3 shows the flow diagram of amine process. In an absorber, the MEA solution is added with the absorber. The MEA is used to get the CO_2 by selectively absorbing from the overall CO_2. Then the selective CO_2 is sent to the stripper. In the stripper, pure CO_2 is made by heating the CO_2-rich with MEA solution. After that the generated the CO_2 lean with MEA solution is out from the stripper. This gas is again sent to the absorber and recycled [18].

The main principle of post-combustion technology is to separate the carbon dioxide from the flue gasses. In the power generation system, for generation of steam, coal is burnt in the boiler with the help of air. Steam is used to drive the turbine and it generates the electricity. The exhaust from the flue gases contains CO_2 and N_2. For coal power plants this Post-combustion technology is best as compared to other technologies. The leading technology is used to scrub the solvents with an amine. This scrubbing of solvents is based on the chemicals which react with the carbon dioxide to generate a high temperature of CO_2, which is suitable for storage and compression of gas.

In post-combustion technology, the advantage of amine scrubbing is that it can be retrofitted with the industries and existing power plants in the suitable location. It is commercially used in small scale industries. MEA can be used with the low pressure of CO_2 to capture CO_2 in flue gas. Separation of CO_2 from the stream of flue gas is one of the challenging tasks. CO_2 with low pressure, high volume and dilute concentration are treated. Scrubbing plants are used to reduce the concentration of acid gasses such as SO_2 and NO_2. The impurities of flue gasses degrade the solvents.

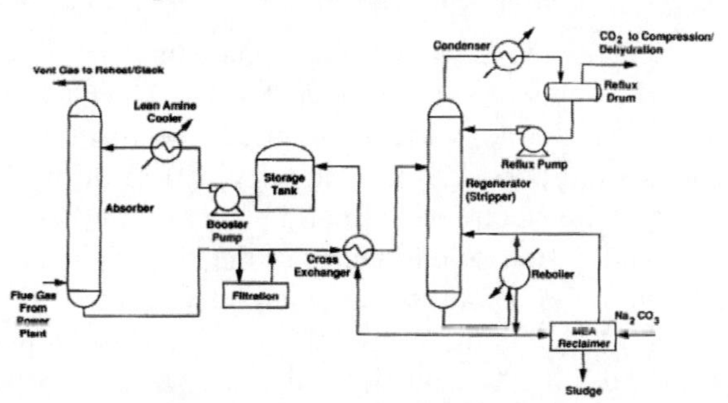

Fig.4.3 Post-combustion process [19]

The degraded items are linked with the problems of corrosion. The limited evolutions are expected for the development of solvents. CO_2 captured by the solvents may use for some other processes. This solvent scrubbing is an existing technology and it has limited experience for the operation of large scale. It has a large operating unit with 800 tons capacity for capturing of CO_2 per day and this power plant produces 8000 ton of CO_2 per day.

5. Research in post-combustion capture process

The research in post-combustion can be divided into two types. Short time research and long-term research. Short term research focuses on the development of existing technologies. It is important to minimize the requirement of heat and temperature.
The improvement of solvents can be divided into three types.

- Concentration of solvents
- Corrosion and degradation of solvents
- Alternative for MEA

The solution of MEA for the absorption of acid gas contains 30 wt% of MEA. To reduce the cost of plant operation and circulation, the concentration and capacity of amine can be increased. Solvent degradation increases the cost of the material, waste disposal and creates a demand for capturing of CO_2. There have been many research conducted to find a superior solvent of MEA as alternative to MEA.

6. Barriers to implement CCS

The adoption of CCS barriers are divided into three categories. They are,

- Financial barriers
- Technical barriers
- Institutional barrier
- Other barriers

The Financial barriers are involving the high capital cost and high energy penalty. The Institutional barriers do not meet the overall development goals. It involves the non-productive expenditure which does not contribute to sustainable development. The Technical barriers are commercially demonstrating the high point sources of CO_2. It is used to capture technologies, but it is not standard for high point sources. It identifies the sinks and its capacities. The technical barriers have some permanence issues and the issues are seismically active in many parts of India. It establishes the EGR/EOR and ECBM in India. The technical barriers are used to determine the potential and cost for the depleted gas and oil. It also optimizes matching and mapping of the sinks and their sources [20].
The other barriers include,

- Financial
- Regulatory
- Storage
- Acceptance

The Storage is used to identify the leakages and their carbon accounting. The regulatory is used to provide the EC establishment and it used to allow the CCS for the European trading.

6.1. Post-combustion

In the world, the CCS is present in the demonstration phase and it gained only one degree of a technology confidence through the deployment of the large scale. The CCS deployment has one major barrier in India. It lacks capture of technology and geological storage side data which can be installed in power plants and their sources are permeability, capacity and location. The CCS implementation has some issues such as increasing the electricity cost and power output. One of the biggest barriers of CCS in India, is electrification and deficit of electricity in the country. In the world, the CO_2 storage uses the Enhanced oil recovery and it is one of the most attractive options for the CCS deployment. In this procedure use, the CO_2 storage cost is offset by accrued revenues. In the petroleum sector, it has been stated by the stakeholders and it has few oil fields that are partially depleted for the enhanced oil recovery [21].
The enhanced oil recovery based on the characteristics of miscibility of the oil that oil is not suitable for all cases. The CCS implementation needs to clarity through retrofit of the capture equipment to the previous plants. The previous plants will modify and change the term of plan references. The barriers are used to access the financial agencies such as Asian development bank and World Bank etc. The previous requirements are verification, monitoring, and measure. The requirements are depending on

the specific clearance of the CCS. The specific clearances of CCS present from the government bodies and Ministry of power to the previous clearance requirement.

The Large scale of CCS deployment requires the best infrastructure and specialized manpower which is not present in India. The CO_2 storage monitors to assure the fulfilment of the CCS implementation and it ensure the rigorous monitor. The monitoring is needed to overlook the techniques and scale development and these are introduced to the Indian stakeholders. The barriers have many legal issues and they are related to the CO_2 leakages, land acquisition and ground water contamination etc. and these are needs to be addressed including the large-scale CO_2 storage and transports. The CO_2 storage and large-scale transports can be permitted via CCS deployment in India.

6.2. Policy amendments by India for CCS implementation

India is world's third largest coal consumer. Coal represents 62% of the nation's power distribution. About 75 percentage of the coal delivered in India which helps to produce an electricity, the remaining 25 percentage is utilized as a part of the steel, binding material, and manure manufacturing. Given the plenitude of coal in India, joined with quickly developing power request, the Indian government is supporting activities, to create up to 9 Ultra-Mega Power Projects. This will include nearly 36 GW of coal driven quantity while suggesting essential chances to examine CCS. The emission of CO_2 is increased up to 1300 Mt, about a portion of which is from expansive point sources that are reasonable for the CO_2 attack. The 25 biggest emitters provided around 36% of aggregate national CO_2 discharge in 2000 which demonstrating the possible presence of various valuable CCS opportunities. As a non-Annex I country to the United Nations Framework Convention on Climate Change (UNFCCC), India has consented to finish Greenhouse Gas emission (GHG) inventories that are still not necessary to meet a binding target for GHG emission under the Kyoto Protocol. India confronts various specialized and administrative obstructions to the use of CCS and cleans coal innovations as a major aspect of a bigger environmental change methodology. In order to acknowledge these problems, the legislature has built up a Clean Coal Technology (CCT) Roadmap with the intention to support the aim of clean coal improvement and strategy (policy) inventions. A CCRD (Clean Coal Research & Development) centre has additionally been set up by industry. CBP (Capacity Binding Programs) has been projected to generate more technology advancement in CCS. Moreover, India has joined various global endeavour to propel the improvement and dispersal of CCS advances. India has become one of founding member countries of the CSLF (Carbon Sequestration Leadership Forum).

6.3. Policy amendments by India for CCS implementation

In 2007, the Department of Science and Technology and Technology Bhawan has started the Indian CO_2 Sequestration Applied Research organizes in New Delhi, which encourages exchange with partners and to build up a system for actions. The Indian CCS research incorporates Carbon Dioxide Enhanced Oil Recovery (CO_2-EOR) perusing thinks about in developing oil fields. Hazira power plant is decided to inject an acid gas. The expenses of CO_2 catch have additionally been surveyed. For instance, emission capture is evaluated to be 21 percentage costlier from Integrated Gasification Combined Cycle (IGCC) and high-ash coal plant than from pulverized coal fired power plant and 12% costlier than from the Ultra Super Critical plant. The Fertilizers Corporation of India has introduced two CO2 attacks plants with 450 t quantity for every day at its Aonla and Phulpur edifices. The study, based on the Deccan Basalt Province in Western India that is one of the biggest surge basalt regions in the world has started in cooperated with Pacific Northwest National Laboratory (PNNL) in the United States [4].

6.4. COP-$_2$1 and outcomes

The United Nations Climate Change Conference COP-21 was held in Paris on November 30, 2015. This agreement explains, there are minimum 55 countries that need to represent 55 percent of greenhouse gas emissions. 174 countries have signed this agreement on April 22, 2016. The main aim of the conference is to reduce the increase in global temperature by greenhouse emissions. In the previous climate conferences, a global agreement pact was signed by the countries to agree to an outline action. These agreements are called as INDCs (Intended Nationally Determined Contributions).

The INDCs used to reduce the global warming and also reduce the emissions [22]. The outcome of this conference holds the binding on reduction of green gas emissions. As per this document, the members agreed to reduce the carbon emissions globally to reduce the global warming. In this agreement, the island states such as the Seychelles, the Pacific and also the non-island state the Philippines strongly mentioned that they will reduce the sea level rise of GHG emissions. They were set the goal as reduce the warm from 2 °C to 1.5 °C [23].

7. Barriers to implement CCS

7.1. CO_2 emission reduction scenario (without CCS)

CO_2 emissions reduction without CCS reduce the CO_2 emissions. BAU has a target of more than 40% time series reduction for CO_2 emission. The CO_2 emission maximum reduction occurs in the power sector by using the clean coal technology deployment as shown in fig 7.1, 7.2 and 7.3 for year 2011, 2021 and 2031 respectively. It has the highest accuracy measurements of the CO_2 concentration atmospheric and documented and it's changing the atmosphere composition with the time series. In CO_2 emissions use the Fossil fuel because it is the CO_2 primary sources. Basically, it involves the deforestation, degradation of soils and land clearing for agriculture. It improves the soils and their activities [24]. The greenhouse gasses emission focuses on the minimizing the emissions. It increases the environmental awareness and regulations. The below diagram explains the sector wise CO_2 emissions in 2011, 2021 and 2031.

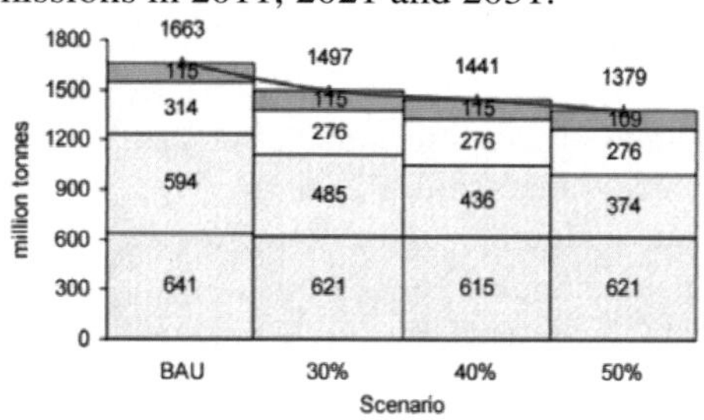

Fig. 7.1 CO_2 emission level in 2011 [20]

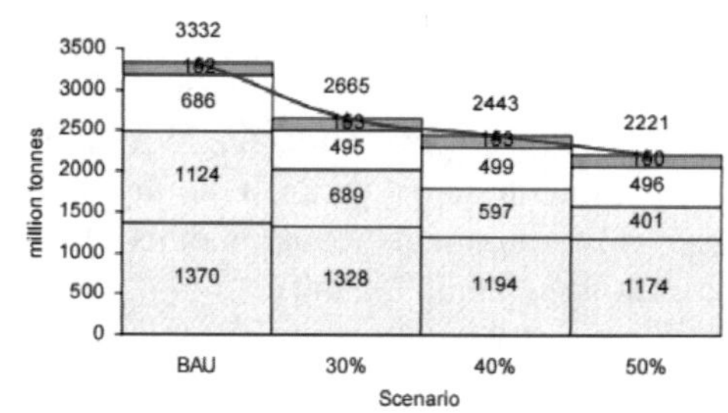

Fig. 7.2 CO_2 emission level in 2021 [20]

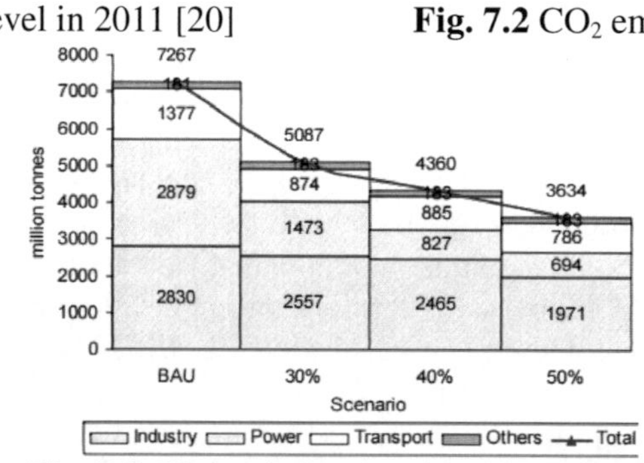

Fig. 7.3 CO_2 emission level in 2031 [20]

7.2. CO_2 mitigation using power generation technology (without CCS)

The power sector has the best opportunity for CCS implementation and it is used to increase the clean coal technology uptake from 10GW in the BAU. The power energy technologies also increase the CCS development opportunities. The power and energy division comprises of various units such as energy policy, petroleum, coal, and power. It reviews, the rising environmental and energy balance concerns and it ensures the energy security. The power energy technologies evolving the energy policy and it covers the commercial and non-commercial energy sources. It includes the natural gas, petroleum, power, and coal [25]. The fig. 7.4, 7.5, and 7.6 explains the power generating technologies in year 2011, 2021 and 2031 respectively.

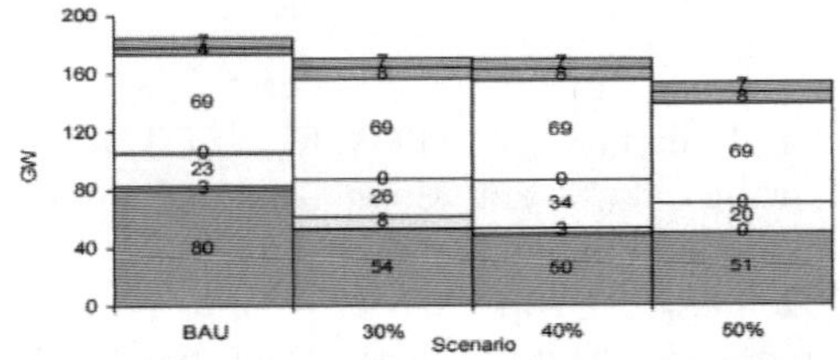

Fig.7.4 Power generation technologies in 2011[20]

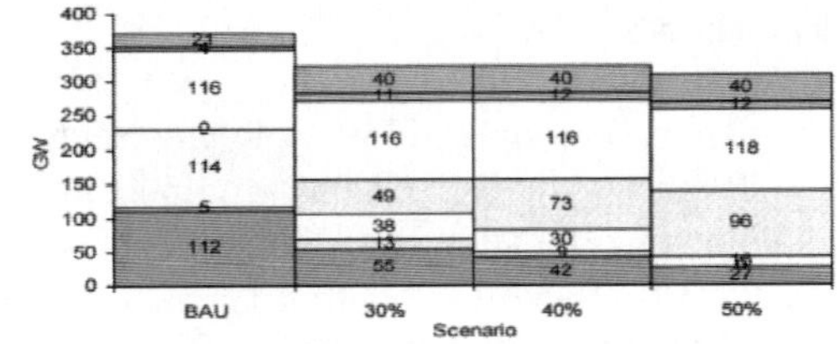

Fig.7.5 Power generation technologies in 2021[20]

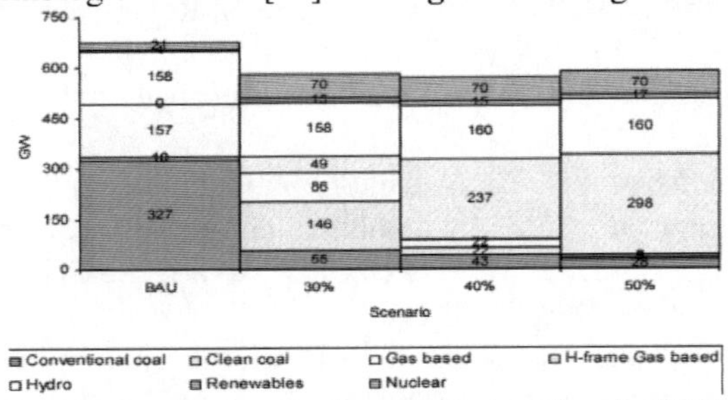

Fig. 7.6 Technologies for Power Generation in 2031 [20]

7.3. Energy related CO_2 emissions (with CCS)

The energy-related CO_2 emissions also reduces the carbon emissions. It only considers the energy-related CO_2 emissions and it analyses the CO_2 emission distribution related to the energy sector. The CCS in energy sector predict the energy consumption. It uses the baseline scenarios, blue map scenarios and the emission level is more than 40% [26] as shown in fig 7.7 the BLUE map scenario offers 14 Gt CO_2/yr if CCS is implemented other else the projected will be 62 Gt CO_2/yr.

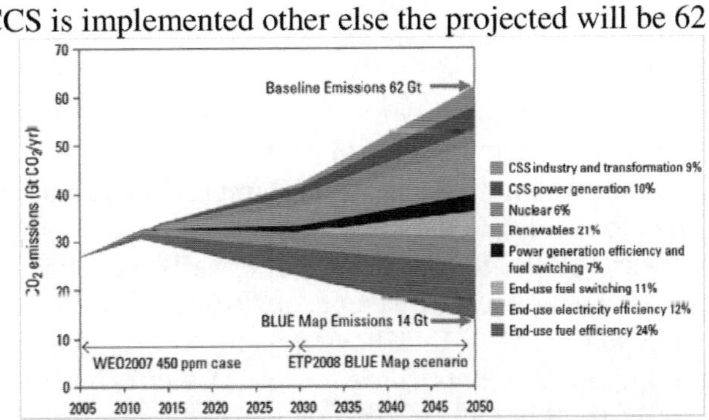

Fig.7.7 CO_2 emission based on energy level [27]

8. Conclusion

The contribution of Indian government in the accumulation of CO_2 and other GHG's from the atmosphere due to the improper fossil fuel combustion with significant amount of fuel procurement with its maintenance is appreciable. The combined infrastructure and domestic reserves are based on the energy sources like wind, solar energy and biomass are needed to be promoted at first level along with integration of CCS with coal fired plants. The clean coal technologies and CCS are the prime media which can be useful to achieve the carbon target and implement the de-carbonization manufacturing processes. The CCS has completed the CO_2 emission reduction in the power sector and it measures the energy efficiency. The CO_2 emissions are projected to reach the more than 55% of the global aggregate in India. The CCS uses the three technologies such as pre-combustion capture, post combustion capture and oxy-fuel combustion capture, among which pre-combustion and post-combustion capture are the most viable options in regards to India due to their flexibility and retrofitting capability with existing fossil fuel fired plants. The technology holds one major barrier of energy penalty which can also be reduced by integrating solar thermal power generation system with post combustion process.

IOP Conf. Series: Earth and Environmental Science **83** (2017) 012024 doi:10.1088/1755-1315/83/1/012024

9. References

[1] Anon Overview of Indian Power Sector | Power Sector in India | INDIAN POWER SECTOR
[2] Raghuvanshi S P, Chandra A and Raghav A K 2006 Carbon dioxide emissions from coal based power generation in India *Energy Convers. Manag.* 47 427–41
[3] Central Electricity Authority 2016 *All India Installed Capacity of Power Stations* vol I
[4] Patware P, Thakur G, Rawat P and Sudhakar K A Roadmap For " Carbon Capture And Sequestration " In The Indian Context: A Critical Review *Int. J. ChemTech Res.* 514 974–4290
[5] Rao A B 2010 CO 2 Capture and Storage (CCS) -Relevance for India *Natl. Res. Conf. Clim. Chang. IIT Delhi*
[6] Russo M E, Olivieri G, Marzocchella A, Salatino P, Caramuscio P and Cavaleiro C 2013 Post-combustion carbon capture mediated by carbonic anhydrase *Sep. Purif. Technol.* 107 331–9
[7] Sood A and Vyas S 2017 A REVIEW: CARBON CAPTURE AND SEQUESTRATION (CCS) IN INDIA *Int. J. Mech. Eng. Technol.* 8 1–7
[8] Viebahn P, Höller S, Vallentin D, Liptow H and Villar A 2011 Future CCS implementation in india: A systemic and long-term analysis *Energy Procedia* 4 2708–15
[9] IPCC|WMO|UNEP *CARBON DIOXIDE CAPTURE AND STORAGE*
[10] Parikh J 2010 Analysis of Carbon Capture and Storage (CCS) Technology in the Context of Indian Power Sector SUBMITTED BY Integrated Research and Action for Development (IRADe)
[11] IPCC 2001 IPCC Third Assessment Report (TAR) *IPCC* 995
[12] Rubin E, Meyer L and de Coninck H 2005 IPCC Special Report on Carbon Dioxide Capture and Storage: Technical Summary *Rite.or.Jp*
[13] Maroto-Valer M M 2010 *Carbon dioxide (CO2) capture, transport and industrial applications* (CRC Press)
[14] Walker J L and Hernandez A *Emissions control in electricity generation*
[15] Anon 4.3 Oxyfuel Combustion | Global CCS Institute
[16] Anon Oxy-fuel combustion systems – The Carbon Capture & Storage Association (CCSA)
[17] Anon CO2 separation before combustion
[18] Viebahn P, Vallentin D and Höller S 2014 Prospects of carbon capture and storage (CCS) in India's power sector – An integrated assessment *Appl. Energy* 117 62–75
[19] Moazzem S, Rasul M G and Kh M M K 2012 A Review on Technologies for Reducing CO2 Emission from Coal Fired Power Plants *Thermal Power Plants* (InTech)
[20] Dadhich K P Potential for CCS in India: Opportunities and Barriers
[21] Anon 8. Barriers to CCS implementation in India | Global CCS Institute
[22] Anon Paris Agreement - Status of Ratification
[23] Anon COP 21 Paris France Sustainable Innovation Forum 2015 working with UNEP
[24] Ramachandra T V, Aithal B H and Sreejith K 2015 GHG footprint of major cities in India *Renew. Sustain. Energy Rev.* 44 473–95
[25] Taylor P 2010 Energy Technology Perspectives 2010: Indian Power Sector
[26] Nataly Echevarria Huaman R and Xiu Jun T 2014 Energy related CO2 emissions and the progress on CCS projects: A review *Renew. Sustain. Energy Rev.* 31 368–85
[27] Anon Climate Change | Schlumberger

IOP Conf. Series: Earth and Environmental Science **83** (2017) 012025 doi:10.1088/1755-1315/83/1/012025

A Fast Evaluation Method for Energy Building Consumption Based on the Design of Experiments

Hocine BELAHYA[1], Abdelghani BOUBEKRI[1] and Abdelouahed KRIKER[2]

1 University Kasdi Merbah Ouargla, Laboratory of Development of New and Renewable Energies in the Arid and Saharan Zones, Faculty of Applied Sciences, Ouargla 30000, Algeria
2 University Kasdi Merbah Ouargla, Laboratory of Exploitation and Valorization of Natural Resources in Arid Zones, Faculty of Applied Sciences, Ouargla 30000, Algeria
E-mail: bhocine30@gmail.com

Abstract. Building sector is one of the effective consumer energy by 42% in Algeria. The need for energy has continued to grow, in inordinate way, due to lack of legislation on energy performance in this large consumer sector. Another reason is the simultaneous change of users' requirements to maintain their comfort, especially summer in dry lands and parts of southern Algeria, where the town of Ouargla presents a typical example which leads to a large amount of electricity consumption through the use of air conditioning. In order to achieve a high performance envelope of the building, an optimization of major parameters building envelope is required, using design of experiments (DOE), can determine the most effective parameters and eliminate the less importance. The study building is often complex and time consuming due to the large number of parameters to consider. This study focuses on reducing the computing time and determines the major parameters of building energy consumption, such as area of building, factor shape, orientation, ration walls to windows ...etc to make some proposal models in order to minimize the seasonal energy consumption due to air conditioning needs.

1. Introduction

Nowadays, Building is one of the most consuming energy sectors, it is responsible for approximately 40% of annual total energy consumption in worldwide [1], [2]. Both residential and commercial, has steadily increased between 20% and 40% in developed countries [3]. In Algeria this sector ranks first by 42% of total energy consumption [4] and it is growing by exaggerated way because the Algerian state has launched a huge plan of housing construction without taking into account the legislation of energy performance, in the absence of awareness and lack of culture on the energy control the energy consumption achieved remarkable level especially for HVAC system, the most reasons that led this increase are: low prices of conventional energy and increase number of air conditioners in each house to maintain the comfort of occupant, knowing that in Algeria there is big variety climate it is one of the largest solar potential in the world. It is valued at more than 3000 hours of sunshine per year and 5 kWh of daily energy received on a horizontal surface on most part of the country and more than 35000 hours in the Sahar region. Buildings face problems of discomfort, mainly due to the overheating phenomenon and the facades exposure to solar radiation. These lead to a large amount of electricity consumption by using air conditioning. Bioclimatic design can minimize the seasonal energy consumption; it refers to the reduction of energy consumption without causing a decrease in the level of comfort [5]. Some published simulation results show that protection against solar radiation and glare has a great influence on the arid climate [6]-[9]. The objective is to optimize the HAVC system, and to reduce the need of the seasonal energy consumption through air conditioning and analysis of the total energy consumption. Many studies have been proposed Design Of Experiment for optimization energy envelop building consumption Romani and al [10] developed and validated metamodels of heating and cooling energy needs for single family houses in six Moroccan climate zones. Xuejun and al [11] enveloped including wall, window, roof and door were selected as the interesting factors regarding to the different layers, materials and insulations. Response factors were: (1) annual building energy consumption, (2) cooling energy consumption, and (3) heating energy consumption in the small-sized commercial building respectively. Jaffal and al [12] evaluate the heating demand used model with a small number of dynamic simulations using the design of experiments.

2. Methodology

This study attempts to develop a new methodology for residential building, by using the co-simulation and investigate the influence geometric envelope building (surface walls, surface roof, ration wall windows, ceiling height, ration shape… etc.) And thermal properties on the energy performance of buildings. We divided our parametric into two types; (1) thermo physical properties and geometric parameters building envelope, also we can reduce envelope building parameters from 10 to 7 parameters, by using the relationship between geometric envelope, the total number of parameter in this study will be 13, Walls and roofs of the building envelope play an important role in the heat transfer between the exterior and the interior spaces, also the windows to wall ration interact between thermal properties and geometric, DoE can lead to modelling this complex problem.. The application field considered in this work is represented by residential buildings, Energy consumption could be divided into two types; the first one is the continuous energy consumption of equipment which is working during the year as lighting, fridge, TV and other equipment to reduce this consumption we should use the height efficiency equipment. The second one represents the seasonal equipment which is used for air conditioning (HVAC system). For many years improving the thermal performance of the building envelope meant predominantly keeping the thermal transmittance values of the opaque and transparent elements as low as possible those building based on high insulation levels of envelope, this simple approach till the easiest way to reduce the energy consumption but doesn't give the optimum solution. During the last years the evolution of computing and software gives the opportunity to study complex systems as building envelope, which the number of parameters are very important such as climatic, architectural, social constriction and material… etc [13].

2.1 Design of experiments method

Design of Experimental is a tool for engineer and scientists to use for product design and development, this tool can reduce development lead time and cost, leading to processes or simulations, and have high reliability than other approaches. The main objective of the experiment is to determine which variables are most influential on the response, even you can set the influential factors that response is near than the desired value with its variety, and reduced the effects of less influence factors [14]. The obtained mathematical model has quadratic/second-order polynomial, the equation resulting (1) from statistical regression analysis or response surface regression,

$$y(x) = a_0 + \sum_{i=1}^{k} a_i\, x_i + \sum_{i=1}^{k} a_{i,i} x_i^2 + \sum_{i=1}^{k-1} \sum_{j=i+1}^{k} a_{i,j}\, x_i\, x_j + \in \quad (1)$$

Here, $y(x)$ is the predicted response variable a_0, a_i and $a_{i,i}$ are the regression coefficients of the intercept, linear, quadratic and interaction effects, respectively, while x_i and x_j are independent input variables, and $\in$ is a random error. Moreover, Eq. (2) can be analyzed graphically by using three-dimensional response surface plots or contour plots in order to evaluate the optimal values [14], [15]. An analysis of these plots is providing a method to reveal the interactions between two significant factors, all their combinations in one studied response variable in order to determine the most efficient conditions for an optimal target response variable[8]

3. Data collection and case study

Ouargla city is located 800 km southern from. Algiers the capital Fig.1. It is one of the most important cities of Algeria and the main source of wealth. It is known as the oil and the oasis capital. It is difficult for people to deal with a lot of natural obstacles, particularly the hard climate that is characterized by high temperatures, low rainfall and low humidity, especially in summer. The values of the average temperature are shown in Fig. 2.

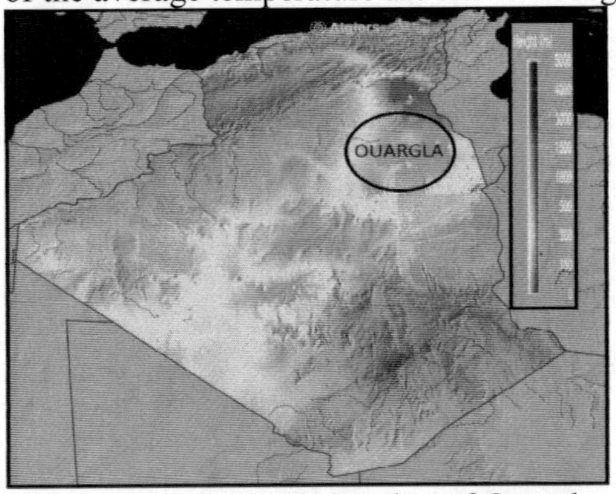

Fig. 1. Geographical situation of Ouargla city

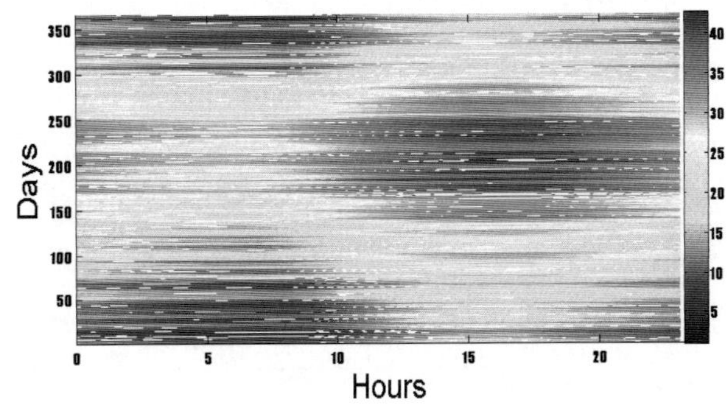

Fig. 2. Yearly temperature degree C° in Ouargla

3.1 House designe characteristics and optimization

Generally the cover surface of house (used HVAC system) is between 60 to 90 m², To estimate yearly energy consumption, 100 houses are taken from 5 different zones to be more representative in the real sample, in order to improve the yearly average energy consumption; we depend the selection on a reference type house which has all the characteristics of the buildings in this region. , we noticed that the maximum load penetrates the building envelope through its roof and walls. Air conditioning system consumes a large proportion of the generated electrical energy and it increases the fuel consumption which causes more environmental pollution. We depend in our study on two types of electrical energy consumption; the first is seasonal consumption which is the value of the seasonal consumed energy (air conditioning in summer and heating in winter), while the continuous consumption is the value of the yearly consumed energy. The historical energetic data analysis of one year shows that the cooling system of the building counts more than an average of 63% of the total energy consumption. The following figure shows the continuous and seasonal consumption of electric energy in the 5 zones of Ouargla.The function of building envelope is to separate physically the building interior from the external environments. Therefore, it serves as an external protection to the indoor environment while facilitating as climate control at the same time. The continuous and seasonal electric consumption per subscriber. The proportion of the continuous electrical energy consumption stays almost constant at five zones with a value of 30% and seasonal consumption of electrical power of 70%.

4. Co-simulation and optimization

The parameter selection is difficult operation and expensive to perform all experiments or simulations. The DOE method can be employed as an efficient technique to accomplish the suitable and necessary simulations with high accuracy. To investigate main and multiple interactions between parameters in this study, a fractional design was employed with two levels for each parameter +1 for high level and –1 for low level, we choose 13 parameters are: Thermal transmission through the vertical walls (Uwall),Thermal transmission through the roof (Uroof), Thermal transmission through the ground (Uground), Thermal transmission through the windows (Uwind), Windows to Wall Ration North wall (WWR_N), Windows to Wall Ration South wall (WWR_S), Windows to Wall Ration East wall (WWR_E), Windows to Wall Ration West wall (WWR_W), Absorption coefficient of the solar radiation of the wall (Awall), Absorption coefficient of the solar radiation of the roof (αroof) Ceiling Height (CH), Area floor A Factor Form (FF) Table(1).

IOP Conf. Series: Earth and Environmental Science **83** (2017) 012025 doi:10.1088/1755-1315/83/1/012025

Table 1: Lower and upper levels for each physical parameter

Parameter	Uwall	Uroof	Ufloor	Uwind	WWR_N	WWR_S	WWR_E	WWR_W	Awall	αroof	CH	A	FF
	W/m² K	W/m² K	W/m² K	W/m² K	%	%	%	%			m	m²	
N	1	2	3	4	5	6	7	8	9	10	11	12	13
Lower level	0.338	0.339	0.354	2.95	5	5	5	5	0.1	0.1	2.5	60	1
Upper level	2.325	2.346	3.351	5.74	10	10	10	10	0.9	0.9	3.2	90	2

We used TRNSYS 17 [16] to simulate building envelope and Minitab for DoE, in this dynamic simulation we use Python as processor Fig. (3), we need Excel to converter dimensional matrix to less dimensional matrix that's can reduce the number of parametric in our study, this process reduced geometric factors from 10 to 7, the complex co-simulation can reduce dramatically the time and avoid the error when the user introduce the data by hand from software to another.

Fig.3 Co-simulation and optimization of cooling and heating energy demands

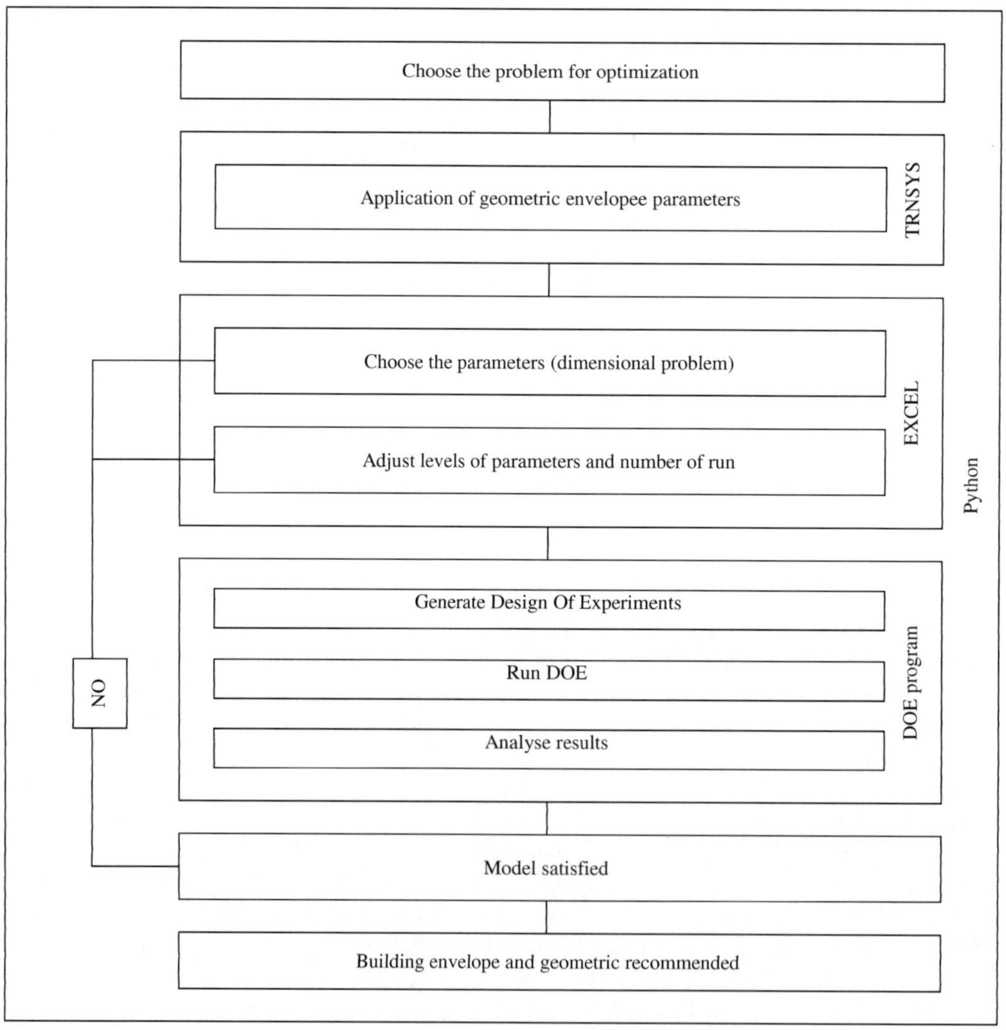

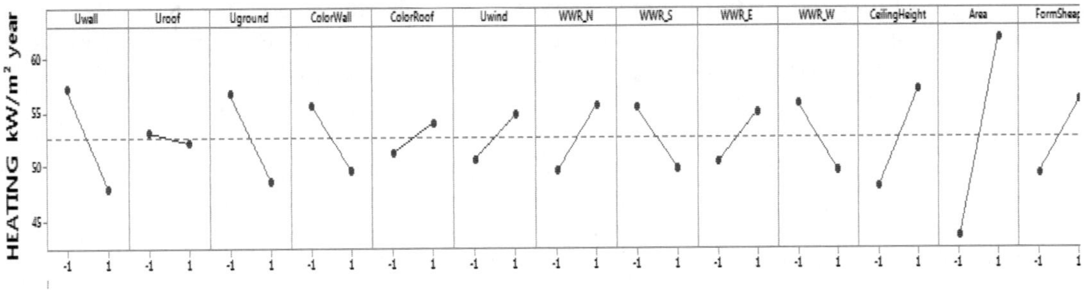

Fig.(4) The graphs of main parameter effect Heating case

The predictor model of Heating energy consumption

(2)

$$HEAT = 52.64 - 4.70\ Uwall - 0.46\ Uroof - 4.08\ Uground - 3.02\ CWall + 1.36\ CRoof + 2.14\ Uwind + 3.06\ WWR\_N + 2.90\ WWR\_S + 2.31\ WWR\_E - 3.15\ WWR\_W + 4.51\ CH + 9.19\ Area + 3.40\ FS$$

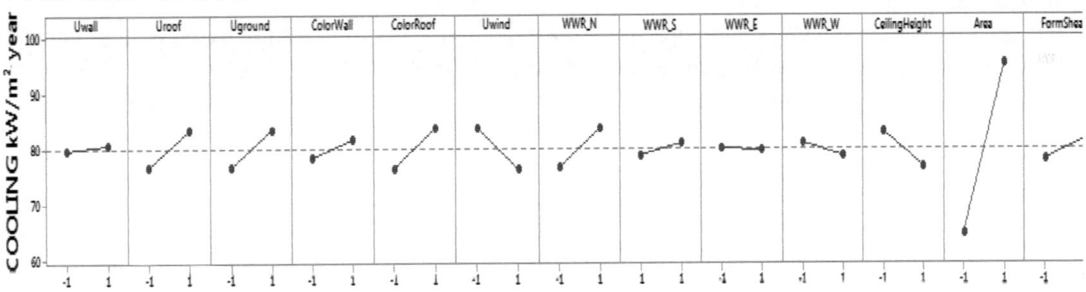

Fig.(5) The graphs of main parameter effect Cooling case

The predictor model of Heating energy consumption

(3)

$$COOL = 80.12 + 0.53\ Uwall + 3.32\ Uroof + 3.45\ Uground + 1.59\ CWall + 3.69\ CRoof - 3.57\ Uwind + 3.52\ WWR\_N + 1.18\ WWR\_S - 0.15\ WWR\_E - 1.10\ WWR\_W - 3.08\ CH + 15.49\ Area + 1.81\ FS$$

5. Results

The simulation results obtained shows the factor which most influence is the surface in both cases cooling and heating, the selection of this parameter is very important. The simulation shows the cooling energy consumption is height than the heating energy consumption, that's returned to the hot climate of Ouargla city, this result gives the cooling consumption the priority for optimization in the case of inverse influence factors, The surface influenced in heating case more than the cooling, the second factor is the factor form, it has big influence in heating case and less important in cooling case, so the square ship is the best for both cases. The influence of thermal transmission through the vertical walls can reduce energy consumption for heating 10 Kw/m² year, but in the case for cooling their influence is so limited, the influence of thermal transmission through the ground has inverse influence in the case of cooling, but its influence is important in the heating case it can reduce this consumption about 8 Kw/m² year, absorption coefficient of the solar radiation of the roof has the same influence for both cases.

6. Conclusinon

The DoE methodology has gained importance for a wide range of industrial applications, pharmaceutical, biotechnology...etc. the advantages of DoE are various its allowing a rapid evaluation of the effects of different parameters or important factors on the selected response variables and their possible interactions, thus factors can be simultaneously changed and optimized, DoE approach provide the possibility of studying a large number of parametric as the case of the envelope building and the feasibility to operate as a promising and efficient optimization technique. In this study a new methodology of building performance evaluation and tool could be developed based on co-simulation and parametric design, to reduce the need of the seasonal energy consumption through air conditioning and analysis of the total energy consumption. plots of this curves give the choose between many factors as thermal property, insulation wall, windows wall ration for each face ceiling high, type of window …ect quickly and accurately according to the performance requirements of designers. The developed polynomial model have simple form and can calculate rapidly the energy needed, which allows to study different factors for all energy envelope building design. Moreover, the design of the experiments reduces significantly the number of dynamic simulations required to determine the coefficients of the parametric models. In fact, a multidisciplinary design tool can integrate all aspects of a building's performance. This type of co-simulation combined with four software can reduce the time of simulation and contribute to the design of low building energy consumption.

7. References

[1] H.-x. Zhao and F. Magoulès, "A review on the prediction of building energy consumption," *Renewable and Sustainable Energy Reviews,* vol. 16, pp. 3586-3592, 2012.

[2] M. W. Ahmad, M. Mourshed, D. Mundow, M. Sisinni, and Y. Rezgui, "Building energy metering and environmental monitoring – A state-of-the-art review and directions for future research," *Energy and Buildings,* vol. 120, pp. 85-102, 2016.

[3] L. Pérez-Lombard, J. Ortiz, and C. Pout, "A review on buildings energy consumption information," *Energy and Buildings,* vol. 40, pp. 394-398, 2008.

[4] Hocine BELAHYA, Abdelghani BOUBEKRI, Abdelouahed KRIKER "A comparative study about the energetic impact of dryland residential buildings with the integration of photovoltaic system," *Energy Procedia* vol. 00 (2016) 000–000, 2016.

[5] D. Popescu, S. Bienert, C. Schützenhofer, and R. Boazu, "Impact of energy efficiency measures on the economic value of buildings," *Applied Energy,* vol. 89, pp. 454-463, 2012.

[6] S. Hoffmann, E. S. Lee, A. McNeil, L. Fernandes, D. Vidanovic, and A. Thanachareonkit, "Balancing daylight, glare, and energy-efficiency goals: An evaluation of exterior coplanar shading systems using complex fenestration modeling tools," *Energy and Buildings,* vol. 112, pp. 279-298, 2016.

[7] B. S. Matusiak, "Glare from a translucent façade, evaluation with an experimental method," *Solar Energy,* vol. 97, pp. 230-237, 2013.

[8] E. J. L. Eriksson, N. Kettaneh-Wold, C. Wikström, S. Wold, " Design of Experiments, 3rd ed. MKS Umetrics AB, Umeå, 2008.."

[9] J. Parasonis, A. Keizikas, A. Endriukaitytė, and D. Kalibatienė, "Architectural solutions to increase the energy efficiency of buildings," *Journal of Civil Engineering and Management,* vol. 18, pp. 71-80, 2012.

[10] Z. Romani, A. Draoui, and F. Allard, "Metamodeling the heating and cooling energy needs and simultaneous building envelope optimization for low energy building design in Morocco," *Energy and Buildings,* vol. 102, pp. 139-148, 2015.

[11] S. W. L. Xuejun Qian, "The Design and Analysis of Energy Efficient Building Envelopes for the Commercial Buildings by Mixed-level Factorial Design and Statistical Methods."

[12] I. Jaffal, C. Inard, and E. Bozonnet, "Toward integrated building design: A parametric method for evaluating heating demand," *Applied Thermal Engineering,* vol. 40, pp. 267-274, 2012.

[13] N. Aste, A. Angelotti, and M. Buzzetti, "The influence of the external walls thermal inertia on the energy performance of well insulated buildings," *Energy and Buildings,* vol. 41, pp. 1181-1187, 2009.

[14] D. C. Montgomery, "Design and Analysis of Experiments," *8th ed. Wiley, New Jersey,2013*

[15] J. E. J.S. Lawson, "Modern Statistics for Engineering and Quality Improvement, 1st ed. Duxbury, California, 2000.."
[16] TRNSYS, "TRNSYS 17 Reference Manual," *Solar Energy Laboratory, University of Wisconsin-Madison;* , 2012.

Evaluation of a School Building in Turkey According to the Basic Sustainable Design Criteria

H D Arslan

1 Architecture Department, Necmettin Erbakan University, Konya, Turkey
E-mail: deryaarslan@konya.edu.tr

Abstract. In Turkey, as well as many other developing countries, the significance of sustainable education buildings has only recently become recognized and the issue of sustainability issue has not been sufficiently involved in laws and regulations. In this study, first of all architectural sustainability with basic design criteria has been explained. After that selected type primary school project in Turkey has been evaluated according to the sustainable design criteria. Type project of school buildings significantly limits the sustainability performance expected from buildings. It is clear that type projects shorten the planning time as they include a designing process that is independent of settlement and they are repeated in various places with different characteristics, indeed. On the other hand; abundance of disadvantages such as the overlook of the natural physical and structural properties of the location mostly restricts the sustainable design of the building. For sustainable buildings, several factors such as the environment, land, climate, insolation, direction etc. shall be taken into consideration at the beginning stage. Therefore; implementation of type projects can be deemed to be inappropriate for sustainability.

1. Introduction

The human being has been in interaction with nature and environment since its existence and they have changed each other during the process. While in ancient ages, this change could be identified as cultural or social change only, with the introduction of the concepts of industry and technology, the venues for a living have been forced to change, too. Thanks to the ever-advancing science, technology, automation and computer technologies, new occupational areas have emerged and the economy has grown fast in this parallel. Along with these; uncontrolled industrialization, the amount of waste, pollution and greenhouse gas on earth and in the atmosphere significantly increased [1]. These intense pressures almost ruined the biochemical cycles and the ecosystem, and consequently, nature has become unable to regenerate itself. The environment designed through the utilization of available resources to provide the individuals with a place to live aims to fulfill the biological, psychological, socio-cultural and economic needs of the individuals by meeting the conditions for comfort. In designing environments that are suitable for the conditions of comfort, this purpose shall be taken into consideration when taking location decisions based on the natural air conditioning, lighting, heating and ventilation systems and when taking planning and designing decisions based on the building and place. A primary principle to take into consideration during the designing process of a structure is that; the structure shall meet the physical, social, psychological and economic needs of the people while being in unity with the natural environment conditions [2]. In order to ensure unity with the environments, designers shall prioritize the preservation of natural resources, environmental impact, and utilization of reproducible materials. Natural resources, the sun, materials, water and climatic data shall be used in a most efficient way; re-utilization and recycling processes shall be considered.

Architectural designs shall be sensitive to the environment and shall be based on the minimization of the consumption of resources and maximization of the use of natural power resources [3]. Education buildings, which are among the elements forming the structured environment, are important both for the society and for the individuals. Individuals form the basis of their life within the framework of the knowledge they have acquired in these buildings. Therefore; education buildings are more significant for children than other buildings, to raise awareness about, to implement, perceive, live and experience sustainability. On the other hand, it is also known that, millions of students on earth receive education in buildings that are insufficient in terms of air conditioning, heating, lighting and acoustic and which have negative impact on the stimulation of the exploring and creativity features of the students, which are required for the learning process, indeed [4].
In order to ensure sustainable life standards on earth for the future generations, provision of efficient education is one of the primary issues to be focused on, as well as the assessment and control of environmental impact. Particularly due to the ever increasing young population in Turkey, the design of education buildings shall consider the environmental sensitivity as an extension of education and education buildings shall be designed in parallel with the sustainability criteria.
Today, most of the total energy consumption is belongs to the construction industry. It is possible to minimize the impact of buildings by making use of technology, materials and current methods without the need to make a further research. School building industry itself is a big industry. Dealing with sustainability practices at schools in an integrated manner would lead to significant improvements in energy and resource consumption. In this sense also, it is important to determine sustainability criteria for school buildings and to put these into practice. In many countries like Turkey, standard type of school buildings provided with the state budget is built ignoring the climatic conditions and environmental factors. The significance of sustainable education buildings has only recently been recognized in Turkey, and the issue has not been sufficiently involved in laws and regulations, yet this research aims to make literature reviews on sustainability, sustainable architecture, and sustainable design and to determine the basic criteria regarding the designing of a sustainable education building. From this motivation, in this study, selected type primary school project has been evaluated according to the sustainable design criteria.

2. Sustainability and architectural sustainability

With an ecological point of view, Foster [5] defines "sustainability" in the simplest and clearest way as; "less is more", which means the ability to manage the most by using the resources the least. The concept of sustainability developed in 1972, through the discussions within the framework of the "eco-development" concept included in the World Environment Conference made in Stockholm with The Club of Rome. Though sustainability is studied on several platforms in many occupational fields, it takes attention that researchers have developed a common perspective when defining the components that form sustainability, based on the ecologic, economic and social aspects of the concepts. For example; some researchers [6], [7] define three categories to classify the data that are influential on the emergence and change of the sustainability concept.

- Ecologic Data; Climatic changes, Difference mitigation, Ecologic factors (environmental sensitivity), Food chain and transformation, Development of the concepts of resource and waste, Energy consumption
- Economic Data; Development of the concepts of agriculture and transportation, Industrial development
- Social cultural Data; Individual and public consciousness-raising (human rights), Investigation of population, intensity and carriage capacity relationships, Historical awareness and development

As the building industry has a significant impact on the global environment today, it must be a priority to recover the environmental impact created by all negative activities about a building. As well as other fields, the current situation shall be analyzed first and discussions shall be made concerning the measures to be taken in the architecture field. Definitions intended for enabling and sustaining healthy living conditions on earth for future generations shall be readdressed in parallel with the conditions that change in time. Sustainable architecture is the main body covering previous architectural approaches and is a comprehensive, strategic and planned way of the building supported by global environmental problems and development problems. Thus; architectural practice is projected to be environment-friendly not only with its morphological properties but also with the contribution it makes to the social, cultural and economic infrastructure. Sustainable architecture is a controversial issue that is often dealt with in international research. Utilization of a rich terminology in the classification of buildings indicates the broadness of the issue and complication of concepts, even the presence of a chaos. Several terms such as environmental design, green architecture, ecologic architecture, environment-friendly architecture, architecture sensitive to the environment, smart architecture, energy-efficient architecture, energy-conscious architecture, climatic architecture address to complicated, conflicting and competing practices [8]. Sustainable design can integrate the life cycles of building systems with the ecologic systems in the biosphere. Components of a building and energy systems shall work in harmony with ecologic systems from resource to the smallest equipment in the building. A successful ecologic building shall have minimum destructive and maximum positive impact on natural systems.

3. Basic criteria in the design of sustainable education building

Sustainable education building can describe as affect the ecosystem and near-environment, have low construction and maintenance cost, have recycling construction materials, having high social and ecological efficiency. According to Commercial International Bank (CIB) [9] and Sakınç [10], main targets of sustainable buildings are reducing using sources, preserving ecological and nature environment, getting human health and comfort top level, taking into account the socio-economic, cultural and political realities of the place. High goals for sustainable buildings can only be achieved through detailed analyses, cost and performance calculations and through dynamic and integrated design methods requiring returns and changes. According to Sakınç [10], the basic steps used in the design of a sustainable building are 1) Basic goals are identified 2) Basic decisions are taken 3) Preliminary design is made 4) Design 5) Enhanced design

Sustainable design of education buildings and integration of sustainable practices would have a high positive impact on energy and resource consumption. It is particularly important to offer social sustainability to students as a philosophy of life. To achieve this, buildings occupied by students shall be carefully designed in terms of sustainability criteria.

The sustainable design shall be discussed in a way that will ensure a healthy cycle in nature-human/society as a whole. In sustainable design; physical criteria such as climatic properties, positioning of buildings, building design, building form, spatial organization, material selection, sanitation system hardware and suitable plant cover are important. Turkish Green Building Council is an institution that contributes to the development of the building industry in parallel with sustainable principles in terms of green building and sustainability issues. There are several green building certification systems in the world intended for sustainable building surveillance. Building Research Establishment Environmental Assessment Method (BREEAM), Leadership in Energy and Environmental Design (LEED) and Comprehensive Assessment System for Built Environment Efficiency (CASBEE) are among the most important of these. The general objective of these systems is to reduce the environmental impact of building and building activities through life cycle approach. Models that are initially developed in parallel with the conditions of the country they have originated from are also in developing countries in time, either directly or after modification. Certification systems were assessed in terms of sustainable lands, water conservation, internal air quality, selection of appropriate material and building components, efficiency in water conservation, energy and atmosphere. Sustainable buildings can be detailed in terms of ecologic, economic, social and cultural sustainability in a way that will involve the common issues included by certification systems.

4. Basic criteria in the design of sustainable education building: sample type project in Turkey

The school building conducted in the scope of study selected from Konya province of Turkey. The school building constructed in 2014 by government as a type project with 24 classrooms. The population of the school is 535 with 6-11 aged children. The buildings locations and general façade view is given in Figure 1.

Figure 1. The buildings locations and general façade view

4.1. Design criteria about ecologic sustainability

Ecologic sustainability includes the economic use of resources, preference of renewable energy sources and preservation of ecosystems. Economic sustainability is divided into two groups as investment and utilization cost. It is significant that building processes, building components, and materials are cost efficient, highly resistant and reusable. By renewing and reusing buildings, "long-term efficiency of resource" is ensured. Low utilization costs are achieved through efficient energy consumption of the building and through simple operation and maintenance. Social and cultural aspects of sustainability are the preservation of health and comfort as well as the values which is the main goal of preservation projects. As ecologic sustainability includes the economic use of resources, preference of renewable energy sources and preservation of ecosystems, selection of location, water conservation, energy and atmosphere, transportation, natural lighting, natural air conditioning and the criteria of distance between buildings are discussed under this heading.

4.1.1. *Selection of location.* Education buildings shall be assessed in terms of topography, land form, underground and above ground riches, settlement pattern, climatic regions and protection from or benefiting from the sun. In this sense, selection of location is the primary and preliminary parameter for sustainability. According to Kayıhan (2016) [11], the main goal in the selection of a sustainable land is the protection of the users of the education building from external pollutants and preferring central locations with complete infrastructure, to which, users can safely access without transportation problem and which have minimum negative impact on the natural habitat and vegetation. The selected school building is located east-west. Daylight in the south should be under control. The classes in the north can experience warming problems because they cannot benefit from daylight. As seen in the school building site plan, the garden has settled on the ground as it will be in front. The building is in semiarid climate zone in Turkey.

4.1.2. *Water conservation.* Water becomes more valuable day by day. It is significant to keep water consumption at an optimal level in education buildings and to reuse the consumed water after purification. Controlling rainwater is also vital for the safety of the school area and ecology as well as water resources on earth and natural habitats. The main goal for water conservation is to protect undesirable excessive water from impermeable layers and to prevent pollution caused by rainwater accumulation. When rainwater is collected and used, low energy consuming installations and tools are used in buildings, water consumption, which is one of the main problems in certain regions, is reduced and the building becomes more ecologic. In this sense, covering materials to be used in the landscape shall be selected from permeable materials that would allow rainwater to reach the ground water. There is no controlling for rainwater in the school.

4.1.3 *Energy and atmosphere.* A general feature of a sustainable education building is that it will enable the use of renewable energy resources instead of exhaustible resources. In parallel with this, the following methods shall be primarily preferred: utilization of wind and solar energy, obtaining solar energy through active solar system, obtaining electric energy through photoelectric transformation, obtaining solar energy through passive solar systems, utilization of Trombe wall, water wall, roof pool systems, metal solar wall systems, controlled double glass facade system, utilization of isolated profit systems in solar energy to create places (greenhouses, solar chambers) for heat accumulation and storage, utilization of separate profit systems for solar energy and water heater collectors etc. Besides these methods, which enable direct acquisition of energy, indirect benefit systems shall be considered. For example Lakot [12], it involves a glass surface and a behind that surface, a thermal mass suitable for heat storage, such as concrete, solid brick, stone or mud brick, with a selective or black-dyed surface for maximum absorption of heat. As the sunlight is absorbed by the surface of the thermal mass and is transformed into heat, it is delivered to the surface of the thermal mass through transmission, and then to the interior through transportation and radiation. These systems are named as; Trombe wall, water wall, roof pool systems, metal solar wall system and controlled double-glass facades. The solar beams coming in at different times. For the evaluation of the day when the school building was received, the sunsets were viewed during the year, at 21:00 on 21 March, 21 June, 23 September and 21 December. The fact that the school building receives direct sunlight all year round reveals some problems in the use of classrooms. In addition, when the buildings around the school were examined, it was found that there was no high building and therefore the heat of the school and the heat to be taken from the sun are not interrupted. On the other hand, when we examine the illumination of the structure, it is observed that the sun rays thought to be used are not sufficient along the long corridors of the plan typology and even daytime they have to be supported with artificial light sources. This is only provided by the windows in the classrooms. From the standpoint of ecological sustainability components, it is seen that natural enlightenment is inadequate in general.

4.1.4. *Transportation* In sustainable school education buildings, transportation lines shall be determined, alternative vehicle transportation, roads, and car parks shall be considered, simple and safe mass transportation options shall be prioritized, car park areas shall be minimized, the number of pavements between residences and school shall be increased, roads shall be sufficiently illuminated to ensure safety of the users, alternative transportation (bike etc.) or pavements shall be projected and bike park areas shall be considered. Furthermore; it is necessary to design windows to enable the observation of areas where mass transportation vehicles wait, drop-off and pick-up, to provide sufficient illumination, to install security cameras and to minimize the amount of hidden areas that could facilitate potential crimes. In the selected school building, alternative vehicle transportation, roads, and car parks were considered.

4.1.5. *Natural lighting and natural air conditioning*. Sunlight is one of the most significant factors that play role in the formation of interior environment conditions. The contribution of sunlight to energy efficiency shall be clearly indicated for the content of sustainability concept and for smart use of energy in buildings. It is also necessary to develop efficient methods intended for the assessment of the contribution of sunlight to the building in terms of energy. Concerning natural lighting and air conditioning criteria for sustainable preschool buildings, the direction should be taken into consideration when sizing window spaces manually-controlled shading elements should be used in front of the windows. Artificial lighting should be supporting the natural lighting as much as possible. Top windows and lights should be used in design and venues should be painted in light colors. With the design of window spaces and directions of the openings and the use of controlled air conditioning system, the quality of the interiors should be improved. The control system should be able to switch off the air conditioning system when the doors and windows are opened. Users should be informed about the use of electric control systems and the amounts of energy consumption.
In the school building, artificial lighting is used throughout the school corridor throughout the day, while there is no problem with lighting in the classrooms (Figure 2 and Figure 3).

Heating in the school is provided by natural gas. There is no heating problem. However, different methods have not been developed to reduce energy consumption. The school does not have solar energy. Solar energy will contribute greatly in order to benefit from the warming of the school and hot water.

Figure 2. General view of classrooms and corridor

The school roof is quite suitable for the installation of solar panels. Insulation material used in the school is Expanded Polystyrene (EPS) which is an oil derivative material. Insulation is beneficial in terms of heating but it is a health hazardous material. There is no cooling system in the school. In the summer months, windows are opened and cooling is done. This causes noise to be affected during the course. In addition, dirty weather and the entrance of various classes into the classroom can disturb the students.

4.2. Design Criteria about Economic Sustainability

According to Cole [13], economic sustainability is divided into two groups as investment and utilization cost. It is significant that building processes, building components, and materials are cost efficient, highly resistant and reusable. By renewing and reusing buildings, "long-term efficiency of resource" is ensured. Low utilization costs are achieved through efficient energy consumption of the building and through simple operation and maintenance. Building form, material selection, venue organization, building the shell and building positioning criteria are discussed under economic sustainability heading.

4.2.1. *Building Form.* A form of the building is one of the design criteria that have an impact on heating and air conditioning energy preservation. Building form can be defined through building related geometric variables such as the building shape (ratio of the building length to the building depth as stated in the plan), building height, roof type, inclination, the inclination of the facade surface. Ratio of the size of the building shell surface - which limits the venues to protect them from external factors- to building area, plays a significant role in energy loss and gain issues significant criteria that should be kept in mind when designing education buildings as sustainable buildings are the following; the width/length of the building should be designed by taking climatic data into consideration, minimization of building's external surface area, selection of optimal building forms by taking climatic data into consideration, determination of the height of floors in a way that would benefit most from natural lighting and designing of the waves of buildings for shading purpose.

4.2.2. *Selection of appropriate materials and building components.* In ecologic architecture, selection of materials is important for the sustainability of buildings. Selection of natural materials that would not harm the nature can come into minds for the first stage. Sustainable education buildings require the utilization of resistant materials, renewable and recyclable materials. Furthermore; utilization of energy efficient materials, from the acquisition of raw materials to their destruction in nature would be useful. It should also be kept in mind that, the basic ecologic design involves the critical selection of natural and environment friendly materials. The materials used throughout the school are generally the most commonly used materials today. But sustainable materials have not been selected. The rate of green space at the school garden is very low. There is a large hard floor in the garden. Garden floor material asphalt. This is not a suitable material for children. Asphalt material is dangerous for children because it is hard and slippery. In cold and rainy weather it is all caused by ice on the uneven floor.

4.2.3. *Spatial organization.* Spatial organization is the combination of user requirements and preferences and aesthetic decisions. Each of these components provides significant input concerning the energy performance and the environmental impact of the building. In this sense, the use of outdoor, semi-outdoor and indoor areas, positioning of these buildings, heat loss and gain on internal and external surfaces limiting the borders of the place determine the impact of the spatial organization on energy use.

Figure 3. General view of toilets and entrance

For a spatial organization in education buildings, gallery systems should be used as much as possible, it should be determined how much heat the places need and for which purposes such places will be used. In parallel with these, places that are desired to be continuously kept warm should be surrounded with places that are temporarily heated and the system should be embraced with a vertical and horizontal buffer zone. For interior comfort, it is important to use mobile materials such as exchangeable mobile panels to ensure a flexible interior design, to select an appropriate positioning to ensure the best insolation and to design appropriately positioned pass ways (such as a greenhouse, glasshouse etc.). Enabling the use of places together with the surrounding community and selection of modular designs that could easily adjust to future needs are particularly important for social sustainability.

4.2.4. *Building shell.* Building shell is the component separating the interior building environment from the exterior building environment, which consists of all horizontal, vertical and inclined building components and it is the most significant variable for the energy conservation and the provision of climatic comfort. One of the criteria that should be kept in mind when designing the shell of education buildings is that, on facades which can make the best use of natural lighting, wide windows should be preferred and on facades exposed to dominant wind, the fewer amount of windows should be preferred. If possible, the use of natural green facades and roof, transparent insulation, roof and wall insulations should be at optimal level. Aesthetic needs, impact on the life cycle of the building and maintenance-repair needs should also be considered in selections.

4.3. Design criteria about social and cultural sustainability

Redclift [14] discusses sustainability within the framework of the continuity of the relations between individuals and social institutions regulating the natural, economic and politic contexts. In the most general meaning, social and cultural sustainability means; "preservation and development of social conditions that would help the fulfillment of human needs and environmental sustainability to ensure that natural resources are used efficiently by current and future generations". According to CIB [9] and Sakınç [10], one of the basic goals of sustainable buildings is the preservation of the socio-economic, cultural and political requirements of the place. Basic principles of this goal are; understanding of the social and economic realities of the society, preservation of social and cultural diversities, understanding the social needs and demands, ensuring active participation of societies to the creation of their own living spaces. Among the socio-cultural sustainability components in sustainable education buildings are; design of building in harmony with its environment, the presence of symbols around the building, recovering the building from a ruined look, prevention of visual pollution, preservation of culture and heritage, improvement of life standards. As within the scope of sustainability it is required to raise the consciousness of people regarding the preservation of natural resources and handing those down the next generations, the issue should start with the children in education buildings [15].

5. Conclusion

In order to be able to hand down a healthier environment to future generations, environmental consciousness of the people shall be raised first. Consciousness can be raised by educating school-age children with this sensitivity. When children grow up with ecologic consciousness starting from the school-age, they will be more sensitive to environmental problems in the future. When children practice and experience the theoretical information they have gained, such information becomes permanent. An education building that is designed in parallel with sustainability criteria must be sufficient in all subheadings like land selection, utilization of natural resources, generation of energy through renewable methods, optimization of water consumption, recycling control, renewal, waste control, building maintenance, and repair. Sustainable education building design shall necessarily include certain parameters such as energy control and efficiency, the sufficiency of interior quality, user comfort and sensitivity of the building to the exterior environment. In Turkey, as well as many other developing countries, the significance of sustainable education buildings has only recently become recognized and the issue has not been sufficiently involved in laws and regulations, yet. A single type of school buildings significantly limits the sustainability performance expected from buildings. Standard type projects shorten the planning time as they include a designing process that is independent of settlement and they are repeated in various places with different characteristics, indeed. On the other hand; abundance of disadvantages such as the overlook of the natural physical and structural properties of the location mostly restricts the sustainable design of the building. For sustainable buildings, several factors such as the environment, land, climate, insolation, direction etc. shall be taken into consideration at the beginning stage. Therefore; implementation of standard type projects can be deemed to be inappropriate for sustainability. In a world, where the resources are rapidly being consumed, the sensitivity shown by developed countries to sustainable architecture shall be shown by developing countries, too. In order to hand down a healthier world to future generations, sustainable building techniques shall be included among the basic design criteria.

7. References

[1] Aytıs S and Polatkan I 2010, *Basic Principles and Structure in Design and Community Appraisal*, YTU, 151-163, Istanbul.
[2] Akın A 2010 *Balanced Design Criteria for Settlement and Building Scale Identification*, YTU, 233-240, Istanbul.
[3] Baysan O 2013 *Concept of Sustainability and Design in Architecture*, Msc thesis, ITU, Istanbul.
[4] Murphy C Thorne A 2010 *Health and Productivity Benefits of Sustainable Schools: A Review*, Brepress, Watford.
[5] Foster N 2001 *Lord Foster of Thames Bank, Architectural Design*, 32.
[6] Williamson T Radford A Bennetts H 2003 *Understanding Sustainable Architecture* Spon Press, London
[7] Kohler N 1999 The Relevance Of The Green Building Challenge: An Observer‘s Perspective *Building Research & Information*, 309-320.
[8] Arsan Z D 2008 Sustainable Architecture in Turkey Architecture Journal, 340, Ankara.
[9] CIB(1999) http:// www.cibeg.com
[10] Sakınç E 2010 Balanced Design Criteria for Settlement and Building Scale Identification, YTU, 151.
[11] Kayıhan K 2006 Sustainable Architecture in Semi-humid Marmara Climate Examination of the Basic Education Buildings and a Methodology, PhD Thesis, YTU, Istanbul.
[12] Lakot E 2007 Energy Effective Double Shelled Building Façade with Respect to Ecology and Sustainability MSc thesis, KTU, Trabzon.
[13] Cole R J 1999 Building Environment Assessment Methods: Clarifying Intensions Building Research&Information, 230-246.
[14] Redclift M and Woodgate G 1997 Sustainability and Social Construction, in The International Handbook of Environmental Sociology", Edward Elgar Pub., 55-68, United Kingdom.
[15] Çahantimur A 2007 A social cultural aspects on sustainable urban development : Burca City Sampe, PhD Thesis, ITU, Istanbul.
[16] Tonguç B 2012 Evaluation of sustainable design for pre-schools Msc thesis, Kocaeli Üniversity, Kocaeli.

IOP Conf. Series: Earth and Environmental Science **83** (2017) 012027 doi:10.1088/1755-1315/83/1/012027

Climate Change Adaptation Practices in Various Countries

A Tanik[1], D Tekten[1]

1 Environmental Engineering Department, Istanbul Technical University, Maslak, 34469, Istanbul-Turkey
E-mail: tanika@itu.edu.tr

Abstract. The paper will be a review work on the recent strategies of EU in general, and will underline the inspected sectoral based adaptation practices and action plans of 7 countries; namely Germany, France, Spain, Italy, Denmark, USA and Kenya from Africa continent. Although every countries' action plan have some similarities on sectoral analysis, each country in accordance with the specific nature of the problem seems to create its own sectoral analysis. Within this context, green and white documents of EU adaptation to climate change, EU strategy on climate change, EU targets of 2020 on climate change and EU adaptation support tools are investigated.

1. Introduction

Today, it is no longer an inevitable fact that consequences of climate change are irreversible. If all greenhouse gas emissions are stopped, existing gases in the atmosphere will continue to warm the earth for next 50 years. For this reason, adapting to climate change and minimizing its possible effects has become a real necessity in today's world. Therefore, experiences gained and lessons-learnt via the application of adaptation policies and practices of some developed countries contribute well to especially the developing countries; in a way by examining their practices, time and money will be highly saved in setting the adaptation policies and strategies of each independent country. It is for sure that adaptation policies and strategies are private and special for each of the countries examined; however, it is seen that so many similarities exist among them regarding their policies set for different sectors. Experiences of 7 selected countries; namely Germany, France, Spain, Italy, Denmark representing European countries, USA as a developed country example from America continent, and finally Kenya from Africa continent together with a trans-boundary catchment of Downstream Basin of Danube will be addressed in this study. Each of the inspected countries and the catchment has already managed to make significant progress towards adaptation, especially their choice of renewable energy, cleaner production, climate-resistance systems, energy conservation, recycling, use of environmentally-friendly technologies such as recycling through implementing an effective action plan and strategy has been determined. Although every country's action plan has some similarities on sectoral analysis, each country in accordance with the specific nature of the problem seems to create its own sectoral analysis. Prior to review the recent practices of the selected 7 countries, initially European Commission's working documents and official web portal were investigated thoroughly to reveal why there is a need for an effective climate change action plan at watershed scale in order to become prepared for climate change and to adapt living with this concept. Within this context, Green Paper and White Paper of EU adaptation to climate change, The EU Strategy on adaptation to climate change, the new general Union Environment Action Programme to 2020 and EU adaptation support tools are investigated. Human-animal-plant adaptation, forest adaptation, infrastructure adaptation, marine adaptation green infrastructure strategy, and blue growth strategies are those significantly referred adaptation tools of EU oriented efforts towards developing EU policies on climate change. The paper will be a review work on the recent strategies of EU in general, and will underline the inspected sectoral based adaptation practices and action plans of 7 countries.

2. Adaptation strategies of EU

The development of strategies by EU on adaptation practices will be briefly summarized in a chronological order.

2.1. Green paper

This document is known to be the initial official document that forms a basis for the adaptation policies of Europe. It had been prepared in 2007 by the European Commission (EC). It generally mentions the probable risks of climate change and further underlines the related regulation, application of the regulation and the funds that may be used meanwhile [1]. 4 main headings come out of this document; namely, necessity for developing adaptation policies over Europe, necessity to communicate with the neighbourhood countries to identify the probable global risks, necessity to enlarge the present database through public consultation and to decrease the uncertainty, and a discussion of the findings and developed strategies in the Adaptation Advisory Council of EU.

2.2. White paper

As a result of thorough discussions following the publication of the initial document, EC declared the second document which is mainly based on a conceptual framework regarding the reduction of the negative effects of climate change. The document explains the fundamental ways of dealing with the problem of concern in order to increase the resistibility of certain fragile sectors like agriculture, biodiversity and coastal ecosystems [2]. Accordingly, this document refers that the priority issue is to reduce greenhouse emissions by 20% till the end of 2020 compared to emissions of 1990. As such, it aims to support the strategies heading the conservation of natural resources in the most sustainable manner.

2.3. The EU Strategy on adaptation to climate change

This strategy has been published in 2013 by EC covering 3 main targets; namely, encouraging adaptation actions by the member states, supporting more conscious decision making steps, and supporting the adaptation of sensitive sectors [3]. The application of this strategy includes 8 actions.

2.4. The new general Union Environment Action Programme to 2020

EC has determined the adaptation targets to be achieved by 2020 through the declaration of 'New and General Environment Action Program' in 2013 that covers 9 priority targets among which protection, safe and effective use of natural resources, initiation of low carbon economy, putting into force new related regulations, enlargement of the present environmental database, encouraging safe, environmental and effective investments, merging environmental problems into politics, promoting sustainability of EU cities and accelerating the adaptation actions all throughout the member states can be addressed as the leading efforts [4].

2.5. Adaptation support tool

The main objective of this tool is to develop adaptation strategies and plans together with providing guidance to users during the application period [5]. With this aim 6 consecutive steps are determined that are listed below.

- Preparing the region to adaptation,
- Determining the sensitiveness and probable risks of climate change,
- Determining region-specific adaptation options,
- Commonly evaluating the selected actions,
- Practicing the adaptation actions, and
- Monitoring and improving the actions though receiving feedbacks from the applications.

In this regard, it is clearly obvious that a successful adaptation process may be initially formed by perfect identification of key issues. Meanwhile, it is also certain that a sufficient financial resource need to be allocated to realize the determined adaptation strategy and the comprehensive researches on the present and future climate change risks and effects should be continuing without any delays.

3. Sectoral analysis

3.1. Sectoral policies and adaptation alternatives

There appears 62 adaptation actions within the context of adaptation activities against climate change in the first official web portal of EU [6]. These actions are categorized under 10 main headings which are namely; Agriculture and Forestry, Biodiversity, Coastal Areas, Mitigating with Natural Hazards, Financing, Health, Infrastructure, Marine Environment and Fishery, Water Management, Urban Areas. While distributing the actions among the basic sectors, it is seen that many of the actions are repeated in different sectors putting forth the reality that there are lots of common adaptation actions valid for various sectors. Thus, another categorization came forth. In this one there exists only 4 main headings; namely, infrastructure (energy, and construction), marine (coastal issues, hazards, fishery, erosion), living bodies (human-animal-plant nexus) and forestry [7]-[10].

3.2. Recent adaptation practices in various countries

The EU strategies and documents published so far act as a road map for the member states. As such, various EU countries try to generate their adaptation policies and actions based on the fundamental concepts processed by the EC. Germany, France, Spain, Italy and Denmark are selected as those countries representing different regions of Europe. Apart from Europe, USA from America continent and Kenya from the Africa continent are also nominated in this study with the basic aim of demonstrating the current situation of adaptation studies in various countries of the world.

Table 1 gives a summary on their demographic and economic characteristics together with information on adaptation policies and actions. It is important to note at this stage that most of the selected countries that are developed countries in general have already declared their national adaptation strategies as well as their vulnerability assessments. Every gained data and information gathered is shared at their portal sites to make the information public with the only exception of Kenya which is expected.

On the other hand, Table 2 is a brief summary of the sectors considered in the selected countries utilized for country specific sectoral analysis within the framework of adaptation studies. It is clearly seen that many of the sectors (shaded in the table) are common ones valid for almost all the countries; however, there is an addition of some other self-specific sectors for different countries indicating that each country may also bear some specific sectors underlined within the context of climate change effects. The 13 sectors stated in Table 2 common to all the examined countries may be accepted as the priority sectors attracting the interest worldwide. As seen and as expected health, water resources and agriculture are shared by all the selected countries indicating that these 3 sectors are the leading sectors that might be referred to as the highly vulnerable sectors regarding the effects of climate change. Forestry and biodiversity/ecosystem, energy/industry, infrastructure/transportation, urban planning/built environment and tourism are the rest of the ones that follow the priority sectors. Coastal areas also are equally important as the other ones; however, this sector is a specific one that recounts a coastal country. Of course, mitigation with climate change effects and application of adaptation policies and activities will be requiring additional budgets to all the countries. In this respect, the realization of cost-benefit analyses is of utmost importance to estimate the probable incidences and occurrences that might take place in the future. Therefore, another sector, funding and insurance is addressed as a common one. While dealing with the probable risks and effects, equally significant consideration must be to supply necessary funding and insurance for the damages on properties to save and protect humans as much as possible.

3.3. Germany

The adaptation action plans in Germany prepared by the federal government has been initially stated in 2008 where there were 14 sectors considered. The revised version of the action plans are declared and published in 2014 where only 8 sectors were reflected. In each of the prepared versions, the probable impacts and the corresponding measures to be taken for each of the sectors concerned are explained [18]. The 14 sectors involved in the adaptation studies are seen in Table 2. Germany is one of the leading countries taking place in the Carbon Disclosure Project (CDP) that aims to form a cooperation platform among investors and companies interested in climate change issues. The project brings together more than 300 firms. Another outstanding activity of the country is that it accelerates insurance companies on dealing with climate change risks and mitigation activities. For realizing active and efficient actions, the insurance companies, reassurance firms, environment associations and scientists have incorporated under the Munich Climate Insurance Initiative to help insurance firms during coping with climate change risks. Another initiative that has been formed is Finance Forum: Climate Change (FFKw) that mainly focuses on the management of research and innovation activities regarding reduction of effects and adaptation. This forum has been established by the Federal Ministry of Education and Research.

Table 1. Information gathered on the selected countries

Country	Population (2016)	Area (km^2)	GDP (€) (2016)	Action Plan	National Adaptation Strategy	Impact, Vulnerability & Adaptation Assessment	Web Portal	Reference
Denmark	5.457.415	49.094	257.444 billion	√	√	√	√	[11]
France	65.856.809	632.833	2.132 trillion	√	√	√	√	[12]
Germany	80.780.000	357.340	2.904 trillion	√	√	√	√	[13]
Italy	61.680.122	301.263	1.616 trillion		√	√	√	[14]
Spain	46.507.760	505.970	1.058 trillion	√	√	√	√	[15]
Kenya	45.846.000 (2015)	569.259	53,5 million (2015)	√				[16]
USA	327,263.800 (2014)	9,629,091	1,695 billion (2014)	√	√	√	√	[17]

3.4. France

The adaptation action plans cover 20 sectors in general and the precautions to be taken are listed by the Ministry of the Ecology and Sustainable Development [19]. These sectors appear Table 2. In this country, a more detailed sectoral analysis is managed by taking into account research, governance, and international cooperation as equally important sectors like the other common sectors that will most probably be affected by climate change. Like is the case in other countries, France is trying to investigate the relationship of human health and climate change. In that sense, it has formed expert teams composed of multidisciplinary scientists. The expectations from these teams is to maintain successful development of health-climate researches, identify of risk causing factors, increase monitoring activities on risk factors, evaluate health and risk relationship under extreme weather conditions, improve and develop protective health measures and increase public awareness through communication and education. Similarly, other sectors also involve certain predetermined mitigation actions. In general, it has been estimated that in case of taking no precautions against reducing the probable effects of climate change, the cost of this situation to the country will be almost around 5-20%

of the national gross product (NGP); however, this cost may be reduced up to 1-2% of the NGP by taking precautions. It is important to conduct such cost-benefit analysis prior to taking adaptive actions. ECONADAPT is an EU 7th Framework Project that is still ongoing that deals with search of the cost-benefit analysis of the worldwide countries. The draft version of the first report in 2015 shows the overall situation of the world in terms of cost-benefit analysis [20].

Table 2. Sectors considered in selected countries for sectoral analysis

SECTORS	Denmark	France	Germany	Italy	Spain	Kenya	USA
Health	√	√	√	√	√	√	√
Water Resources	√	√	√	√	√	√	√
Agriculture	√	√	√	√	√	√	√
Forests	√	√	√	√	√		√
Biodiversity/Ecosystem	√	√	√	√	√		√
Funding and Insurance	√	√	√		√		
Fisheries and Aquaculture	√	√	√		√		
Energy and Industry		√	√	√	√	√	√
Infrastructure and Transport Systems		√	√	√	√		√
Urban Planning and the Built Environment		√	√	√	√		
Tourism		√	√	√	√	√	
Coastal Areas		√	√	√	√		
Soil			√	√	√		

3.5. Spain

The Ministry of Environment has prepared the adaptation policies and strategies of the country in which there are 13 main sectors defined that will be under the effect of climate change [21]. As seen from Table 2, the sectors covered in Spain are quite similar to the other countries. The initial action plan has been published by the government in 2008 and then revised in 2014. As the sectoral based studies have not been translated into English, detailed information cannot be obtained on the realization of the actions.

3.6. Denmark

The prominent issues in the country are addressed as increase in precipitation, rise of sea level, temperate winters, comparatively more hot summers, severe storms and increase in the extreme weather conditions. Thus, the Denmark Government has prepared adaptation action plans and strategies in 2008 [22]. Apart from the 7 common sectors of health, water resources, agriculture, forests, biodiversity, funding and insurance, fisheries and aquaculture, 2 additional specific sectors are considered. These are natural hazards and land-use planning.

3.7. Italy

The initial action plan has been prepared in 2008 that covered 4 main categories. This plan has been valid till 2015. The revised action plan includes short term (till 2020) and long term (after 2020) action plans. However, details of this new version have not been translated into English hindering to provide more information. Table 2 shows that apart from the common sectors covered like the other countries, Italy considers 2 specific sectors of hydro-geological risks and cultural heritage. As such, one can easily understand that this old and historical coastal country has already faced problems related to these two sectors. Therefore, the country being aware of the future probable climatic fluctuations added these sectors to their adaptation strategy in order to reduce the negative impacts of global climate change effects.

3.8. USA

In this developed country, there appears 2 action plans; one stated by the Government whereas the other one is by the President. As the country owns large land quite different from the comparatively smaller Europe continent, different regions are under the effect of different climatic conditions. In this respect, adaptation measures and applications are considered state and/or region based. As seen from Table 2, similar to EU countries, USA also pays attention to some common sectors [17]. Adaptation actions deliberated by the Federal Government initially started to direct public entities in 2009 especially on reducing greenhouse emissions, disposal of wastes, increasing the performance of energy and water consumptions, using of clean production technologies and on utilizing environmentally-friendly products. Further onwards, in October 2010, a series of adaptation targets and remedial suggestions are forwarded to the President. A year later in October 2011, a new report has been published and Adaptation Task Force has recommended a list of adaptation items and has underlined the significance of dissemination activities to strengthen and to better understand the adaptation policies and requirements. The integrated adaptation actions of the President have been initially published on July 2013. These actions emphasize on the termination of carbon pollution, preparation of the country for the probable impacts of climate change and on the importance of internationally and inter-institutional dealing with the main topics related to emission reductions. A few of these inter-institutional examples can be accounted as recovery from disasters as was the case in Sandy Hurricane, coordination of disaster management, management of heavy fires, taking precautions against sea level rise, protection of biodiversity, understanding the impacts of climate change, protection of human health, management of natural resources, supporting agriculture sector, better building transportation networks, preparation for the coming energy requirements, development of tools to support local decisions, supporting public awareness, protection of governmental bodies, designing infrastructure for future, accelerating the work of adaptation initiatives at local, state region and/or tribe basis.

3.9. Kenya

Adaptation plan for the country has been made ready in 2013. Its application period has been determined as 2013-2017. It is considered that the expected targets will be achieved by 2030. The country has already focused on 5 sectors; namely, health, water resources, agriculture, energy & industry, and tourism. Cost analysis has also been done for the sectors and it is seen that health and sanitation issues are of high importance followed by transportation infrastructure especially on arid lands [16].

4. Cost-benefit analysis for adaptation practices

ECONADAPT Project (2013-2016) relates to Cost and Benefits of Adaption. The first draft report of this project that has been published in July 2015 refers to cost estimations of adaptation practices in both Europe and other continents. It is expected that at the global scale adaptation investments will be around 50-170 billion USD$/year till 2030. A majority of this amount will be allocated to the developed countries for their infrastructural needs whereas the developing countries will be using the minority (30-70 billion USD$/year) [20]. Figure 1 illustrates the status of the worldwide countries regarding their activities on cost-benefit analysis. In this respect, many countries in the America continent led by USA have already managed to cost-benefit analysis regarding climate change. Europe continent seems to be highly lacking similar trials and efforts. Still many countries in the world have not initiated any cost analysis so far regarding climate change impacts.

Sectoral based global scale cost analyses exert some interesting results. A few of these striking findings are as follows;

- Protection of coastal areas against disasters like sea level rise and overflows may technically be achieved through construction of seawalls along coastlines. According to the RCP2.6

scenario, it will annually cost 12-31 billion USD$ till 2100 whereas based on RCP8.5 scenario, this cost will be around 27-71 billion USD$.

- Protection of important coastal structures like ports and harbours against sea level rise will charge for an annual cost of 0.5 billion USD$ till 2050 in the developing countries.
- It is estimated that additional budget for improving water supply systems at global scale will be 9-11 billion USD$/year at global scale.
- Greece confronts high water losses (60%) via technical reasons and illegal uses in the water transmission systems. In order to decrease this loss to 10% in a stepwise manner till 2100, it is estimated that the annual cost will be 68 million USD$ whereas the benefit derived from this reduction will be 380 million USD$. This single estimation proves that the benefit gained through application of adaptation activities is far more above the related cost required to realize these actions.

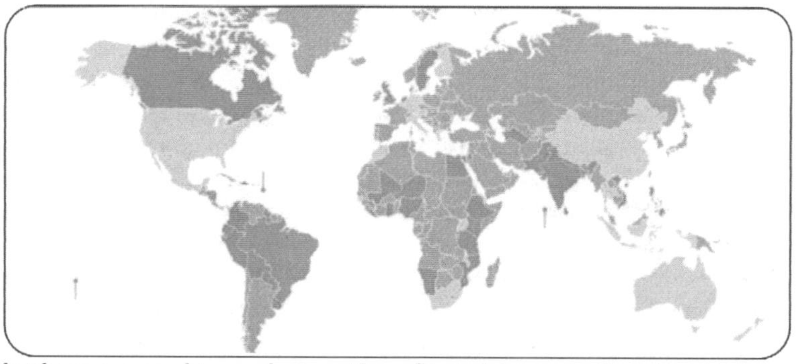

Countries that have managed to conduct cost/benefit analyses regarding climate change,
Countries with ongoing national efforts on cost/benefit analyses,
Countries that have not started any studies on cost/benefit analyses

Figure 1. The current worldwide situation on cost/benefit analyses of adaptation practices [20]

Table 3 shows the condition of cost-benefit analyses at the global scale.

Table 3. Condition of cost-benefit analysis at the global scale [20]

RISK / SECTOR	Cost Analysis	Benefit Analysis
Coastal areas and coastal storms	+++	+++
Infrastructure development against flood risk	++	++
Water Management	++	+
Other infrastructure services	+	+
Agriculture	++	++
Overheating (building, energy, health)	++	+
Health risks	+	+
Biodiversity / ecosystem services	+	
Industry, trade and public services	+	+

+++: Comprehensive analyses of different geographical conditions. These analyses include uncertainty studies.
++: Medium-level sectoral case; +: Low-level sectoral case.

- According to a study conducted in England and Scotland, it is found out that the cost of adaptation activities between 2006-2080 will be around 6-39 million £ /year; whereas the cost in case of taking no protective action against climate change effects is expected to be around 41-388 million £ /year.

- According to World Bank, the cost of infrastructure renewal/repair based on the worst (wet) scenario will be 13.5-27 billion USD$/year in the developing countries between the period of 2010-2050. 54% of this amount corresponds to urban infrastructure. The rest is shared among railways and highways [23].
- According to World Bank, approximate cost of the agriculture sector in the developing countries at the global scale is expected to vary between 2.5-3 billion USD$[23].
- Adaptation activities in the health sector at EU is estimated to be around 1-10 million € [23].

It may easily be observed from Table 3 that coastal countries meditate that they will suffer from the probable risks posed on their coastal regions either due to flooding/overflow and sea level rise or to strong storms. Therefore, the most comprehensive cost/benefit analyses have so far been performed on the coastal areas. Medium level analyses are done on water management, agriculture, infrastructure and on overheating as expected.

5. Concluding remarks

The aim of this study is to compare the current climate change adaptation practices and action plans of 7 countries. Lessons-learnt and the initial attempts towards mitigating with climate change effects are outlined. Uncontrollable climatic factors, extreme weather conditions and, unexpected natural occurrences threaten almost all the countries of the world no matter what the development level of the nation is. By inspecting the today's condition, it can be stated that such efforts are still not at a desirable level. Conducting such analyses is so important in understanding the reality that '*doing nothing'* will cost more than the protective actions that has to be taken. This review study puts forth another point that countries urgently have to provide additional funding to their budget regarding the cost required to realize adaptive actions. Otherwise, the consequences of not taking immediate/urgent actions may in the short and long run will cost them a lot. Moreover, through this study, it is once more seen that the three pioneer sectors to be involved in adaptation studies are health, water management and agriculture.

6. References

[1] European Commission 2007 Communication from the Commission to the European Parliament The Council The European Economic and Social Committee and The Committee of The Regions adapting to climate change in Europe – options for EU action Green Paper
[2] European Commission 2009 White Paper Adapting to Climate Change: Towards a European Framework for Action Commission of the European Communities
[3] European Commission 2013a The European Economic and Social Committee and The Committee of The Regions An EU Strategy on Adaptation to Climate Change Communication From The Commission to The European Parliament
[4] European Commission 2014 The Economic Impact of Climate Change and Adaptation in the Outermost Regions Final Report (Luxembourg: Publications Office of the European Union)
[5] http://climate-adapt.eea.europa.eu/adaptation-support-tool. Accessed on 29.09.2015.
[6] http://climate-adapt.eea.europa.eu/adaptation-measures. Accessed on 01.02.2016.
[7] European Commission 2013b The European Economic and Social Committee and The Committee of The Regions An EU Strategy on Adaptation to Climate Change Adapting Infrastructure to Climate Change Communication From The Commission to The European Parliament The Council
[8] European Commission 2013c The European Economic and Social Committee and The Committee of The Regions An EU Strategy on Adaptation to Climate Change Coastal and Marine Adaptation Communication From The Commission to The European Parliament The Council
[9] European Commission 2013d The European Economic and Social Committee and The Committee of The Regions An EU Strategy on Adaptation to Climate Change Commission

Staff Working Document, Adaptation to Climate Change Impacts on Human, Animal and Plant Health Communication From The Commission to The European Parliament The Council
[10] European Commission 2013e The European Economic and Social Committee and The Committee of The Regions A new EU Forest Strategy: for Forests and the Forest-based Sector Communication From The Commission to The European Parliament The Council
[11] http://climate-adapt.eea.europa.eu/en/web/guest/countries/denmark. Accessed on 29.10.2015
[12] http://climate-adapt.eea.europa.eu/en/web/guest/countries/france. Accessed on 29.10.2015
[13] http://climate-adapt.eea.europa.eu/en/web/guest/countries/germany. Accessed on 29.10.2015
[14] http://climate-adapt.eea.europa.eu/en/web/guest/countries/italy. Accessed on 29.10.2015
[15] http://climate-adapt.eea.europa.eu/en/web/guest/countries/spain. Accessed on 29.10.2015
[16] http://www.ekonomi.gov.tr/portal/faces/home/disIliskiler/ulkeler/ulke-detay/Kenya. Accessed on 11.04.2016.
[17] United States Climate Action Report 2014 First Biennial Report of the United States of America 6th National Communication of the USA Under the UN Framework Convention on Climate Change (US Department of State)
[18] Nature Conservation, Building and Nuclear Safety The German Government's Climate Action Programme 2020 Cabinet decision of 3 December 2014 Federal Ministry for the Environment
[19] Transport and Housing, French National Climate Change Impact Adaptation Plan 2011 The French Republic Ministry of Ecology Sustainable Development
[20] Costs and Benefits of Adaptation 2015 EU7th Framework Project ECONADAPT First Draft Report (2013-2016)
[21] Spanish National Climate Change Adaptation Plan 2008 Gobierno De España Ministerio De Medio Ambiente Y Medio Rural Y Mari
[22] http://denmark.dk/en/quick-facts/facts/. Accessed on 11.04.2016.
[23] The Costs to Developing Countries of Adapting to Climate Change: New Methods and Estimates 2010 The Global Report of the Economics of Adaptation to Climate Change Study Synthesis Report World Bank (Washington)

Development of Landscape Architecture through Geo-eco-tourism in Tropical Karst Area to Avoid Extractive Cement Industry for Dignified and Sustainable Environment and Life

Pita A. B. Cahyanti[1] and Cahyono Agus[2]

1 Department of Architecture, Faculty of Engineering UGM Yogyakarta Indonesia
2 Faculty of Forestry, UGM Yogyakarta Indonesia
E-mail: pita.asih.b@mail.ugm.ac.id

Abstract. Karst areas in Indonesia amounted to 154,000 km^2, potentially for extractive cement and wall paint industries. Exploitation of karst caused serious problems on the environment, health and social culture of the local community. Even though, karst region as a natural and cultural world heritage also have potential environmental services such as water resources, carbon sink, biodiversity, unique landscapes, natural caves, natural attractions, archaeological sites and mystic areas. Landscape architectural management of in the concept of blue revolution through the empowerment of *land* resources (soil, water, minerals) and biological resources (plant, animal, human), not only have adding value of economy aspect but also our dignified and sustainable environment and life through health, environmental, social, cultural, technological and management aspects. Geo-eco-tourism offers the efficiency of investment, increased creative innovation, increased funding, job creation, social capital development, stimulation of the socio-entrepreneurship in community. Community based geo-eco-tourism in Gunung Kidul Yogyakarta rapidly growing lately due to the local government banned the exploitation of karst. Landscape architecture at the caves, white sand beaches, cliffs in karst areas that beautiful, artistic and have special rare natural architecture form of stalactite and stalagmite, become the new phenomenal interested object of geo-eco-tourism. Many hidden nature objects that had been deserted and creepy could be visited by many local and foreign tourists. Landscape architectural management on hilltops with a wide view of the universe and fresh, sunset and sunrise, the clouds country are a rare sight for modern community. Local cultural attractions, local culinary, home stay with local communities will be an added attraction, but the infrastructure and human resources should be developed. Traveler photographs that widespread rapidly through social media and mass media became a great and effective promotion. With geo-eco-tourism, people can empowering natural resource to gain harmonization of economic, environment and social-culture aspect, without destroy it.

1. Introduction

Indonesia is located in equator and represents a series of equatorial emerald with huge potential of natural resources [1]. Over explotation of natural resources that are deposited deep in earth by open mining causes serious degradation of living and environmental things on the earth [2]-[5]. Karst area in Indonesia covers huge natural landscape of about 154,000 km2. The karst area in Java Island is 11,000 km2 and it becomes the target of extractive industry such as cement factory [6]. It has vital economic, environmental, social and cultural functions because it produces calcium carbonate, natural water reservoir, climatic change mitigation, habitat of swallows, bats and other flora and fauna and also human being. Therefore, it is necessary to rearrange green karst area by putting the emphasis on the harmonization of all of the existing aspects.

Gunung Sewu that has been officially announced as UNESCO Global Geopark in September 2015 stretches across several regencies (Gunungkidul, Wonogiri, and Pacitan) and provinces (Yogyakarta, Central Java, and East Java) [6]. The Gunung Sewu Geopark is a classic tropical karst landscape in the south central part of Java Island that is dominated by limestone, which is well-known in the world. Tectonic activity still occurs in the region because it is situated in front of an active subduction zone between the Indian Ocean, Australian and Eurasian plates. Active uplifting has been taking place for 1.8 million years and results in the emergence of river terraces at Sadeng dry-valley and also coastal terraces along the southern coast of the Global Geopark [6], [7]. Approximately 805,000 people inhabit the area., with the economy of the local people is driven by agriculture and service sectors [6]. In addition to its aesthetic and recreational values, the area is also rich of biodiversity, archeology, history and culture.

It is necessary to develop the Gunung Sewu Geopark in the effort of venerating earth legacy for the purpose of the prosperity of the local people. Geopark represents a geographic area in which geological legacy sites are parts of the concept of conservation, education and sustainable development that puts three elements in harmony, which are geodiversity, biodiversity and cultural diversity [7]-[9]. The synergy among the natural diversity should be managed in an integrated and sustainable manner by empowering local people in a sustainable development that gives economic, environmental, social and cultural added values [9].

2. Materials and methods

The study uses descriptive method and qualitative approach with naturalistic paradigm. Data are collected using in-depth interviews, observation and documentation. The objective of the triangulation data collection technique is to examine the validity of the data using other objects in comparing the results of the interviews about the object of the study. The data are analyzed inductively and qualitatively and the results of the study put more emphases on meaning than on generalization. The concept of integrated bio-cycles system is implemented through holistic and sustainable landscape ecological management in the development of the geopark that is based on education for sustainable development.Other paragraphs are indented (BodytextIndented style).

3. Results and discussions

3.1. New paradigm in karst management

The massive transformation necessary to do in the management of karst is to change extractive economy that causes environmental degradation into productive conservation activity that improves the prosperity of the local people in a holistic and comprehensive development. The karst area should be managed in harmonious, holistic, integrated and sustainable manner for qualified (i) biomass production, (ii) living environment, (iii) biological habitat, (iv) infrastructure, (v) mineral resources, and (vi) aesthetics and culture [3], [5]. Each of the elements of the landscape should not compete for its own sectoral interest, but they should be in mutual and harmonious supporting relation. The output and the outcome of the system are given more emphasis than the output of each constitutive elements [1], [2], [3], [5].

The management program of the karst area should be implemented on the basis of three main pillars of personality, community and institutional empowerment. It is expected that emphaty, care, multidisciplinary cooperation, personality, the contribution of local/national competitiveness and learning community/society. The empowerment program should also be implemented in co-creation, co-finance, and in sustainable and flexible ABCG (academician, business, community, government) network [4]. The empowerment program of 6M (men, money, material, machine, method, and management) should be implemented in a synergetic and optimal manner that all of the existing stakeholders have the ability, the willingness, the chance and the authority to really contribute and to gain optimal merit [5].

Gunung Sewu geopark has been successful in improving the prosperity of the local people, though it is still necessary to make further effort to gain maximal results. The lack of the cooperation among the stakeholders and the lack of the regional zoning are the main obstacles in the development of the area. It is necessary to make a zoning of the geopark into three zones, including main zone, supporting zone, and service delivery zone [9], [10]. The main zone (Blue) is managed by putting the emphasis on the development of the core tourist attraction of geo-tourist objects of flora and fauna. The supporting zone (Green) is managed by putting the emphasis on the construction and the maintenance of supporting facilities that prevent the tourists from doing any harm to the nature and its beauty and set the threshold of the supporting capability of the environment. The service delivery zone (Yellow) consists of the areas with the facilities necessary for the tourists, which are developed by considering the requirements of the geo-tourism.

It is necessary to formulate different strategies in developing the geopark area that are specific for each area considering that there are many emerging tourist destinations. People participation-based development strategy for the geo-tourism area should include economic, social, customary and cultural, environmental and political aspects. The managers of the geo-tourism may organize themselves in association that enabled them to improve their managerial capability and their capability in delivering tourist services in addition to the development of the integrated facility in the geopark area.

Gunung Sewu contributes only about IDR 800 millions to the local original income in the area of tourism in 2011. However, a year after the official announcement of UNESCO Global Geopark (UGG) its contribution increases to IDR 22.5 billions [8].Tthe Gunung Sewu Geopark is undergoing the tourism development phase that involves local people and even the managers of some geo-tourism objects that have been at overcapacity level begins to explore other areas with huge potential of the geo-tourism objects in the hope that they could be as successful as the prior locations [7], [8]. Therefore, it is necessary to accelerate the consolidation phase to open access for external investors while keeping the strong roots of the local potentials that must be managed in an integrated and sustainable manner. According to United Nation [10], the development of the sustainable natural tourism should follow the following principles: (a) the participation of a community, (b) community goals, (c) stakeholder involvement, (d) lokal ownership, (e) establishing lokal business linkages, (f) cooperation, (g) sustainability of the resource base, (h) carrying capacity, (i) monitoring and evaluating, (j) accountability, (k) training, and (l) positioning. The contribution of the UUG in the sustainable development goals (SDG's) includes some goals, which are goal 1 (poverty alleviation), 4 (education), 5 (gender quality), 8 (sustainable economic growth), 11 (sustainable cities), 12 (sustainable consumption and production), 13 (climate change), 17 (global partnership) for sustainable development.

The development of the cultural landscape represents the reflection of the relationship of human being and its culture and natural environment in a wide and integrated time and spatial unity. The natural environment may include mountain, mountains, forest, dessert, and rivers, while the culture may include the outcomes of reason, emotion, and intention and human works such as traditions, beliefs, the way of life, and so on. The results of the engineering of physical landscape by human being include among others settlements, roads, houses, rice fields and non-irrigated land that are formed on the basis of geomorphologic condition and its ecological values. The Gunung Sewu Geopark in Indonesia has many outstanding cultural landscape heritages with strong historical values, heirloom resources, typical geographical conditions, natural system and biogeophisical transformation process that continuously occurs.

Considering the unique principles and locality, Indonesian tourism should be based on the national philosophy of life, which is the concept of harmonious and balanced life, meaning that there must be a good balance relationship between human and God and also between human and its natural environment. The concept teaches us to uphold the noble religious values and to actualize the values, to respect humanity values, tolerance, equality, togetherness, brotherhood, and the importance of maintaining natural environment. It also stimulates the awareness of the balance between material and spiritual needs, the balance between the use of the resources and the conservation of them in order to prevent greedy behavior.

3.2. Integrated bio-cycles system

Integrated bio-cycle system (IBS) manages the resources of land (including land, water, air, temperature, etc.), biological resources (flora, fauna, human and other living things) and environmental resources (the relationship among living things, etc.) in an optimal manner [3], 4, [5]. The program pays special attention to the increase in economic value, environmental conservation, social justice and culture in a synergetic and optimal manner that the regional unity can produce food, fodder, shelter, fertilizer, water, herbal medicine, tourism and so on. The IBS program in the karst area will be very useful in improving the quality of the environment and the life through the development of living environment and livelihood for the households of the local people. It is expected that the program would continuously increase the income of the people. The continuously increasing income will in turn cause the improvement of the prosperity of the people and finally the poverty would decrease. It will help people lead independent life and manage local natural resources in a wise manner. The IBS enables the local people to earn daily, monthly, and annual and decade incomes in short, medium and long terms. It is useful for those with small, medium and big capitals and very prospective in continuously establishing sustainable economy, environment, and social and culture that subsequently serve as leverage and locomotive of the prosperity of the local people [4], [5].

It is necessary for the managers of the natural resources to manage the natural resources in an integrated and sustainble mananer with 9R principles (reuse, reduce, recycle, refill, replace, repair, replant, and reward) to make optimal use of the natural resources for all of the living things and our living environment. The integration of the upper stream and the lower stream of the land should be established from input, process, output and outcome with 9A (agro-production, -technology, -business, -industry, -infrastructure, -marketing, -management, -structure and infrastructure, and –tourism) [3], [5].

The agro-production intends to produce multiple products in an entity of land and the products represent the real "golds" that has been being ignored and given less value, including "brown gold" (timbers), "yellow gold" (grains rich of carbohydrate necessary for human life), and "black gold" (organic fertilizer, compos, etc.) in addition to "blue gold" (biomass and biogas energies), "green gold" (green vegetables, fodder, environment, temperature, and humidity), "white gold" (milk, fish, food), "red gold" (animal protein of cattle meat, pork, chicken, ducks, etc.), "transparant gold" (water for life and oxygen) and "colorful gold" of herbal medicine that plays very important role in maintaining human health and dignity human life [5].

3.3. *Geo-eco-tourism*

Caves in the karst area that have been long being considered as desolate and spooky places actually have special architecture of stalactites and stalagmites that are rare, beautiful, artistic and phenomenal and formed naturally by drops of water and calc solution. All of the caves may be managed as very interesting natural geo-eco-tourism objects that are very attractive for both domestic and international tourists. White sand beach and limestone ravines represent beautiful and virgin natural architecture with huge attraction for the tourists [5].

Tops of high rising hills with huge, beautiful and fresh natural sceneries are very rare for modern people who begin to be bored with modernity. New spectracular and interesting icons with wooden viewing places, tree bridges, tree houses or other artistic and typical constructions such as wooden canoes, bamboo rafts with artistic ornamentation of butterflies, birds or wild animals and gigantic sculpture as background for selfie photography will provide the tourists with beautiful and rare place to take memorable pictures [5]. Even when the situation is foggy and cloudy the tops of the hills still provide the tourist with amazingly beautiful sceneries that give them a sensation of flying somewhere over the cloud. Beautiful sunrise and sunset sceneries become the most wanted ones with premium beauty of natural phenomena that provides the tourists with the most beautiful moment to take photographs. The selfie photographs of the tourists that are rapidly distributed in social media and become viral are special attraction for people to visit the emerging natural tourist objects. The selfie photographs in the geo-eco-tourism become a must in the present life style and even people feel very strong eagerness to experience the nature and take their own selfie photographs and they consider the experience as the proof that they have conquered nature and no need to cause any harm to the nature and no need to unearth something from it.

The dramatic increase in the number of the tourists of the karst area for weekend in Gunung Kidul give the local people many blessings that has positive impact on the economy of the people [5]. However, it seems that the infrastructure and the human resources have not been ready to meet the increasing demand of the tourist for the tourist objects. There has not been any sufficient network of wide and smoothly paved roads to the tourist objects and traffic jams still represent serious problem in addition to parting areas and good lodging houses with good facilities such as clean and decent bathrooms. Additionally, the local people seems to react to the increasingly demand for the tourist object in a wrong manner by drastically increasingly the renting rate of the basic facilities for the tourists. The more demand for the tourist objects, the higher they set the renting rate and the price of the items necessary for the tourist during their visits. Consequently, many tourists get disappointed because of the crowded situation during certain visiting season despite of those with strong eagerness to experience it themselves [5].

Even, seriously damaged ex-limestone mining area may be managed for the geo-eco-tourism by establishing certain buildings with artistic carvings, such as breccia ravine in Yogyakarta and Garuda Wisnu Kencana (GWK) in Bali that become natural tourist attraction. The geo-eco-tourism provides human with the change to enjoy the nature and no need to cause any harm to the nature. The attraction of local cultural events, local culinary, home stay in the family of the local people plays an important role in the implementation of the concept of the geo-eco-tourism.

It is expected that the integrated land management based on the local people and environment would be able to establish a harmonious relationship among economic value, environment and social aspects. Blue Earth revolution, a revolution in the paradigm of karst area management, may be national and international reference in improving the quality of the environment and the prosperity of the local people that they may lead dignity and sustainable life [5].

3.4 Acknowledgment

The authors express their greatest gratitude toward Directorate of Research & Community Services Ministry of Research, Technologi & Higher Education RI and UGM Yogyakarta for their research funding. This work is the revision and extension of the paper in 2017 8th International Conference on Environmental Engineering and Applications (ICEEA 2017). The opportunity is gratefully acknowledged.

4. References

[1] Agus, C., Karyanto, O., Kita, S., Haibara, K., Toda, H., Hardiwinoto, S., Supriyo, H., Na'iem, M., Wardana, W., Sipayung, M., Khomsatun and Wijoyo, S. 2004. Sustainable site productivity and nutrient management in a short rotation Gmelina arborea plantation in East Kalimantan, Indonesia. *New Forest J.* 28: 277-285

[2] Agus, C. Sunarminto, B.H., Suhartanto, B., Pertiwiningrum, A., Setiawan, I. Wiratni and Pudjowadi, D. 2011. Integrated Bio-cycles Farming Ssystem for production of Bio-gas through GAMA DIGESTER, GAMA PURIFICATION AND GAMA COMPRESSING. *Journal of Japan Institute of Energy* 90 (11) : 1086-1090.

[3] Agus, C. 2013. Management of Tropical Bio-geo-resources through Integrated Bio-Cycle Farming System for Healthy Food and Renewable Energy Soverignty: Sustainable Food, Feed, Fiber, Fertilizer, Energy, Pharmacy for marginalized communities in Indonesia. Proceeding of 2013 IEEE GLOBAL HUMANITARIAN TECHNOLOGY CONFERENCE (GHTC). www.ieeeghtc.org. San Jose, California USA October 20 - 23, 2013

[4] Agus C., Putra PB, Faridah E., Wulandari, D. and RNP Napitupulu. 2016a. Organic Carbon Stock and Their Dynamics in Rehabilitation Ecosystem Areas of Post Open Coal Mining at Tropical Region. *Procedia Engineering* 159: 329–337.

[5] Agus, C. 2017. Revolusi Jagat Biru Rahayu pada Lahan Konflik Tambang Kapur Industri Semen. Majalah Indikator Fakultas Ekonomi dan Bisnis Universitas Brawijaya. Nomor 49/Tahun XXX/2016 (In Indonesian)

[6] Anonim. 2017a. GUNUNG SEWU UNESCO GLOBAL GEOPARK (Indonesia). http://www.unesco.org/new/en/natural-sciences/environment/earth-sciences/unesco-global-geoparks/list-of-unesco-global-geoparks/indonesia/gunung-sewu/ access 27 Mei 2017.

[7] Anonim. 2017b. Memuliakan Warisan Bumi untuk Mensejahterakan Masyarakat. http://gunungsewugeopark.org/INA.html . access 27 Mei 2017

[8] Anonim. 2017c. Geopark Bakal Menjadi Wisata Alam Andalan Indonesia. http://kanalwisata.com/geopark-bakal-menjadi-wisata-alam-andalan-indonesia . access 27 Mei 2017

[9] Darsiharjo, Upi Supriatna and Ilham M. Saputra. 2016. Pengembangan Geo-park Cileuh Berbasis Partispasi Masyarakat sebagai Kawasan Geo-wisata di Kabupaten Sukabumi. *Jurnal Manajemen Resort dan Leisure*. Vol.13, No.1, 55-60. April

[10] Utama, IGB R. 2017. PRINSIP-PRINSIP PEMBANGUNAN PARIWISATA BERKELANJUTAN. https://tourismbali.wordpress.com/2013/03/10/prinsip-prinsip-pembangunan-pariwisata-berkelanjutan-2/

IOP Conf. Series: Earth and Environmental Science **83** (2017) 012029 doi:10.1088/1755-1315/83/1/012029

Explorations of Public Participation Approach to the Framing of Resilient Urbanism

Wei-Kuang Liu[1], Li-Wei Liu[2], Yi-Shiang Shiu[2], Yang-Ting Shen[1], Feng-Cheng Lin[3], Hua-Hsuan Hsieh[1]
1 Architecture, Feng Chia University, Taiwan
2 Urban Planning and Spatial Information, Feng Chia University, Taiwan
3 GIS Research Center, Feng Chia University, Taiwan
E-mail: wkliu@fcu.edu.tw

Abstract. Under the framework of developing resilient and livable cities, this study was aimed at engaging local communities to achieve the goal of public participation. Given the prevalence of smart mobile devices, an interactive app called "Citizen Probe" was designed to guide users to participate in building resilient and livable urban spaces by enabling users to report the condition of their living environment. The app collects feedback from users regarding the perceived condition of the urban environment, and this information is used to further develop an open online index system. The index system serves as a guide for the public to actively transform their city into a resilient and livable urban environment. The app was designed for the reporting of flood incidents with the objective of resilient disaster prevention, which can be achieved by enabling users to identify disaster conditions in order to develop a database for basic disaster information. The database can be used in the prevention and mitigation of disasters and to provide a foundation for developing indices for assessing the resilience and livability of urban areas. Three communities in Taichung, Taiwan, participated in the study. Residents of these communities were requested to use the app and identify local environmental conditions to obtain spatial data according to four stages in disaster response: assessment, readiness, response, and recovery. A volunteered geographic information database was developed to display maps for providing users with current reports of predisaster risk assessment, disaster response capacity, real-time disaster conditions, and overall disaster recovery. In addition, the database can be used as a useful tool for researchers to conduct GIS analyses and initiate related discussions. The interactive app raises public awareness on disaster prevention and makes disaster prevention a daily norm. Further discussion between the public and experts will be initiated to assist in policy management pertaining to the ongoing development of cities in addition to improving disaster prevention and response measures.

1. Introduction

This study examined a system that was developed to facilitate public participation in building urban resilience through the use of a digital platform and basic environmental data constructed from feedback from students and community residents through specially arranged educational courses. The collected environmental information can be used as a basis for further public participation and discussion. Recently, the concept of urban resilience has received increasing attention in the field of spatial planning. Discourse on urban resilience has included not only environmental and ecological resource management but also the structural effects on urban space governance, particularly in the context of disaster prevention (Dudley, 2010) [1]. In the field of spatial planning, the concept of urban resilience is mainly expressed in the institutional and implementation aspects of urban policies, particularly in city-wise adaptive strategies for addressing climate change. The present study used information technology to facilitate public participation in resilient disaster prevention by enabling urban communities to collectively make proactive urban development decisions and create urban livability and resilience.

Manuel Castells (1989a) indicated that information technology has altered the material basis of society as well as the processes through which new spatial forms and meanings are created [2]. Such technologies are particularly effective in realizing the concept of volunteered geographic information (VGI), which can facilitate public participation in developing social organizations (Castells, 1989b) [3]. Given the prevalence of smart mobile devices, the present study was aimed at using an interactive mobile app to enable public participation in providing local environmental information to build resilient and livable urban environments. Public feedback was collected to construct a database of information on the perceived condition of urban cities, and an open index system was developed to guide the public in actively transforming their urban environment into a more livable environment. With the trend of public participation in urban development and the development of networked societies, identifying and solving problems pertaining to urban living environments is no longer the sole responsibility of governments and scholars in democratic societies. Therefore, encouraging public participation in issues pertaining to the public sphere and providing appropriate communication platforms have become critical tasks that must be faced in this new era. Through the use of mobile terminals (e.g., smart mobile devices), participatory sensing can be realized, which would change the nature of traditional communication modes, in which communication is generally limited to members within the same community, enable the public to provide immediate feedback on the condition of their neighborhood, thereby assisting in collecting data for resilient city indices though a bottom-up approach.

Castells (2000) noted this contemporary social trend 17 years ago by stressing that, after the emergence of networked societies with the advent of the information age, social governance would transition from a traditional vertical hierarchy into various forms of information flows and sharing [4]. The Internet enables users to share or comment on information, which indirectly affects how other users interpret information; this was unachievable in the traditional mode of governmental information transmission or media user experience. By using such an online community system, the present study asserted that the public could use their own smart mobile devices to provide environmental information. Through the use of mapping technologies, the feedback can be imported into a geographic information system (GIS) analytical framework for further analysis of regional problems or public participation.

2. Resilient disaster prevention

In recent years, the concept of resilience has received considerable attention in addressing natural disasters in urban areas. This concept implies that a city must be able to restore its living functions and order as soon as possible following a disaster. Disaster prevention requires improving forecast and monitoring technologies, managing all basic environmental data, detecting anomalies to minimize loss of life and property, and issuing warnings in advance. Therefore, applying information communication and Internet technologies to developing effective solutions has become a global trend in disaster prevention and rescue, particularly in moving away from the traditional top-down approach to disaster prevention governance. Thus, this study focused on discussing the use of the Internet in assisting various communities in participating in disaster prevention through a bottom-up approach.

Foster (2006) proposed a framework comprising four stages for assessing a city's resilience in disaster prevention: assessment, readiness, response, and recovery. A resilient city must be able to assess and prepare for potential disasters, respond to disasters appropriately, and immediately restore city-operation mechanisms to the normal level. The continuous cycle of these four stages enables a city to respond properly to all types of potential disaster [5].

To develop resilience, a city must change not only its disaster prevention and management strategies but also the residents' mindset and behavior. Thus, this study proposed the mobile app Citizen Probe to incorporate the collective feedback of online communities into a city's disaster management decision-making processes and to gather basic environmental data for analyzing disaster prevention capability. The data are gradually accumulated by repeating the three procedures of public participation: identification (of disaster prevention locations), coconstruction (establishment of disaster prevention locations), and mutual assistance (shelter capacity).

3. Public participation in collecting and managing environment information

French philosophers Deleuze and Guattari (1988) explained the impact of living experience on urban structures and emphasized that an orderly urban system is constructed and organized through a dynamic bottom-up social relationship. This social power from the lower levels forms the foundation of urban structures, whereas a top-down planning order, although rational, cannot fully govern the voice and energy of the social collective [6]. British geographer Massey (1997) considered this as a process of place making. The form of a space cannot govern local cultures and the sense of a place is represented through the subjectivity and experience of human life (Massey, 1997; Pred, 1984; Tuan, 1977) [7]-[9]. Public participation serves as a mechanism for coproducing a sense of place. Sociologist Habermas (1989) indicated that the public sphere must involve public participation to shape society and politics, and he emphasized the importance of social communication platforms because communication ensures the preservation of modern values (e.g., social rationality, consensus, liberation, and solidarity), which can be used as a foundation for generating critical perspectives and ensuring social reconstruction [10]. Therefore, identifying the ideal method for building an appropriate decision-making platform through public participation is a major objective of the present study.

The integration of global positioning systems (GPSs) and the Internet with mobile devices enables users to have easy access to spatial information and to share information with the public. Thus, convenient access to posting and receiving information in addition to advancements in mobile device technologies have given rise to VGI apps. Previously, information on maps, traffic accidents, and disaster reporting has been provided by official sources, and information releases can be delayed and unlikely to contain local information. However, VGI transforms information typically released in a top-down manner, under authoritative structures, or by professionals, into information that can be continually accumulated and disseminated by the public through various methods, hence reducing the distinction between information providers and users (Goodchild, 2007) and rendering VGI as a potential resource for collecting official geographic information. American geographer Goodchild (2007) suggested that, when employing participatory-sensing technology, citizens effectively become sensors for their city by marking the geographic locations of incidents and providing relevant information [11]. For instance, a platform must allow the dynamic registration of geographic information to increase user willingness to participate and to maximize the platform's effectiveness. If the public can participate in disaster reporting by marking the location of a disaster via GPS and autonomously providing information of predisaster risk assessment, disaster damage, and postdisaster recovery, the collected information can be used for GIS analysis and enhance current developments in disaster prevention.

Figure 1: Reporting on surface water permeability on Citizen Probe

The present study applied the assessment framework for urban resilience proposed by Foster (2006) to the identification of flood incidents and developed a mobile app called Citizen Probe. The app enables users to participate in the following stages of disaster management and prevention: (1) Assessment: Photos can be taken and uploaded to report the surface water permeability of a living area (Fig. 1) and imported in an online platform to determine the overall distribution of surface water permeability and thereby measure its capacity for coping with flood incidents. (2) Readiness: Potential disaster risks can be reported in advance to confirm whether an area is ready and well prepared for disaster response (e.g., ensuring drain cover blockages that may lead to poor drainage do not occur). (3) Response: Flood conditions can be immediately reported by users to inform others or rescue personnel of the actual flood height (Fig. 2) and enable them to understand overall flood conditions, thereby reducing water damage and increasing response efficiency. (4) Recovery: Flood water level decrease in flood areas can also be reported (Fig. 3) to calculate the total recovery time required for individual areas and to identify environmental problems that require further improvement actions.

Figure 2. Reporting flood height on Citizen Probe

Figure 3. Reporting flood recovery on Citizen Probe

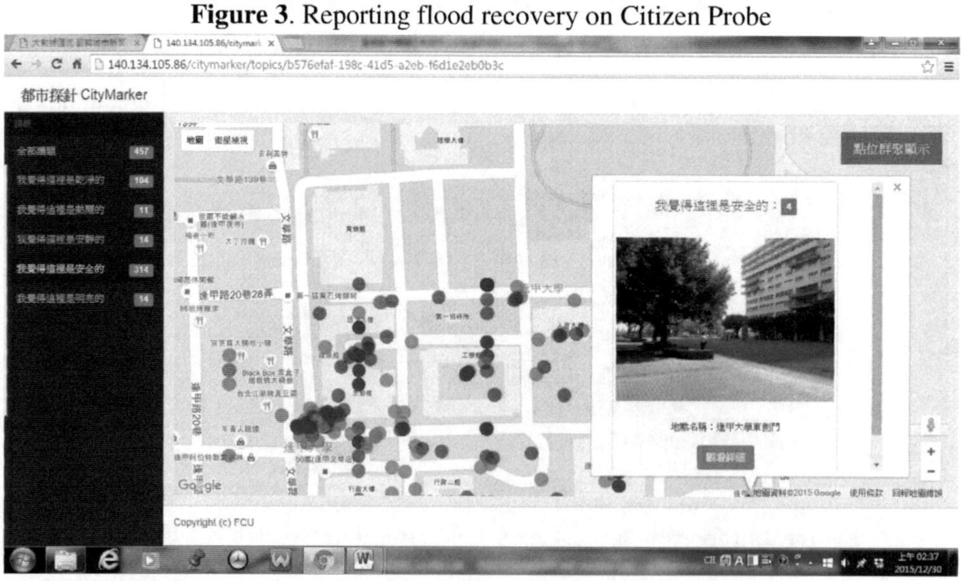

Figure 4. Spatial distribution of safe locations in the campus of Feng Chia University as reported by Citizen Probe users

To inform the public about the distribution, conditions, preparation, and response to flood incidents in their immediate area, this study constructed an open cloud database for users to connect to via

Facebook to upload photos taken with their mobile devices. The use of GPS enables the uploaded files to be integrated with geographic information on the flood location. The identified content is then accumulated to form a VGI database, which is visualized in the form of a map (see Fig. 4), providing immediate public access to current reporting on predisaster risk assessments, disaster response capacity, real-time disaster conditions, and overall disaster recovery. In addition, the VGI database can serve as a tool for researchers conducting GIS analyses and discussing disaster prevention and management. To display information, the online platform has a user-friendly interface that allows both the public and experts to discuss policies regarding disaster prevention developments in urban areas. The present study employed community volunteers to facilitate public participation and provided regular training on disaster prevention and mobile device usage.

4. Community participation

According to the following three criteria, three communities were selected from Taichung City to guide local residents in identifying and reporting local flood-related information to develop local disaster resilience. In the selection of participants and research regions, the following were considered: (1) past history of severe flood incidents, (2) convenience for observing flood levels, and (3) fully developed and comprehensive community operation. The selected communities were Wukuang (Wuri District) and Chungho (Nantun District). All locations are close to rivers and have a history of severe flooding due to heavy rainfall, and this continues to be a problem despite several attempts at river rectification. The community heads agreed to participate in the present study on a long-term basis. The present stage involves education courses designed to enroll students and community residents as volunteers to participate in experiments to test the Citizen Probe app (see Fig. 4). However, since the commencement of this study (August 2016), no severe flood has occurred in the selected communities; consequently, no data were collected to evaluate flood severity in the areas. Cooperation with these communities will be ongoing with regular disaster prevention drills to test Citizen Probe to ensure its effectiveness during disaster events.

5. Conclusion and future research directions

At present, this study is mainly focused on developing the Citizen Probe app and establishing a system for community participation. However, the following limitations were observed during the test drills and require further improvement: (1) the app cannot be operated with the mobile device in offline mode, (2) GPS accuracy is poor, and (3) lack of diverse presentation modes for displaying user-required information on the GIS maps. By continuously improving and adjusting the proposed app and system, this study suggests that a functional identification and interaction platform can be developed to achieve the goals of resilient disaster prevention. Moreover, Citizen Probe was applied only in the identification of flood incidents. It could also be used for other disasters or its user base could be expanded to determine its capacity for developing urban resilience.

In addition to the ongoing improvement of the app and system, this study suggests the following topics for future research: (1) integrating Internet-of-things technologies to strengthen the current disaster prevention and monitoring network, (2) establishing a sharing and cooperation system between public and private sectors, and (3) applying big data to disaster damage analysis. These efforts would improve the effectiveness of the current approach and transform the present feedback data from individual users into big data, which would have considerable analytical value and may serve as reference for future developments in resilient urbanism.

6. Acknowledgements

This work is supported by the Ministry of Science and Technology of Taiwan, under grant No. MOST105-2627-M-035-006 and 105-2221-E-035-012. This manuscript was translated into English by Wallace Academic Editing.

7. References

[1] Dudley, M. (2010). Resilience. In *Green Cities: An A to Z Guide*, N. Cohen, ed. Sage Publications.
[2] Castells, Manuel. (1989a). 'Social Movements and the Informational City'. *Hitotsubashi Journal of*

Social Studies 21: 197-206.

[3] Castells, Manuel. (1989b). *The Informational City: Information Technology, Economic Restructuring, and the Urban Regional Process*. Oxford; Cambridge, Mass.: Blackwell.

[4] Castells, Manuel. (2000). *The Rise of the Network Society*. 2nd edn. Vol. 1, The Information Age: Economy, Society and Culture. Oxford; Cambridge, Mass.: Blackwell.

[5] Foster, Kathryn A. (2006). 'A Case Study Approach to Understanding Reginal Resilience'. Working Paper prepared for the Building Resilient Regions Network and presented at the Annual Conference of the Association of Collegiate Schools of Planning.

[6] Deleuze, Gilles, and Fâelix Guattari. (1988). *A Thousand Plateaus: Capitalism and Schizophrenia*. London: Athlone Press.

[7] Massey, Doreen. (1997). 'A Global Sense of Place'. In *Reading Human Geography: The Poetics and Politics of Inquiry*, T. J. Barnes and D. Gregory, eds. London: Arnold.

[8] Pred, Allan R. (1984). 'Place as Historically Contingent Process: Structuration and the Time-Geography of Becoming Places'. *Annals of the Association of American Geographers* 74 (2):279-297.

[9] Tuan, Yi-Fu. (1977). *Space and Place: the Perspective of Experience*. London: Edward Arnold.

[10] Habermas, Jurgen. (1989). *The Structural Transformation of the Public Sphere*. Cambridge, Mass.: MIT Press.

[11] Goodchild, M. (2007). 'Citizens as sensors: the world of volunteered geography'. *GeoJournal* 69 (4): 211-221.

IOP Conf. Series: Earth and Environmental Science **83** (2017) 012030 doi:10.1088/1755-1315/83/1/012030

The Role of Soil Amendment on Tropical Post Tin Mining Area in Bangka Island Indonesia for Dignified and Sustainable Environment and Life

C Agus[1], D Wulandari[2], E Primananda[1], A Hendryan[1], V Harianja[1]

1 Faculty of Forestry, Universitas Gadjah Mada, Jl. Agro - Bulaksumur Yogyakarta 55281, Indonesia.
2 Scientist and Research Coordinator in Southeast Asia Minister of Education Research Centre for Tropical Biology (SEAMEO BIOTROP), Jl. Raya Tajur, Bogor 16720, Indonesia.
E-mail: cahyonoagus@gadjahmada.edu

Abstract. Openly tropical tin mining in Bangka Island Indonesia expose heavy metal that had been buried became a part of our environment and life. This has become a major cause of land degradation and severe local-global environmental damages. This study aims to accelerate reconsolidation of degraded ecosystems on the former tin mine land, to increase land productivity and dignified environment through appropriate rehabilitation technology on marginal land that is inexpensive, environmentally friendly and sustainable. This study is a part of a roadmap research activities on the rehabilitation of degraded land in tropical ecosystem, that consist of (a) characterization of degraded tin mining lands through the determination of chemistry, physics, biology and mineral soil properties, (b) introducing multi-function pioneers plant for acceleration of peak pioneer plant in the reestablishment of degraded tin mining ecosystem (c) management of natural soil amendment (volcanic ash, organic waste materials and legume cover crop as a material for soil amelioration to increase land productivity, (d) role of biotechnology through the application of local bio-fertilizer (mycorrhizae, phosphate soluble bacteria, rhizobium). Soil from post tropical tin mining acid soil (pH 4.97) that dominated by sand particles (88%) with very low cation exchange capacity, very low nutrient contents (available and total-N, P, K, Ca, Mg) and high toxicity of Zn, Cu, B, Cd and Ti, but still have low toxicity of Al, Fe, Mn, Mo, Pb, As. Soil amendment of biogas and volcanic ash could improve soil quality by increasing of better pH, high available-P and cation exchange capacity and maintained their low toxicity. The growth (high, diameter, biomass, top-root ratio) of exotic pioneer plant of Kemiri sunan (*Reutealis trisperma*) increased in the better soil quality that caused by application of proper soil amendment. The grand concept and appropriate technology for rehabilitation of degraded tin-mining land ecosystems in tropical regions which are the lungs of the world have a high contribution for development of our dignified and sustainable environment and life.

1. Introduction

Over exploitation of tropical forest ecosystem is drastically leading to land degradation and damages [1]-[5]. Open mining become one of anthropogenic disturbance on tropical forest ecosystem drastically leading to land damage and degradation. Nevertheless, mining of natural resource is one of important industries because of its contribution to economic development in Indonesia. Approximately 30% of global tin production took place in two islands of Indonesia, which were Bangka and Belitung in which tin mining has been being operated since 1668 [6]. Indonesia was the second biggest mining producer in 2013 (95,200 tons) and 2014 (84,000 tons) after China (110,000 tons in 2013 and 125,000 tons in 2014). Global tin reserve was predicted to be about 4.8 million tons and China ranked the highest (1.5 million tons) followed by Indonesia (800,000 tons) [7].

Tin mining process using simple purification process and sluice box was common method in Indonesia. The method was based on the difference in the weight of tin and quartz sands. Tin ore would be left in the sluice box, while the quartz sands would be carried by the water flow [8]. The tin mining process in huge tailing number resulted in very toxic metal concentration at low soil pH [9]. The removal of upper layer of ground in the mining process caused loss of important macro and micro nutrients useful for plants [10]. One of several soil health parameters was soil nutrient concentration and traditionally considered as important parameter in mining area.

The mining process by removing soil changes soil formation, texture and structure. The mining soil texture would be dominated by coarse quartz in ex-tin mining [10] that was prone to erosion [11]. The removal of vegetation and the lessening soil in steep slope resulted in the high risk of watershed degradation and sedimentation [10], [12]. Huge quantity of overburdened soil, dumping, reclamation of mining area and large scale cavity in the mining area and other phenomena of the impact of the mining has led to ecosystem destruction [12]-[19]. The condition caused unsporting environment for natural vegetation and other organisms to grow.

The high net primary production (NPP) of tropical forests are influenced more by nutrient cycling rate than by amounts of nutrient availability in soils [1]-[5]. The use of soil ameliorant is one of the alternatives in the rehabilitation of degraded land of post-mining area [15]. It was expected that organic and inorganic soil ameliorant materials would be able to improve physical, chemical, biological and mineral properties of the degraded soil [16], [17]. Initial step to improve the quality and the health of the ex-mining area was to increase the organic material content [18]. The organic materials were very useful in improving the physical and chemical properties of the soil that became the determinant factors of the fertility of the soil. The organic materials played an important role of soil conditioner. Soil organic material compounds increased main particles in chemical and physical aggregates that would in turn improve the aggregate stability and hamper the breaking of the aggregates. The organic materials also played an important role of providing nutrients, food and energy sources for soil microorganisms and also maintaining soil moisture [19].

Rehabilitation of extremely degraded areas through re-vegetation by fast growing species is expected to speedily recover their dynamic of organic-carbon stocks [5]. Kemiri sunan (*Reutealis trisperma*) was a pioneer plant able to grow in marginal land. It was a fast growing species and produced useful oil. Also, it was able to adapt to acid soil with low pH and to tolerate various types of soil, including saline, sandy, heavy clay soil, stony soil and inundated soil. It was expected that the introduction of soil ameliorating materials such as biogas waste, compost and volcanic dust in the ex-tin mining soil to boost the growth of the R. trisperma could give useful information to wisely rehabilitate the land in after the tin mining period has been over, especially in forestry areas [11].

2. Materials and method

The study used tailing soil obtained from the ex-tin mining area in Bangka Island in November 2016-March 2017. It was conducted in the Laboratory of Intensive Silviculture, Faculty of Forestry, Universitas Gadjah Mada (UGM), Yogyakarta. The test of the physical and chemical properties of the soil was carried out in the Soil Laboratory of the Faculty of Forestry UGM, the Laboratory of Soil-Faculty of Agriculture UGM and Integrated Laboratory of LPPT UGM.

It used completely randomized design with single factor, consist of 5 treatments of soil amendments, namely: Control (C), volcanic ash 5% (A5), volcanic ash 10% (A10), biogas waste 5% (B5) and biogas waste 10% (B10) with 6 replications and 1 tree-plot. The 5 months age of *Reutealis trisperma* seedlings were grown.

Data was analyzed using ANOVA complete randomized design to find out the impact of the treatments. Duncan's median value test was carried out to find out the best treatment at the confidence interval of 95% and 99% toward the median value that had significant impact. The R. trisperma was grown by introducing the ameliorant materials according to its respective treatment proportion in the

media of the ex-mining soil in polybags and then maintenance, pest and disease control, and regular watering were conducted once a day.
The chemical and physical tests of the soil were carried out before harvest after the introduction of the soil improving materials at the beginning of growing period. The height and the diameter of the plant were observed once in two weeks after the cultivation for 12 weeks. The physical properties of the soil observed/tested as parameters included soil texture by groping, pH H2O, EC and EH were measured by electrometric methods, N was observed using Devarda's alloy method, P was observed using Bray II method, K was observed using ammonium acetate method, and Cation Exchange capacity (CEC) with the ammonium acetate method; the total macro nutrients of N, P, K and secondary elements and micro nutrients of Ca, Mg, S, B and metals and other elements were observed using SEM (Scanning Electron Microscopic) + EDX method. The harvest of the plant was conducted in the week 12, along with photographs and dry weight of biomass were analyzed using tissue analysis with SEM method. The data was analyzed using SPSS 16.

3. Results and discussion

Mining activity caused ex-tin mining tailing with bad physical and chemical properties (Table 1). The texture of the ex-tin mining tailing soil was classified into loamy sand dominated by 70-90% sand. The results of 50x and 1000x magnifications using SEM showed that the ex-tin mining tailing soil was dominated by big size single grain with big macro pore and many fissures. It resulted in very low capability of retaining water and nutrients that the water and the nutrients quickly released and were not retained for the growth of plants. The water loss per location and the nutrients dissolved in the water was excessive because of earth gravity, while the existing reserve of the water and the nutrients was very low. However, there was inundation in the land with low groundwater and it resulted in anaerobic condition and intensive soil reduction that had negative impact on land productivity and plant growth. The impacts of the tin mining on the soil physical properties were the loss of top soil and the change of soil structure [20]. The mean loss of the top soil was 27.6 cm and the texture of the soil became coarser, while the sand content increased from 70% to 90% in tin tailing soil.
The ex-tin mining soil was not productive soil because of its lack of macro elements necessary for the growth of plants. The pH of the soil was 5.34 (acid), electronic conductivity (EC) was 28.1 mS (very low), electronic hydrostatics was 147 mV (medium reduction), CEC was 3.61 me/100g (very low). C-total in ex tin mining was 18.65%, that categorized as very high, because of accumulation of $CaCO_3$. It had very small essential macro nutrients of N that was 23.18 ppm, P was 0.38 ppm and K was 0.07 me/100g (Table 1). The proportion of nitrite (NO_2) in extracted-N content of water logged soil was very high. Such condition resulted in very low productivity of the ex-mining land because it had very low capability to provide plants with necessary nutrients for growth.
The results of the test using Scanning Electron Microscopic (SEM) + EDX showed that the tailing soil of the ex-tin mining had very low total essential macro nutrients (Mg, P, Ca, and Mo). Some micro elements that played only minor role for plants to grow were available in excessive quantity or beyond normal condition, including B, Cu, Zn and Cd, so that they became toxic for the plants. Some micro elements were in fact in normal range (Al, S, K, Mn, and Fe) that they did not cause any significant disturbance for the plants to grow.
The use of soil amendment of volcanic dust to the tailing soil (control) changed into the same texture of the loamy sand because the quantity of the volcanic dust was not enough to give significant impact to the texture of the soil. Meanwhile, the most optimal treatment by introducing biogas waste was able to slightly reduce the sand content. The texture of the loamy sand of amended soil that had 50-70% sand, 15-20% clay and 10-40% loam, was quite ideal condition for some species of plants [21].
The treatment of biogas waste caused the increase in the pH from acid to neutral condition. However, the treatment of volcanic dust decreased the pH into more acid (4.9-5.1) compare to the control. It was consistent with the results of the study by [22], which was 4.7-5.6. It was because the main source of acidity was dominated by Al^{3+}.
The micro element of Al^{3+} in acid soil was highly soluble in excessive quantity so that it became toxic for plants and even it fixed the element of P into Al-P and hence the P was not available for the plants. The treatment of the soil ameliorant of the volcanic dust caused the reaction of the soil that released

H+ ion and then it caused the increase in the soil acidity, while the treatment of the biogas waste increased the pH of the soil.

Table 1. Soil chemical properties of tailing ex-tin mining soil in Bangka Island and their treatment of soil ameliorant of volcanic ash and biogas waste at dosage of 5% and 10%.

Element	Control	Volcanic 5%	Volcanic 10%	Biogas 5%	Biogas 10%
pH (H_2O)	5.34	5.05	4.87	6.425	6.57
EC (µS)	28.1	125	154.5	413	614.5
EH	147	99.5	110	17.5	13.5
CEC (me/100g)	3.61	5.61	4.41	3.61	4.82
Macro nutrient					
C-total (%)	18.65	16.92	12.64	13.97	16.01
N-total(%)	6.28	7.06	5.2	5.73	5.62
N-extracted (ppm)	23.18	20.38	17.56	72.17	35.07
P total (%)	0	0	0.01	0	0.09
P-available (ppm)	0.38	0.22	1.78	6.8	19.61
K –total (%)	0.19	0.28	0.23	0.14	0.09
K-available (me/100g)	0.07	0.08	0.08	0.17	0.15
Secondary element					
Ca-total (%)	0	0.26	0.42	0.11	0.1
Mg-total (%)	0	0	0.2	0.01	0.01
Na-total (%)	0	0	0	0	0
S-total (%)	0.16	0.02	0.02	0	0
Micro element					
Fe-total (%)	0.67	0.67	1.01	0.57	0.42
Mn-total (%)	0.06	0.01	0	0	0
Zn-total (%)	0.39	0.14	0.02	0.15	0.21
B-total (%)	4.35	1.03	4.71	0	8.84
Cu-total (%)	0.35	0.24	0.07	0.54	0.4
Mo-total (%)	0	0	0	0.32	0
Cl-total (%)	0	0.07	0.03	0.04	0
Heavy metal					
Pb-total (%)	0	0.28	0	0.09	0.1
Cd-total (%)	0.02	0	0	0	0
Si-total (%)	20.54	23.39	21.92	27.41	17.2
As-total (%)	0	0	0	0.04	0.02
Ti-total (%)	0.39	0.27	0.67	0.15	0.16
Sn-total (%)	0	0.04	0.2	0	0
Pt-total (%)	2.73	3.13	1.96	3.7	1.36

At the neutral condition of pH the fertility status of the soil was good in which it was not deficient of the elements of N, P, and K. It was also the case of the micro nutrients such as Ca, Mg, S, Fe, Zn, Mn, Co, Bo, and Mo. At the condition of the pH value in the range of 6.0-7.5, all of the necessary elements were available for the majority of plants, except for the plants that grew in acid land [23].

IOP Conf. Series: Earth and Environmental Science **83** (2017) 012030 doi:10.1088/1755-1315/83/1/012030

The treatment of volcanic dust caused the increase in the electronic conductivity from 0.28 dS/m (very low) to 1.25-1.55 dS/m (low), whereas the treatment of biogas waste could increase 4.13-6.15 dS/m (very high) (Table 1). The condition of very low to low EC indicated that actually the salt content of the soil did not have any significant impact on the availability of the nutrients in the soil, but the high EC value would result in osmosis in which plants should use higher energy to absorb water from the soil and finally it hampered the growth of the plants and decreased their productivity [24]. The roots of the plants had semi-permeable membrane that faced difficulties in absorbing water and nutrients from the soil because of the presence of the dissolved salt. It was because the biogas waste had very high EC value as compared to the ameliorant of the volcanic dust. The range of the Eh value in all of treatment soil were almost the same as medium reduction, related to not so drainage land condition in which there was excessive water [24].

CEC the tin mining soil in were classified into very low, even after treatment by biogas dust, but treatment of 5% volcanic dust increased more significantly compare to the control, although still considered to be low. Soils with higher clay/colloid content or high organic content had higher CEC than those with low clay content (sandy soil) and low organic content [23].

The total-N of the ameliorated soil and control was considered to be very high. It was assumed that it was because nitrate (NH_4^+) evaporated. The N was changed into ammonium (NH_4^+) and then changed into nitrate and the nitrate evaporated because of the change by microorganisms into NH_3 (volatilization) and N_2O, NO, N_2 (denitrification) [23]. The treatment of biogas waste increased N-total significantly from 23.18 ppm to 35.07-72.17 ppm, while the treatment of volcanic dust decreased to be 17.56-20.38 ppm. It was assumed that the available inorganic N content in the form of NO^{3-} and NH_{4+} plus the organic material of the biogas waste was bigger than the tailing soil without any addition of organic material (control). It was indicative of the presence of the mineralization process or the addition of inorganic N resulting from the weathering process of organic materials. The nitrogen value could be influenced by reorganizing the organic materials of the soil into N mineralization.

The available-P in the tailing soil was very low (0.38 ppm) with the total-P that was very low as well. The treatment of volcanic dust caused not any significant addition or remained in low category at 0.22-1.78 ppm. Meanwhile, the treatment of biogas waste was able to increase the available-P to 6.8-19.61 ppm and it was considered to be very high. The available-P in volcanic ash was smaller because the majority of the P was bound by the elements of Al, Fe and Mn that they could not easily be dissolved in soil [17].

The total-K in the tailing soil was 0.19%, it increased with the treatment of volcanic ash to 0.23-0.28%, while the treatment of biogas waste decreased to 0.09-0.14%. Available-K was 0.07 me/100g in the tailing soil, increased to be 0.15-0.17 me/100g, after treatment by biogas waste. Soil available-K was not constantly available, but it was changing slowly and hence it was difficult for plants to absorb it (i.e., slowly available). The sandy soil would be easily dried when the water came out, so that the availability of the K for the plants decreased [17]. The slow exchange of K because the CEC value of the tailing soil or the tailing soil after the treatment was still very low.

Total-Ca in the tin tailing soil was very small, but increased to 0.1-0,11% after the treatment of volcanic dust, and to 0.1%0.26-0.42% with the treatment of volcanic dust. The pH of the tailing soil was sufficiently acid and the Ca was very small that it could not increase its pH. Total Mg was very small, however, the treatment of volcanic dust was able to increase the total Mg that it reached 0.2% (normal range). The small total-Mg in the treatment of the biogas waste resulted from the total number of the Mg that was only 0.3%. The imbalance of the Ca and the Mg contained in the soil and the equally small number of the Mg contained in the CEC soil caused Mg deficiency [23].

There were 7 micro nutrients that played as important role as macro nutrients and secondary nutrients. However, the necessary micro nutrients are not as much as the macro nutrients and the secondary nutrients. The lack of the micro nutrients in the soil might also limit the growth of plants though other nutrients were sufficiently available. The soil acidity (pH) had significant impact on the availability of the micro nutrients, except Mo and Cl that would decrease if the pH increased. If the number of the Ca was bigger in the organic material of the biogas waste than that in the volcanic dust, the decrease in the acidity caused the decreased in the number of B contained in the soil. There were a lot of Boron (B) detected in the tailing soil and according to [25], it was beyond the standard limit of the number of the

B in the soil. After the treatment of the organic materials, the number of the B was still in the category of beyond the limit, except the treatment of 5% biogas waste.

The treatments of biogas waste decreased toxicity of some micro nutrients. Total-Mn, Zn, Fe, Al, Mo, and Cu had the same properties that their solubility would increase the acidity and toxicity. The abundantly available Fe in the tailing soil and the treatment of the volcanic dust and the treatment of the biogas waste related to the decrease in the availability of the P after the treatment. Other metals contained in the soil such as Fe, Mn, and Al influenced the availability of the Cu for the growth of plants. The excessive number of the Cu could suppress the activity of Fe and might cause the symptom of Fe deficiency [23].

Aluminum (Al) represented supporting nutrient that was not required by plants because the Al was toxic. It had the characteristics of providing and absorbing similar to the micro nutrients without any absorbing zone that in the excessive content it was toxic. Meanwhile, if the Al was not available sufficiently in the soil, the growth of the plants would be hampered and their flower and fruits would not be optimal. Total-Al in the tailing soil was 5.71% (in normal range) and the treatment of the ameliorant of the volcanic dust and the biogas waste caused the progressive increase in the number of the Al. Other micro nutrients that had metal properties and were detected in the tailing soil included Ti, Si, Cd and Pt.

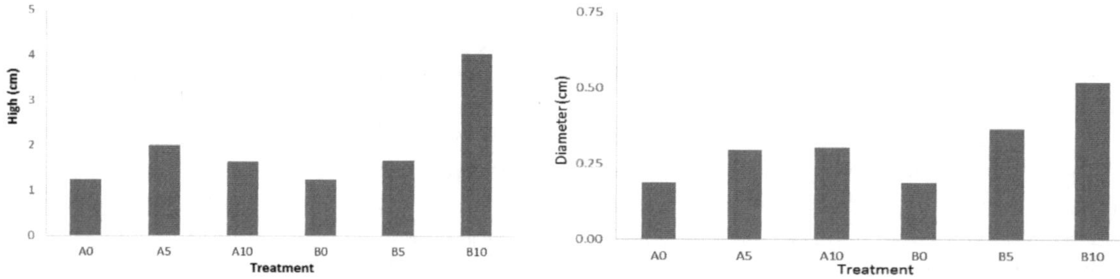

Figure 1. The growth of high and the diameter of *Kemiri sunan* on the ex-tin mining soil with the treatment of the soil ameliorants and the dosages till 12 weeks after cultivation.

The best treatment with high growth rate was the treatment of 10% biogas waste with the increased height of 4.03 cm for 12 weeks after cultivation, that mean 322% than their height on the tailing soil (only 1.25 cm). It also indicated that the increase in the height resulting from the treatment of 5% volcanic dust was 160%. The best treatment with the growth rate in the diameter was correlated to the height resulting from the treatment of the biogas waste at the dosage 10% with the increased diameter of 0.52 cm, that increasing of 274% compare to the tailing site that only 0.19 cm for 12 weeks. The treatment of the ameliorant materials at higher dosage did not cause the increase in the high and diameter (Figure 1).

During the growing period of the *Kemiri sunan* for 12 weeks after cultivation the leaves often fell down, except the plants on the tailing soil with the treatment of 10% biogas waste in which the color of the old leaves was yellow and they were rolled before they fell. The *Kemiri sunan* at all of the treatments tended to be susceptible to plant diseases such as insects and so on. It was clearly observed in the morphology of the leaves with holes as a result of the bite of the insects or the leaves full of with flour as a result of the plant animal disease of *embun jelaga.* The high value of the top-root ratio in the treatment of 10% biogas waste was the most optimal treatment in the growth of the *Kemiri sunan*, whereas the tailing soil gave the lowest top-root ratio value.

The treatment of the organic ameliorant that gave the highest N value was the treatment of ameliorants of both volcanic dust and biogas waste that the growth of the roots would be stimulated and the weight of the roots would increase. However, the high intake of the N resulted in soft leaves and caused the increase in the water content. The high N content would surely cause lower P content in the treatment of 5% organic materials than the treatment of 10% organic materials. The two organic materials at their respective dosages have proven to be able to increase the total dried weight and the top-rood ratio value of the *Kemiri sunan* for 12 weeks. According to [21], non-fertile soil or low soil quality of tin mining condition would cause low top-root ratio, while the application of organic fertilizer would be able to increase the top-root ratio value.

4. Conclusion

Soil from post tropical tin mining acid soil (pH 4.97) that dominated by sand particles (88%) with very low cation exchange capacity, very low nutrient contents (available and total-N, P, K, Ca, Mg) and high toxicity of Zn, Cu, B, Cd and Ti, but still have low toxicity of Al, Fe, Mn, Mo, Pb, As. Soil amendment of biogas and volcanic ash could improve soil quality by increasing of better pH, high available-P and cation exchange capacity and maintained their low toxicity. The growth (high, diameter, biomass, top-root ratio) of exotic pioneer plant of Kemiri sunan (Reutealis trisperma) increased in the better soil quality that caused by application of proper soil amendment. The grand concept and appropriate technology for rehabilitation of degraded tin-mining land ecosystems in tropical regions which are the lungs of the world have a high contribution for development of our dignified and sustainable environment and life.

5. Acknowledgment

The authors express their greatest gratitude toward Directorate of Research & Community Services Ministry of Research, Technologi & Higher Education RI and UGM Yogyakarta for their research funding, and local Government of Bangka and PT Tambang Timah for supporting of field research materials. This work is the revision and extension of the paper in 2017 8th International Conference on Environmental Engineering and Applications (ICEEA 2017). The opportunity is gratefully acknowledged.

6. References

[1] Agus C, Karyanto O, Hardiwinoto S, Haibara K, Kita S. and Toda H 2003 Legume cover crop as a soil amendment in short rotation plantation of tropical forest *J. For. Env.* **45(1)**: 13-19.

[2] Agus C, Karyanto O, Kita S, Haibara K., Toda H, Hardiwinoto S, Supriyo H, Na'iem M, Wardana W, Sipayung M, Khomsatun and Wijoyo S 2004 Sustainable site productivity and nutrient management in a short rotation *Gmelina arborea* plantation in East Kalimantan, Indonesia *New Forest J.* **28**: 277-285

[3] Agus C. Sunarminto B.H, Suhartanto B, Pertiwiningrum A, Setiawan I. Wiratni and Pudjowadi D 2011 Integrated Bio-cycles Farming Ssystem for production of Bio-gas through GAMA DIGESTER, GAMA PURIFICATION AND GAMA COMPRESSING *Journal of Japan Institute of Energy* **90(11)** : 1086-90.

[4] Agus C, Pradipa E, Wulandari D, Supriyo H, Saridi, Dan Herika D 2014 Peran Revegetasi Terhadap Restorasi Tanah pada Lahan Rehabilitasi Tambang Batubara di Daerah Tropika *Jurnal Manusia dan Lingkungan* **21(1)**:60-66.

[5] Agus C, Putra PB, Faridah E., Wulandari D and Napitupulu RNP 2016 Organic Carbon Stock and Their Dynamics in Rehabilitation Ecosystem Areas of Post Open Coal Mining at Tropical Region *Procedia Engineering* **159**: 329–337.

[6] Juwarkar AA and Jambhulkar HP 2008 Phytoremediation of coal mine spoil dump through integrated biotechnological approach *Bioresource Technology* **99**: 4732-4741.

[7] Yulianingsih E 2004 Pengaruh Tingkat Kelengasan Tanah Terhadap Beberapa Sifat Tanah dan Pertumbuhan Kedelai di Lahan Pasir Pantai bugel, Kulonprogo Tesis Program Pascasarjana. UGM, Yogyakarta.

[8] Irawan RR., Sumarwan U, Suharjo B, dan Djohar S 2014 Strategic model of tin mining industry in Indonesia (Case study of Bangka Belitung Province) *International Journal of Business and Management Review* **2**: 48 – 58.

[9] Ashraf M.A, Maah M.J, Yusoff I 2012 Chemical speciation and potential mobility of heavy metals in the soils of former tin mining catchment *The Scientific World Journal* **Vol 2012**: 11 pages.

[10] Sugeng W 2005 Kesuburan Tanah : Dasar Kesehatan dan Kualitas Tanah. Penerbit Gava Media. Yogyakarta.

[11] Herman MM, Syakir D, Pranowo, Saefudin dan Sumanto 2013 Kemiri Sunan (Reutealis trisperma (Blanco) Airy Shaw) Tanaman Penghasil Minyak Nabati dan Konservasi Lahan IAARD Press, Jakarta: 88 hlm.
[12] Clemente AS, Werner C, Maguas C, Cabral MS, Loucao MAM, Correia O 2004 Restoration of a Limestone Quarry: Effect of soil amendments on the establishment of native Mediterranean schlerophyllous shrubs *Restoration Ecology* **12**: 20 – 28.
[13] Hadi H dan Sudiharto 2004 Pengembangan Perkebunan Karet di Daerah sekitar tambang batubara: Kasus di Kabupaten Tabalong Kalimantan Selatan *Warta Perkaretan* **23(3)**:28-36.
[14] Nurcholis M, Wijayani A, Widodo A 2013 Clay and organic matter applications in the coarse quartz tailing material and sorghum growth on the post tin mining at Bangka Island *Journal of Degraded and Mining Lands Management* **1**: 27 – 32.
[15] Hutahaean BP dan Yudoko G 2013 Analysis and proposed changes of tin ore processing system on cutter suction dredges into low grade to improve added value for the company *The Indonesia Journal of Business Administration* **2**: 1946 – 56.
[16] Agus C and Wulandari D 2012 The Abundance of Pioneer Vegetation and Their Interaction with Endomycorrhiza at Different Land Qualities after Merapi Eruption *Jurnal Manajemen Hutan Tropika* **18(3)**: 145-154.
[17] Rosmarkam A dan Yuwono NW 2002 Ilmu Kesuburan Tanah Kanisius, Yogyakarta
[18] Haering KC, Daniels WL, Galbraith JM 2004 Appalachian mine soil morphology and properties: Effect of weathering and mining method *Soil Science Society of America* **68**: 1315 – 25.
[19] Anonim 1991 Studi Evaluasi Lingkungan (SEL) Unit Penambangan dan Unit Peleburan Timah Pulau Bangka. Ringkasan Eksekutif, Vol 1-4. PT. Tambang Timah. Pangkal Pinang.
[20] U.S. Geological Survey 2015 Mineral commodity summaries 2015: U.S. Geological Survey, 196 p., http://dx.doi.org/10.3133/70140094.
[21] Syekhfani 2013 Kriteria Penilaian Sifat Kimia Tanah Jurusan Tanah Fakultas Pertanian Universitas Brawijaya. from : http://syekhfanimd.lecture.ub.ac.id/files /2013/10/Kriteria-Sifat-Kesuburan-Tanah.pdf. Tanggal akses: 6 April 2017.
[22] Anonim 2002 Effect of bio-organic on soil and plan Improvement of post tin mine site at PT. Koba Tin Project Area, Bangka Research Center of Biotechnology IPB, Bogor.
[23] Sumardi 2009 Prinsip Silvikultur reforestasi dalam rehabilitasi formasi gumuk pasir di kawasan pantai Kebumen. *Prosiding seminar nasional Silvikultur Rehabilitasi lahan: Pengembangan Strategi untuk mengendalikan Tingginya Laju Degradasi Hutan. Yogyakarta*, 24-25 November 2008, pp.58-65.Yogyakarta: Fakultas Kehutanan Universitas Gadjah Mada.
[24] Stevenson FJ 1982 Humus Chemistry: Genesis, Composition, Reactions. A Wiley Interscience Publication, New York.
[25] Issac RA dan Kerber JD 1971 Atomic Absorption and flame photometry: Techniques and uses in soil, plant, and water analysis, In L.M. Walsh (ed), Instrumental methods for analysis of soils and plant tissue. Soil Sci. Soc. of Armer., Inc. Ma., Wisc. USA.

Sorption Capacity Measurement of Chlorella Vulgaris and Scenedesmus Acutus to Remove Chromium from Tannery Waste Water

Liliana Ardila[1], Rubén Godoy [2] and Luis Montenegro [3]

1 &2 Department of Chemical and Environmental Engineering Department, National University of Colombia – South America. E-mail: lardilaf@unal.edu.co or lardilaf@ecci.edu.co; rdgodoys@unal.edu.co
3 Biology Department, National University of Colombia – South America.
E-mail: lcmontenegror@unal.edu.co.

Abstract. Tanning process is a polluting activity due to the release of toxic agents into the environment. One of the most important of those toxic chemicals is chromium. Different alternatives have been proposed for the removal of this metal from tanning waste water which include the optimization of the productive processes, physicochemical and biochemical waste water treatment. In this study, the biological adsorption process of trivalent chromium was carried out in synthetic water and tannery waste water through two types of native green microalgae, called *Chlorella vulgaris* and *Scenedesmus acutus* in Free State and immobilized in PVA state. This, considering that cellular wall of microalgae has functional groups like amines and carboxyl that might bind with trivalent chromium. Statistical significance of variables as pH temperature, chromium and algae concentrations was evaluated just like bio sorption capacity of different types of water and kind of bioadsorbent was calculated to determine if this process is a competitive solution comparing to other heavy metal removal processes.

1. Introduction

For the last years, the concern about the pollution of air, water and soil has become bigger due to the side effects that have been seen like climate change, the increase of human diseases and human poisoning with food or beverage. Heavy metal residues in water released by industries, are one of the most relevant pollutants. They may endanger ecosystems and public health because of their high mobility, their no degradability and their high ability to accumulate inside all living being and participate in biological reactions ruining vital processes inside cells, tissues and organs and promoting illness even when metals are in low concentration in environment. [1] Chromium is a heavy metal, which is used in industries of electroplating and tanneries. Those activities produce enriched effluents with trivalent chromium which could oxidize to hexavalent chromium that is more toxic

To avoid this health risk and preserve environment from metal pollution, industry discharges must be treated to separate them. For metals in aqueous solution there are several solutions to remove them such as precipitation, ion exchange or sorption. Those are physical chemical processes which are highly effective in metal removal from water but they are expensive due to their high demand of reagents and they might produce greater amount of other toxic products that must also be eliminated from treated water [1]-[3]

Alternatively, there are biological processes which use live or dead, free or immobilized cells of bacteria, fungi or algae and farming waste as sorbents, because cells have in their walls carbohydrates and polypeptides with hydroxyl, aldehydes, ketones, amines, phosphates and carboxyl groups that are responsible for the sorption and chelation of the metal cations. [4]-[6] The use of bio sorbents overcomes some of the disadvantages of the physicochemical processes and are useful at low concentrations [1], [2], [5], [7]. But many studies have been done using synthetic water done with deionized water, buffer solution and salt with the metal(s) of interest. The result of those studies were high loading capacity of specific metals. [1]-[3], [5], [7]-[10].
In spite of, real condition of waste water is more complex because of the great variability of water composition due to industrial processes. The appearance of many other organic or inorganic compounds, might affect the performance of bioprocess as Murphy exposed [11] Also, Han et al [12] used the green microalgae *Chlorella miniata* for the removal of hexavalent chromium· and found that the presence of other ions affect the adsorption capacity of chromium.
For that reason, in this study it was necessary to understand the behavior of metal´s bio sorption in real waste water. That was done, by definition of relevant variables like pH, temperature, microalgae and chromium concentration, algae particle size or bio sorbent (considering the mass of algae and the carrier), and others, during bioprocess. Other aspect to know is the estimation of bio sorption capacity for each microalgae specie. In addition, some studies have consider the immobilization of bio sorbent in bigger structures constructed in natural or synthetic materials as polymers. The benefits of this procedure is the ease and economical recovery of employed biomass in bio sorption. Due to the greater size of the immobilized biomass, it would be necessary a simple mechanist to separate bio sorbents like sedimentation or filtration chambers. If biomass where free in the bio sorption process, it would be necessary to employ more complex and expensive mechanisms like ultrafiltration for example. However, there are some disadvantages like the loss of active side of biomass cell wall or drain of biomass cell from carrier material with the consequence of loss of bio sorption effectiveness [13]. In this study, a part of microalgae biomass was immobilized in a resistant material called Polyvinyl alcohol PVA and the bio sorption studies proposed for free microalgae where done for new bio sorbent material.

2. Mmaterials and methods

2.1. Evaluation of tannery waste water treatment

Tannery waste water was taken from different tanneries inside Bogotá city and surrounding area. Waste water quality was determined by evaluation of several physical chemical factors that were recommended in literature [14]-[16]. Analysis was done following the Standard Methods for the Evaluation of Water and Wastewater APHA, AWWA and WEF at Environmental Engineering Laboratory at National University of Colombia [17]

2.2. Microalgae culture

The strains *C. vulgaris* and *S. acutus* were obtained from a natural water body in Bogotá. Each strain was isolated and grown in the Algae Cultivation Laboratory LAUN (maximum culture volume reached: 10 L) and Estación La Terraza (maximum culture volume reached: 1 m^3). Both laboratories of National University of Colombia. The following table present the culture conditions for each volume.

Table 1: Culture parameters to produce microalgae biomass

Culture variables	Volume of 10 L*	Volume of 1 m^3
Culture Media	Liquid BBM [18]	MK10-30-10 FERTITEC ®**
Temperature	24 ± 2 ° C	18 ± 3 ° C

pH	6,3 ± 0,3	6,5 ± 0,1
Light source	Artificial	Natural
Culture period	8 d – 10 d	8 d
Illuminance	4500 ± 50 lux	---
Photoperiod	16 h with light	12 h with light
CO_2 supply	From pumped and filtered air (Millipore filter of 0,2 μm)	

* Volume of ten litters was used as a seed for 1 m^3 volume of culture media.

** Massive biomass production was done with a liquid fertilizer called MK 10-30-10 FERTITEC ® (Tecnoquímicas SA). Biomass production results were very similar to those obtained using BBM medium, but at a fraction of the cost. [19]

2.3. *Harvesting of microalgae*

After biomass cultivation, each microalgae was harvested by centrifugation at speed of 4000 rpm for a period of 10-15 minutes. The recovered biomass was washed once with deionized water and separated again by centrifugation. The supernatant was removed and the biomass was dried at a temperature of 60 ° C in ovens until a humidity value of 3% (established with a Sartorius MA 35 humidity balance). Dried biomass was manually macerated and separated by a 32 mesh Tyler screen (average particle size of 0.05 cm)

2.4. *Microalgae immobilization*

Immobilization of microalgae was made by encapsulation according to the methods described by Cheng and Houng [20], Pramanik and Khan [21]. With the result of previous experiments of chemical resistant and compression test, it was selected the following microalgae bead formulation: 15% ± 0,3 WW^{-1} of PVA, 1% ± 0,1 WW^{-1} of sodium alginate and 3,3% ± 0,2 WW^{-1} of microalgae. PVA monomer and sodium alginate was dissolved in water at boiling point. Then, solution was cooled to 40 °C and microalgae powder (*C. vulgaris* or *S. acutus* separately) was dissolved in monomer solution to obtain a homogeneous dispersion. Solution of monomers and microalgae was fed in the form of dropwise in the saturated solution of boric acid and 2% WV^{-1} of calcium chloride that was magnetically stirred at a speed of approximately 600-700 rpm during polymerization and immobilization process (calcium chloride destroys sodium alginate bonds and it could provide more porosity to PVA beads).

The feed rate was 5 drops per minute to prevent polymer agglomeration during formation of spheres. Immobilization time was 2 hours. Then, beads were collected and washed quickly to remove residual boric acid. The spheres were transferred to 2 M sodium orthophosphate solution for 5 hours (more than suggested by Cheng and Houng [20], Pramanik and Khan [21]) at room temperature with magnetic stirring plate to harden the bio sorption beads. Phosphorylated and hardened spheres were filtered, washed with water, allowed to room temperature to remove some of their moisture and then they were stored in a plastic container in the refrigerator at 4°C. (To evaluate if polymer beads would uptake chromium, the same immobilization process was done without algae). Structure of bio sorption beads was observed by electron microscopy SEM.

2.5. *Bio sorption essays*

The removal of chromium from synthetic water and tannery waste water with free an immobilized algae, would be affected by some variables. The choice of them was suggested by some authors. [5], [11], [22], [23] Two Experimental Response Surface design were selected. The rotatable Draper & Lin experimental design was applied for trivalent chromium removal in synthetic and tannery waste water and both free microalgae. Considering the results obtained in the experimental design of Draper & Lin, the factor(s) which did not have statistical significance was (were) eliminated from the following experimental design executed. Box Benhken experimental design was conducted for both species of immobilized microalgae and tannery waste water. Table 2 shows the statistical parameters for each experimental design.

Table 2: Statistical parameters applied on each experimental desing.**

Parameter	Draper & Lin	Box Benhken
Number of experiments	16	14
Number of replicates	6*	6*
Confidence level	95%	95%
F value	3,53	3,02

* Replicates were done at central point of experimental design.
** The distribution of the levels and determination of ANOVA´s response surface was performed using Statgraphics Centurion XV® software (StatPoint Technologies, Warrenton, VA, USA)

The choice of the experimental design was performed under the criterion of the smallest number of experiments needed to obtain statistically reliable results. The relevant of selected factors was determined by ANOVA and the comparison between the amounts of chromium removed by both microalgae was studied using Student's T test and Fisher Test. Table 3 shows the selected factors and their value in which was executed both experimental design.

Table 3: Selected factors and their levels.

Factor *	Code	Levels				
		-2	-1	0	1	2
T	X_{1**}	23,3	25	27,5	30	31,3
pH	X_2	3,6	4,0	4,5	5,0	5,3
MC	X_3	0,6	1,0	1,5	2,0	2,3
CCS	X_{4s}	58,0	70,0	100,0	130,0	153,0
CCW	X_{4r}	11,3	15	20	25	28,3

* Where T is temperature (°C), MC is microalgae concentration (gL^{-1}), CCS (mgL^{-1}) and CCW (mgL^{-1}) is chromium concentration in synthetic water and tannery waste water respectively.
**This factor was proved only on Draper & Lin experimental design

Tannery waste water has trivalent chromium, thus synthetic water was prepared by reduction of hexavalent chromium from potassium dichromate $K_2Cr_2O_7$ with sulfuric acid and ethanol. The quantity of trivalent chromium in solution must reach a similar concentration according to the metal in tannery waste water. Reduction was verified by spectrophotometry measurement of hexavalent chromium [24], [25]. After precipitation and verification were completed, pH adjustment was done with solutions of sodium hydroxide and sulfuric acid 1M and 0.1M according to experimental designs. The supernatant with chromium was separated and its metal concentration was determined by Atomic Absorption Spectrometry AAS [17]. Then, dilutions were done to reach the metal concentration according to experimental designs. For tannery waste water, a volume of 500 ml was treated first by precipitation and determining the concentration of trivalent chromium was done in the same way as for synthetic water.
Experimental designs were done in 100 ml conical flasks with a volume of 30 ml of each solution, for a time of 24 hours and stirring of 120 rpm at different temperature (see Table 3) y using the Heidolph incubator shaker. For pH adjustment it was seen that after washing the microalgae, this retained traces of salts of the nutritive medium that are basic in nature so the microalgae increases the pH of the essay solution, which rather favors precipitation than bio sorption. The maintenance of the pH to the desired values was done using buffer solution of citric acid - disodium phosphate.
The amount of microalgae that should be added to the solution of 30 ml in each flask to meet the predefined concentration in the experimental design was recommended by different authors [21]-[23], [26]. It was calculated and weighted by a four-decimal digit balance Mettler Toledo brand. For essays

of chromium removal by immobilized microalgae, the following calculation was performed to obtain the approximate weight of PVA – Microalgae beads to assure the microalgae quantity according to experimental designs:

$$\frac{X\ g\ free\ microalgae}{L\ solution} * \frac{100\ g\ polymer\ solution}{3{,}3\ g\ microalgae} * 0{,}03L\ solution \qquad (1)$$

Where X is the amount of free microalgae required in Box Behnken experimental design. Besides bio sorption essays, test was done with the negative control (essays with no microalgae and essays with white polymer under the same experimental conditions to evaluate the effect of precipitation and adsorption of chromium by the polymer used in the immobilization of microalgae).

2.6. Sorption capability measurement

In clean and metal free metal flasks of 100 mL, six solutions with different chromium concentration in synthetic water (0 mgL^{-1}; 32,1 ± 0,3 mgL^{-1}; 74,2 ± 0,4 mgL^{-1}; 110,2 ± 0,4 mgL^{-1}; 154,4 ± 0,2 mgL^{-1} and 200 ± 0,2 mgL^{-1}), citric acid - disodium phosphate buffer solution and 2,34 gL^{-1} and 1,5 gL^{-1} of free *C. vulgaris* and *S. acutus* respectively (according to optimal factor values obtained as results in experimental designs) were prepared. Bio sorption essays were executed in Heidolph shaker at 120 rpm and at temperature where maximum chromium removal was presented for each microalgae, and for 24 hours.
Chromium concentration for bio sorption essays in tannery waste water were: 0 mgL^{-1}; 20,1 ± 0,1 mgL^{-1}; 40,2 ± 0,4 mgL^{-1}; 60,2 ± 0,3 mgL^{-1}; 80,3 ± 0,3 mgL^{-1} and 100,4 ± 0,3 mgL^{-1}. Solutions were prepared with pH adjustment and dilution with deionized water. Concentration of free microalgae used for the residual water tests was 2,34 gL^{-1} of *C. vulgaris* and 2,0 gL^{-1} of *S. acutus*. The amount of immobile biomass used for the residual water tests was 40 gL^{-1} of immobile *C.vulgaris* and 46,8 gL^{-1} of immobile *S. acutus*. Stirring, bio sorption temperature and bio sorption time was the same as for synthetic water.

3. Results and discussion

3.1. Observed structure of bio sorption beads

The Figure 1 shows the observed structure of a blank bead. In this, it is seen a cavernous structure in the middle of beads. But outside, it is seen a smooth surface (Part B). Tiny pore may drive chromium solution inside bead. Because of the size of these pores, diffusion possibly is the controlling step.
Figure 2 shows a polymer – microalgae bead. Also in this bead, a cavernous structure is in the middle, but is bigger comparing to the blank bead (Part B) and it could be seen heterogeneous surface and black points which could be the pores of the bead. C and D present the microalgae distribution inside the bead. Some of them are available to be in contact with chromium solutions, but some of them might be totally or partially covered by PVA polymer.

3.2. Evaluation of bio sorption performance

From tannery waste water quality, it was seen that chromium concentration was from 3000 mgL^{-1} to 3200 mgL^{-1}. After precipitation and pH adjustment it was seen that chromium concentration in synthetic water was higher although the initial chromium concentration was similar to tannery waste water. That is the reason for the evaluation of higher chromium concentration in the Draper and Lin experimental design.
Although higher concentration of chromium in synthetic water, it was seen that higher chromium removal percentages was reached by sorption for each free microalgae in synthetic water in comparison with bio sorption of chromium by free microalgae in tannery waste water. This might be because of the presence of many other compounds in tannery waste water that compete with chromium for active site of each microalgae as Murphy and Hang suggested [11] [12]. This is not the case in synthetic water which has less compounds. According to the comparison between chromium removal by free and immobilized microalgae in tannery waste water, it is seen that there is a

significant difference in bio sorption percentages. This can be explained by a decrease in cell wall activity after immobilization. This would happen when immobilization process is so harsh that could break cell wall or cover surface sites during the crosslinking procedure as Wijffels exposed [13].

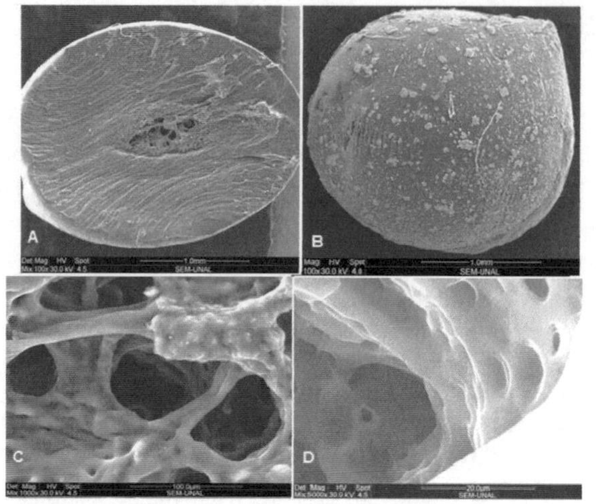

Figure 1: Structure of blank bead obtained by SEM

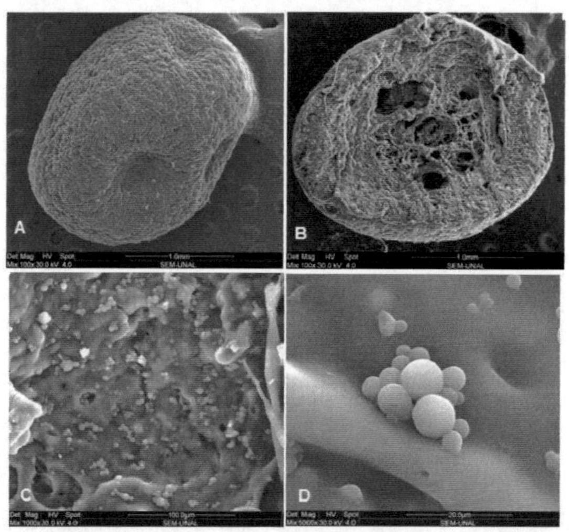

Figure 2: Structure of polymer - microalgae bead obtained by SEM

Table 4 shows the optimal values of factors and the chromium removal percentage. It shows that *C. vulgaris* remove more chromium in synthetic water than *S. acutus*. However with tannery wastewater, free and immobile *S. acutus* made a higher chromium adsorption than *C. vulgaris*. To evaluate this difference in removal response between the two microalgae, there were performed the T-Student and Fisher tests. These tests assume that population means and variances are equal.

Table 4: Factors values at which chromium removal was the highest

Operative parameters	Optimal values					
	Free microalgae				Immobilized microalgae	
	Synthetic water		Tannery waste water		Tannery waste water	
	C.v	*S.a*	*C.v*	*S.a*	*C.v*	*S.a*

T (°C)	27,5	27,5	27,5	25	27,5	25
pH	4,5	5,34	4,5	4,0	4,0	4,5
MC (g L^{-1})	2,34	1,5	2,34	2,0	2,0	2,34
CC (mg L^{-1})	93,9	93,9	20	25	15	20
Stirring (rpm)	120	120	120	120	120	120
Max. removal %	88,2	87,1	61,7	64,1	32,6	35,3

*Where T is temperature (°C), MC is microalgae concentration, CC is chromium concentration.
*C.v is *C. vulgaris* microalgae and S.a is *S. acutus microalgae*

Table 5: Obtained values of t-student and fisher tests.

Removal system*	Fisher test for equal variances		t Student test for equal simple means		standardized kurtosis**
	F value	P value	T value	P value	
FM – SW	0,853	0,719	1,926	0,061	C.v= -1,164 S.a = -1,609
FM - TW	0,787	0,587	0,288	0,775	C. v = -1,155 S.a = -1,067
IM - TW	1,242	0,641	0,771	0,445	C.v= -0,124 S.a= 0,953

*Chromium removal was done by FM-SW free microalgae in synthetic water, FM-TW free microalgae in Tannery waste water and IM-TW immobilized microalgae in Tannery waste water.
** C.v is *Chlorella vulgaris* and S.a is *Scenedesus acutus* microalgae

Population variance for the different type of bio sorption (with free or immobilized biomass and synthetic or tannery waste water) is equal because the relation between variances were less than the critical F value (For FM-SW and FM-TW = 4,35 for IM-TW = 4,41). Also P significant level (α = 0,05) shows higher values that validate null hypothesis. Thus, comparison between population means was done. In all cases is greater than the critical t value (for FM-SW and FM-TW: 2.086; for IM-TW is 2,101). The T value in all cases is lower and the P value for the t test is always higher than the level of significance. Thus, for the three chromium removal systems, null hypothesis is approved. There is no statistical evidence that any of the two species of microalgae as chromium bio sorbents is more effective and can be used both without preference (See values in Table 5)

3.3. Factors affecting bio sorption analysis

Significance of chromium bio sorption variables were evaluated by Statgraphics Centurion XV® software (StatPoint Technologies, Warrenton, VA, USA). Figure 3 and Figure 4 show four diagrams of studied factors which may interact and affect bio sorption process of chromium in synthetic water and tannery waste water for both free microalgae. The vertical line represents the significance level of 5% and the bars which exceed the line have statistically relevance.

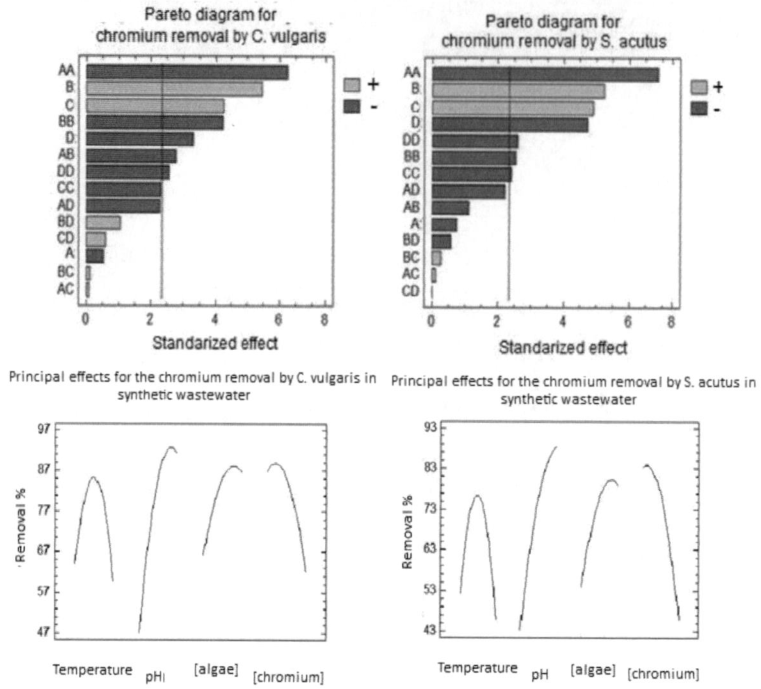

*Factor A: Temperature °C. Factor B: pH, Factor C: Concentration of biomass, Factor D: Concentration of chromium. Other bars are combined effects that may affect chromium bio sorption process.

Figure 3: Factor´s behavior in chromium bio sorption in synthetic water

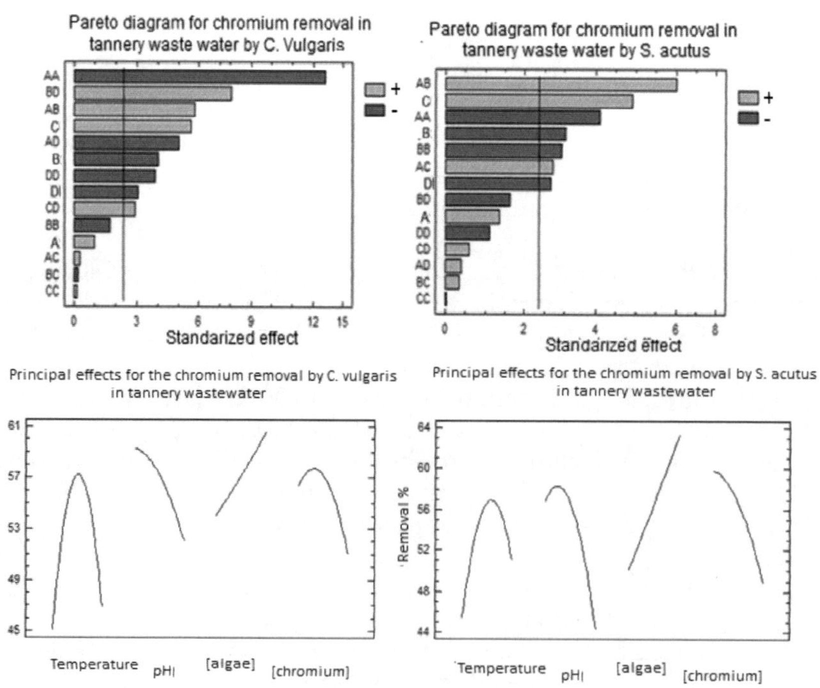

Figure 4: Factor´s behavior in chromium bio sorption in tannery waste water

Therefore, those diagrams express that pH, chromium and microalgae concentration are important factors in bio sorption process in synthetic water and tannery waste water. Concentration of microalgae (both species) and pH have positive effects because when they increase, higher amount of chromium is adsorbed in synthetic water as Onyancha had explained before [23]. But in Tannery waste water, pH has a negative effect on the chromium removal.

This ambiguous behavior might be described for the ionic charge of the active sites on the surface of the microalgae due to pH changes. That can attract or repel the metals in solution. At low pH values, carboxyl, hydroxyl and sulfonates groups, are deprotonated and thereby attracts the metal cations as it was explained by different authors [5], [11], [26], [27]. However, the increase of pH increases the importance of precipitation of metal which reduces the amount of chromium which adheres to the microalgae. Besides, it may rise the amount of sodium and other compounds which are soluble a higher pH and that increases competition for active side in cell wall. The chromium concentration has a negative effect in bio sorption process because when it increase, less amount of chromium is adsorbed, despite the increasing of the driving force [28], [29] which may lead the mass transfer and favors bio sorption only a fraction of them is adsorbed and the rest can repel each other when approaching the active sites.

The temperature in no one of the Draper and Lin experimental designs presented significant effect according to ANOVA analysis. However, the temperature has a negative effect on the chromium removal in synthetic water for both microalgae, while the temperature effect is opposite in chromium removal in tannery waste water. This is due to the increase on the metal diffusivity in the solution and this effect improves the chromium bio sorption if the controlling step is the diffusion [30]. But, up to the 27 °C (25 °C for removal of chromium from waste water by S. acutus), the reduction in adsorption capacity can be caused by a possible tendency of chromium to escape to the aqueous phase, disabling or damaging the active sites of the bio sorbent by breaking bonds discussed before by Sari and Tuzen [9]. Removal of chromium from synthetic water and tannery wastewater by both microalgae had other significant effects. The double effect of temperature (AA), is negative and is the most significant among the effects evaluated in the experimental design. It indicates that the breaking of bonds between chromium and the active site because of the increase in temperature is much more important than the increase in the diffusivity of chromium in the solution.

The dual effect of pH (BB) is negative due to the formation of chromium hydroxide Cr(OH)3 precipitates and cannot be adsorbed by the microalgae. This effect, decrease the concentration of chromium in the solution and let other free and dissolved ions to be adsorbed like sodium, potassium and magnesium which are found in higher proportion in the solution. Consequently those metals can be more likely to the microalgae active sites than chromium as Murphy explained [11]. The dual effect of pH is important in synthetic water because of precipitation but less for metal competition. for both species of microalgae and in chromium removal from tannery wastewater by *S. acutus* as shown in Figure 3 and Figure 4, while for *C. vulgaris* is not significantly important, possibly for a greater selectivity of the functional groups at chromium ions compared to *S. acutus*.

Other important factors in the removal of chromium from synthetic water and tannery wastewater for both microalgae were the combined factors. Combined factor of temperature-pH (AB) may be explained because with the temperature increase and pH values under 6, the hydrogen ions are easily detached from the functional groups of the surface and allow the microalgae to adsorb chromium ions. More microalgae concentration (CC) might lead to higher bio sorption because there is more active sites available. But it also could lead to minor removal of chromium if electrostatic interactions in cells that make them agglomerate with the loosing of bio sorption capability. This behavior was seen by Haeng et al and Vaghetti et al [5], [10] If chromium concentration (DD) upraise, the ratio between chromium and other cations also do, upholding chromium bio sorption. Thereby there is not enough adsorption site which could lead less adsorption capacity.

The temperature and concentration of microalgae effect (AC) for *C. vulgaris* is significant negative on the chromium removal in tannery wastewater, possibly because it stimulates electrostatic interactions between the functional groups of the microalgae. Likewise, with temperature rising, the chromium and compounds solubility and diffusion increase and that promote the rise in repulsion between ions. This hinders the arrival of chromium to the active sites on the surface of the microalgae. The effect between

IOP Conf. Series: Earth and Environmental Science **83** (2017) 012031 doi:10.1088/1755-1315/83/1/012031

pH and chromium concentration (CD) is explained by the proportional increase of other ions with increasing chromium concentrations; however chromium decreases by precipitation with the increasing of pH and lower amount of chromium that adheres to the microalgae.
Figure 3 and Figure 4 show the positive and negative effect of each principal factor evaluated in Draper and Lin experimental design. Bio sorption by free microalgae in synthetic and tannery waste water temperature has reached a maximum value where percentages of bio sorption is the highest.

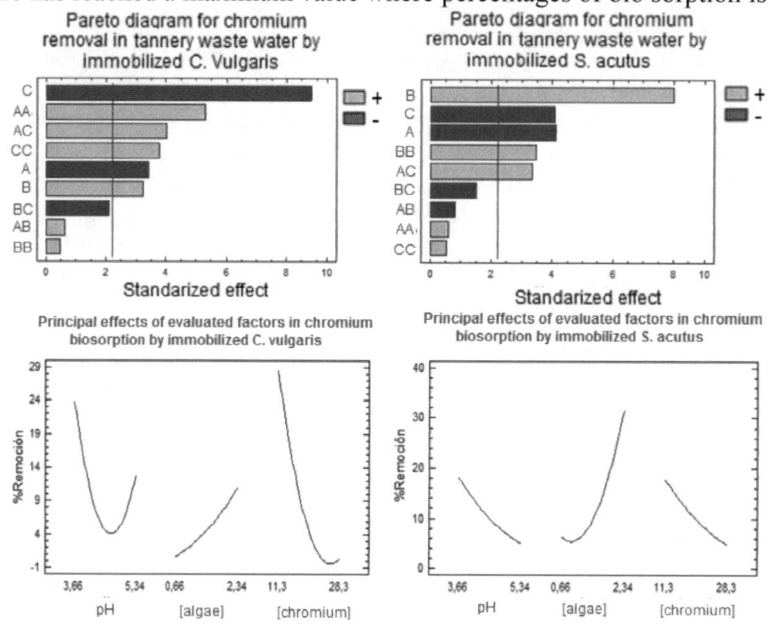

Figure 5: pH. Factor B: Microalgae concentration, Factor C: Chromium concentration.

Figure 5: Factor´s behavior in chromium bio sorption in tannery waste water by immobilized microalgae.
Microalgae concentration has always a positive effect while chromium concentration has negative effect. The ambiguous behavior has seen clearly in this diagrams.
The Figure 5 shows for both immobilized microalgae in polyvinyl alcohol 15% + 1% alginate, that the three principal factors are significantly important. In the case of chromium removal from wastewater and immobile C. vulgaris, the most important variable is the concentration of chromium and this has a negative effect on the bio sorption, while microalgae concentration has a positive effect and this factor is the most important variable in the removal of chromium from wastewater by immobilized S. acutus. By elimination of temperature as a variable, it also eliminate the importance of combined factors associated to temperature that explained most of the variation in the chromium removal by free algae.
This consideration would be important because bio sorption statistical model would not represent the exact behavior. The twin factors of pH (AA) concentration of microalgae (BB), chromium concentration (CC) and pH - concentration of chromium explained before, also appear as statistically significant variables in the biological process of adsorption. Other combined factors are not statistically significant.
From essays of chromium bio sorption in tannery waste water by both immobilized algae, the twice pH factor (AA) has always positive effect, because the increase in pH promotes the deprotonation of the functional groups of microalgae and makes them very reactive and selective to heavy metal but environmental conditions for immobilized microalgae are milder than for free biomass. Therefore the basic precipitation and the presence of other cations have important effect, but functional groups of cell wall would maintain deprotonated, for that bio sorption process would be feasible. For immobilized *C. vulgaris*, the increased chromium concentration has a positive effect, it would be explained partially by the increase of the driving force and the bio sorption of chromium. But, for *S. acutus*, the increase in the concentration of chromium is negative due to the presence of functional

groups that are more selective to other ions presented in the waste water. This increase of the combined driving force with the deprotonation of the active site explains the positive effect of the combined factors of pH - chromium concentration (AC). A high microalgae concentration has a positive effect on the adsorption of chromium, because immobilized microalgae cells does not present the same effect of agglomeration by electrostatic effects as with free microalgae.

3.4. Chromium bio sorption capability determination

The following Figure 6 and Figure 7 shows isotherms for chromium removal from synthetic and tannery waste water with free and immobilized microalgae. To determine the type of adsorption of chromium by biomass, the evaluation of several models presented in the literature as Freundlich, Langmuir, Flory-Huggins, Dubinin - Radushkevich and Temkin was done [2], [5], [23], [31]-[33]
Table 6 presents the summary of models that best fit the experimental behavior of the bio sorption in synthetic water by both microalgae. The regression of the parameters, the coefficients of determination and the sum of squares (mse) for free microalgae and synthetic water are presented. According to Table 6 Freundlich and Dubinin - Radushkevich fitted with a correlation coefficient close to or greater than 90% and a sum of squares smaller than 0.1 Freundlich model is the one that best fits the experimental data. This model assumes that the surface is heterogeneous [31]and that the affinity and the energy of the active sites is different.

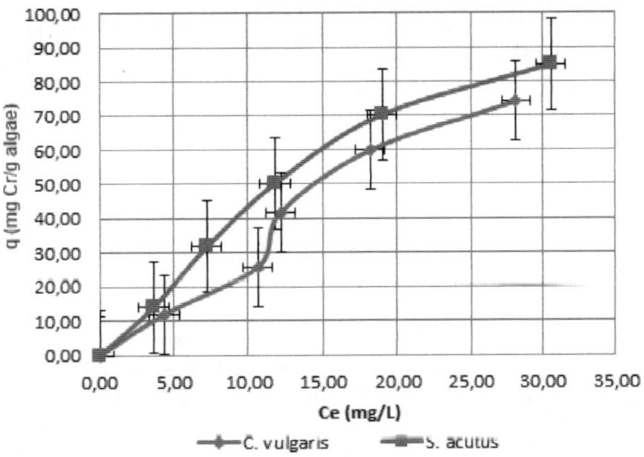

Figure 6: Chromium bio sorption isotherms in synthetic water

Table 6: Isoterm models to determine bio sorption capability in synthetic water

Model*		Unit	C. vulgaris Values	S. acutus Values
FM	n		0,973 ± 0,281	1,173 ± 0,227
	KF	L g-1	1,534 ± 0,876	2,073 ± 0,718
	R2		0,9594	0,9653
	mse		0,0033	0,0027
DRM	QD	mg g-1	57,005 ± 5,513	68,851 ± 8,638
	KD	mol2J-2	6,30E-6 ± 3,68E-6	4,66E-6 ± 3,70E-7
	E	J mol-1	281,73 ± 81,536	327,67 ± 12,104
	R2		0,830	0,913
	mse		0,0725	0,0357

*MF is Freundlich model and DRM is Dubinin – Radushkevich model.

The values of the constant adsorption intensity n for *C. vulgaris* and *S. acutus* are close to one therefore it can be said that the adsorption tends to be linear and the effect of affinity for the solvent is not important on the adsorption of chromium. [2], [34]. For *C. vulgaris*, the value of n less than one. It indicates that adsorption may be chemical, whereas for *S. acutus*, the value of n is greater than one and therefore the adsorption of chromium might be physical according to Nemr Et al [35] and based on the work of Crini et al. [36] The adsorption capacity of Freundlich KF for *C. vulgaris* and *S.acutus* is $1.534 \pm 0.876\ Lg^{-1}$ and $2.073 \pm 0.718\ Lg^{-1}$ respectively and this indicates that *S. acutus* can adsorb 1.35 times more chromium than *C. vulgaris*.
The Dubinin - Radushkevich model adjusts with a relatively good of accuracy and the quadratic mean of the error is low. This model represents the adsorption on nonhomogeneous surfaces as predicted by Freundlich and is carried out on porous surfaces [1]. The adsorption energies obtained by this model are $0.281 \pm 0.081\ KJ\ mol^{-1}$ and $0.327 \pm 0.012\ KJ\ mol^{-1}$. These are low indicating that the joints between the metal and the bioadsorbent material are weak.
The following Figure 7 shows the obtained results of bio sorption capability essays in tannery waste water. With bio sorption in tannery waste water, classical isotherm models were evaluated as multicomponent isotherm models (Lamgmuir model characterized by Equation 1 and the combined model of the multi-component Langmuir - Freundlich Equation 2), taking in count that in real waste water there is many other compounds which could be adsorbed by microalgae [34]

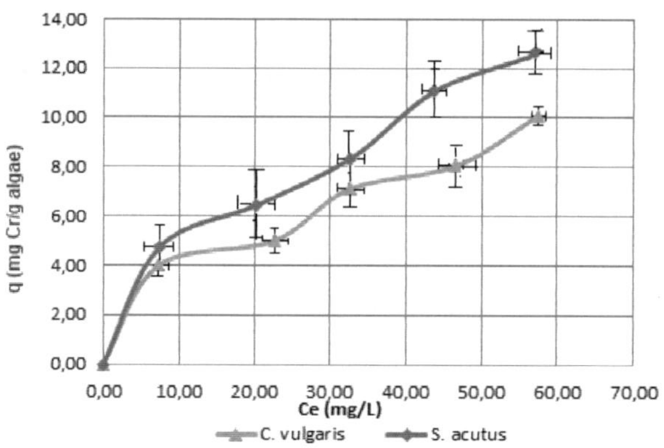

(a)

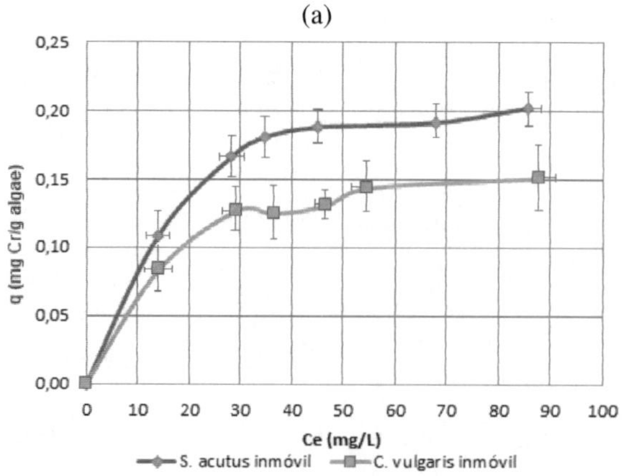

(b)

Figure 7: Chromium bio sorption isotherms in tannery waste water by free microalga (a) and immobilied microalgae (b)

$$q_{e,i} = q_{max,i} \frac{K_{L,i} C_{e,i}}{1 + \sum_{j=1}^{N} K_{L,i} C_{e,i}} \qquad (1)$$

$$qi = \frac{qmi * bi * ci^{1/\overline{n}}}{1 + bi * ci^{1/\overline{n}} + S^{1/\overline{n}}} \qquad (2)$$

Where qmax,i; the maximum metal adsorption, adsorption capability, KF_i, mean adsorption intensity n, adsorption constant bi, for each metal in aqueous solution and sum of concentration and affinities of other metals S. Table 7 shows the isotherms models which better describe bio sorption behavior. The models of Langmuir and Freundlich presented a high coefficient of correlation for *S. acutus* and only Freundlich for *C. vulgaris*, the adsorption of chromium on the surface of the microalgae is a combination of both processes. This implies same affinity for the chromium in certain places of the surface of the microalgae although of the repulsion that can generate cations in solution [35], [36].

Table 7: Isoterm models to determine bio sorption capability in tannery waste water for free and immobilized microalgae

Parameter		Free algae		Immobilized algae	
		C. vulgaris	*S. acutus*	*C. vulgaris*	*S. acutus*
Simple Freundlich model					
n		2,311 ± 0,54	2,083 ± 0,37	3,127	3,129
KF	L g^{-1}	1,214 ± 0,73	1,258 ± 0,59	0,2432	0,2781
R^2		0,9001	0,9513	0,9089	0,835
mse		0,0021	0,0012	6,07E-4	0,0014
Simple Langmuir model					
Qmax	mg g^{-1}	14,78 ± 3,64	20,08 ± 7,42	0,1782	0,240
b	L mg^{-1}	0,029 ± 0,002	0,027 ± 0,01	0,067	0,071
RL		0,21- 0,57	0,11-0,39	0,10–0,36	0,09 – 0,33
R2		0,8429	0,9097	0,9689	0,9385
mse		2,812	1,117	0,00457	0,0093
Temkin Model					
A	L g^{-1}	0,462 ± 0,02	0,385 ± 0,06	0,8135	0,9043
b	J mol^{-1}	911,52 ± 100	662.094 ± 240	6,77E4	5,11E4
R^2		0,831	0,8842	0,941	0,8793
mse		0,801	0,9600	2,65E-5	1,14E-4

For that reason more chromium can be adsorbed in the vacant active sites generating a chromium monolayer. In other places on the surface it is not like that, the adsorption is physical and favorable as predicted by the constant of adsorption intensity n but there were other compounds that can compete with chromium for active sites on the surface of microalgae.

Thus, the effect of the affinity of the solvent should be considerable and n value should be less than one [2], [34]. The higher affinity sites are occupied first and as the surface is saturated with metal and other ions, after that the surface of the microalgae reduces the adsorption energy and its affinity until it stops adsorbing more chromium as it happens on the heterogeneous surface [2], [31]. These models have adequately represented the chromium adsorption model by both microalgae, however, these models have not taken into account the effect of the presence of the other ions present in the wastewater. The multicomponent Langmuir model presents the same adjustment as the simple Langmuir model together with the maximum Qimax adsorption capacity.

The combined model of Langmuir - Freundlich multicomponent has also presented a good fit and the value of n, indicates that the adsorption is physical and favorable [34]. However the values of the maximum capacity of adsorption are higher than those predicted for adsorption of chromium in synthetic water for which these models do not represent correctly the reality in spite of its good adjustment. The Temkin model presents a coefficient of determination close to 90%. This might validate some of the assumptions of the Langmuir model however the decrease of the power of the adsorption no other models can predict it like this model The adsorption energies are 0.911 ± 0.10 $KJmol^{-1}$ and 0.662 ± 0.24 $KJmol^{-1}$, which are low and the prediction that the interaction between the chromium and the cell surface is physical as assumed by Freundlich model [34].
A difference in chromium bio sorption between synthetic water and tannery waste water is that models of Flory Huggins and Dubinin - Radushkevich presented high coefficients of determination. Even the model of Dubinin - Radushkevich is the model that better adjusts in comparison with the others and indicate that the chromium adsorption chromium is carried out on a heterogeneous surface in the pores of the bio sorbent [2], [31], [35]. This is partially true for microalgae immobilized in PVA beads. Flory Huggins model has a high coefficient of determination. However, this model must be discard since the calculation of the free power of the adsorption presents a positive value, therefore the adsorption does not occur spontaneously and is not feasible.

4. Conclusion

Statistical experimental design allowed to identify that pH, microalgae and chromium concentration had statistical importance in the bio sorption of chromium by microalgae in contrast to temperature, but combined factors of last variable were important. For that temperature must be taken in count in the bio sorption process. The experimental design showed that the major chromium removal was done in synthetic water with a percentage of 88,2% with *C. vulgaris* as a sorbent at a pH of 4,5, 2,34 g/L of microalgae, temperature of 27°C and a chromium concentration of 20 mg/L, while for the bio sorption by *S. acutus* was 87,1% and a pH of 5,34, 2,0 g/L of microalgae, temperature of 27°C and a chromium concentration of 20 mg/L.
The removal of chromium of tannery waste water by the microalgae was reduced to 30,01% for C. vulgaris and 26,41% for *S. acutus*. The explanation of this efficiency drop was a major interaction between funtional groups of the microalgae and other compounds present in wastewater which are competitor to the chromium removal. The removal of chromium from tannery wastewater comparing free microalgae and immobilized microalgae is reduced by 57% for *C. vulgaris* and 50% for *s. acutus*. This may be explained by the cover and destruction of a part of the active sites of the microalgae surface because of the process immobilization. Isotherm models which explained bio sorption of chromium for both type of microalga in synthetic were Freundlich model and Dubinin - Radushkevich model, but it changed in tannery waste water where it was found that chemical bonds were weak and deactivation of free o immobilized microalgae would occur.

5. References

[1] S. Lesmana, N. Febriana, F. E. Soetaredjo, J. Sunarso y S. Ismadji, «Studies on potential applications of biomass for the separation of heavy metals from water and wastewater,» *Biochemical Engineering Journal* , 2009.
[2] J. Febrianto, A. Kosasih, J. Sunarso, Y.-H. Ju, N. Indraswati y S. Ismadji, «Equilibrium and kinetic studies in adsorption of heavy metals using bio sorbent: A summary of recent studies,» *Journal of Hazardous Materials,* nº 162, pp. 616 - 645, 2008.
[3] J. Wang y C. Chen, «Biosorption of heavy metals by Saccharomyces cerevisiae: A review.,» *Biotechnology Advances,* vol. 24, pp. 427 - 451, 2006.
[4] W. Jianlong y C. Chen, «Bio sorbents for heavy metals removal and their future,» *Biotechnology Advances,* vol. 27, pp. 195 - 226, 2009.
[5] H. D. Cho., E. Y. Kim y Y.-T. Hung., «Heavy metal removal by microbial bio sorbents,» de *Handbook of Environmental Engineering: Environmental Bioingenieering*, New York, Humana Press, 2010, pp. 375 - 402.

[6] V. Prigione, M. Zerlottin, D. Refosco, V. Tigini y A. Anastasi, «Chromium removal froma real tanning effluent by autochthonous and allochthonous fungi,» *Bioresource Technology,* vol. 100, pp. 2770 - 2776, 2009.
[7] B. Volesky, «Biosorption and me,» *Water Research,* vol. 24, pp. 427 - 451, 2007.
[8] K. Cheung y J. Gu, «Mechanism of hexavalent chromium detoxification by microorganism and bioremediation application potential: A review,» *International Biodeterioration & Biodegradation,* vol. 59, pp. 8 - 15, 2007.
[9] A. Sari y M. Tuzen, «Biosorption of total chromium from aqueous solution by red algae (Ceramium virgatum): Equilibrium, kinetic and thermodynamic studies,» *Journal of Hazardous Materials,* vol. 160, p. 349–355, 2008.
[10] J. Vaghetti, E. Lima, B. Royer, J. Brasil, B. d. Cunha, N. Simon, N. Cardoso y C. Zapato, « Application of Brazilian-pine fruit coat as a bio sorbent to removal of Cr(VI) from aqueous solution—Kinetics and equilibrium study,» *Biochemical Engineering Journ,* vol. 42, pp. 67 - 76, 2008.
[11] V. Murphy, An Investigation into the Mechanisms of Heavy Metal Binding by selectec Seaweed species (PHD Thesis), Waterford Institute of Technology, 2007.
[12] X. Han, Y. S. Wong, M. H. Wong y N. F. Y. Tam, «Biosorption and bioreduction of Cr(VI) by a microalgal isolate,Chlorella miniata,» *Journal of Hazardous Materials,* Vols. %1 de %2146,, p. 65–72, 2007.
[13] R. Wijffels, Immobilized Cells, Springer Lab. Manuals, 2001.
[14] J. Buljan, G. Reich y L. Ludvik, «Mass balance in leather processing,» United Nations Industrial Development Organization UNIDO, 2000.
[15] Centro de promoción de tecnologías sostenibles. CPTS, Guía técnica de producción más limpia para curtiembres, Centro de promoción de tecnologías sostenibles. CPTS, 2003.
[16] M. Aragón y A. Alzate, «Sistema de referenciación ambiental (SIRAC) para el sector curtiembre en Colombia,» Bogotá, 2004.
[17] American Public Health Association; American Water Works Association; Water Environment Federation, «METALS BY FLAME ATOMIC ABSORPTION SPECTROMETRY MEASUREMENT,» de *Standard Methods for the Examination of Water and Wastewater*, American Public Health Association; American Water Works Association; Water Environment Federation, 1999.
[18] R. A. Andersen, «Bold's Basal Medium,» de *Algae Culturing Techniques*, 437, Phycological Society, 2005.
[19] TECNOQUÍMICAS S.A., «Agricultura,» TECNOQUÍMICAS S.A., [En línea]. Available: http://www.tqconfiable.com/detalle-producto?idProducto=fertitec. [Último acceso: 20 mayo 2017].
[20] C. K.H y J. Houng, «Cell immobilization with phosphorilated Polyvinyl alcohol,» de *Immobilization of Enzymes and Cells. Some practical considerations*, New Jersey, Humana Press, 1997, pp. 207 - 216.
[21] S. Pramanik y E. Khan, «Effects of cell entrapment on growth rate and metabolic activity of mixed cultures in biological wastewater treatment,» *Enzyme and Microbial Technology,* vol. 43, pp. 245 - 251, 2008.
[22] M. Carmona, M. Antunes y S. Ferreira, «Biosorption of chromium using factorial experimental design,» *Process Biochemistry,* vol. 40, pp. 779 - 788, 2005.
[23] D. Onyancha, W. Mavura, J. C. Ngila, P. Ongoma y J. Chacha, « Studies of chromium removal from tannery wastewaters by algae bio sorbents, Spirogyra condensata and Rhizoclonium hieroglyphicum,» *Journal of Hazardous materials,* vol. 158, pp. 605 - 614, 2008.
[24] American Public Health Association, American Water Works Association, Water Environment Federation, «3500-Cr B. Colorimetric Method,» de *Standard Methods for the Examination of Water and Wastewater*, American Public Health Association, American Water Works Association, Water Environment Federation, 1999.
[25] P. Pattanapipitpaisal, N. Brown y L. Macaskie, «Chromate reduction by Microbacterium liquefaciens immobilised in polyvinyl alcohol,» *Biotechnology Letters,* vol. 23, pp. 61-65, 2001.

[26] A. Sari, D. Mendil, M. Tuzen y S. Mustafa, «Biosorption of Cd(II) and Cr(III) from aqueous solution by moss (Hylocomium splendens) biomass: Equilibrium, kinetic and thermodynamic studies,» *Chemical Engineering Journal,* vol. 160, pp. 349 - 355, 2007.
[27] V. Gupta y A. Rastogi, «Biosorption of hexavalent chromium by raw and acid treated green algae Oedogonium hatei from aqueous solutions,» *Journal of Hazardous material,* vol. 163, pp. 396 - 402, 2009.
[28] R. E. Treybal, Operaciones de Transferencia de Masa, Ciudad de México: McGraw Hill.
[29] C. Geankoplis, Procesos de transporte y operaciones unitarias, Ciudad de México: Compañia Editorial Continental S.A de C.V , 1998.
[30] B. B. a. H. Benaissa, «Cadmium removal from aqueous solutions by chitin: kinetic and equilibrium studies,» *Water Research,* vol. 36, p. 2463–2474, 2002.
[31] B. Kiran y A. Kaushik, «Chromium binding capacity of Lyngbya putealis exopolysaccharides,» *Biochemical Engineering Journal,* vol. 38, pp. 47 - 54, 2008.
[32] R. D. M., Principles of adsorption and adsorption processes, USA: John Wiley and sons, 1984.
[33] B. Volesky, Biosorption: Application Aspects – Process Simulation Tools, Montreal: McGill University.
[34] A. Delle Site, «Factors Affecting Sorption of Organic Compounds in Natural Sorbent / Water Systems and Sorption Coefficients for Selected Pollutants. A Review,» *American Institute of Physics,* pp. 189 - 192, 2000.
[35] A. E. Nemr, A. Khaled, O. Abdelwahab y A. El-Sikaily, «Treatment of wastewater containing toxic chromium using new activated carbon developed from date palm seed,» *Journal of Hazardous Materials,* vol. 152, p. 263–275, 2008.
[36] G. Crini, H. Peindy, F. Gimbert y C. Robert, «Removal of C.I. Basic Green 4 (Malachite Green) from aqueous solutions by adsorption using cyclodextrin-based adsorbent: Kinetic and equilibrium studies,» *Separation and Purification Technology,,* vol. 36, pp. 97 - 110, 2007.

Institute of Physics
Dirac House, Temple Back
Bristol BS1 6BE UK

ISSN: 1755-1307
ISBN 978-1-5108-4773-6
9 781510 847736

CURRAN ASSOCIATES INC.
proceedings.com